APPLICATIONS OF MOLECULAR SIMULATION IN THE OIL AND GAS INDUSTRY

MONTE CARLO METHODS

IFP PUBLICATIONS

Ph. UNGERER
Professor, IFP • Associate Professor Paris XI University

B. TAVITIAN
Research Engineer, IFP

A. BOUTIN
Research Fellow, CNRS

APPLICATIONS OF MOLECULAR SIMULATION IN THE OIL AND GAS INDUSTRY

Monte Carlo Methods

Foreword by
François Montel • Thermodynamics Expert, TOTAL

2005

Editions TECHNIP 5 av. de la République, 75011 PARIS FRANCE

FROM THE SAME PUBLISHER

- Basin Analysis and Modeling of the Burial, Thermal and Maturation Histories in Sedimentary Basins
  M. MAKHOUS, Y. GALUSKIN
- Sedimentary Geology
  Sedimentary basins, depositional environments, petroleum formation
  B. BIJU-DUVAL
- Geomechanics in Reservoir Simulation
  P. LONGUEMARE
- Oil and Gas Exploration and Production
  Reserves, costs, contracts
  CENTRE OF ECONOMICS AND ADMINISTRATION (IFP-SCHOOL)
- Integrated Reservoir Studies
  L. COSENTINO
- Geophysics for Sedimentary Basins
  G. HENRY
- Basics of Reservoir Engineering
  R. COSSÉ
- Well Seismic Surveying
  J.L. MARI, F. COPPENS
- Geophysics of Reservoir and Civil Engineering
  J.L. MARI, G. ARENS, D. CHAPELLIER, P. GAUDIANI

Manuscript corrected by
Bowne Global Solutions
92-98, boulevard Victor Hugo, 92115 Clichy, France

Printed in France

ISBN 978-2-7108-0858-9

Foreword

François Montel

Thermodynamics is required at any stage of Oil and Gas exploration and production. During the last three decades, empirical Equations of States (EOS) have become essential to extend the small number of reliable experiments available in the open literature. But despite our efforts the physical background of the most efficient EOS remains poor and their predictive nature is not proven except for very simple molecules and thanks to a very large number of specific parameters.

The Development of Oil and Gas technology requires more accurate fluids properties and the application domain spreads out to new area like extreme P&T conditions or high level of acid gas components. Moreover, the search for an optimal development scenario in an unsettled economic context assumes a perfect knowledge of the evolution of effluent composition during production. In the absence of satisfactory control of the thermodynamics aspects, the most profitable scenarios will be discarded because they are often based on more complex compositional evolution.

Surface installations are designed and sized on the basis of the fluid properties. Proper characterisation is of prime importance since the equipment installed must ensure compliance with a given number of product specifications while minimising energy consumption and the required investment. Too much uncertainty regarding effluent behaviour may lead to needless investment, oversize production plant or, conversely, make it impossible to comply with the product specifications using existing plant.

The Petroleum industry has also to face a new challenge: monitoring and controlling the greenhouse gas emissions and the production of final waste. To achieve this goal, new processes have to be developed. Acid gas injection for instance, requires many experiments at high pressure with toxic materials like hydrogen sulphide.

The laboratories devoted to data acquisition cannot face the increasing demand and molecular simulation appears to be the only valuable alternative to get these data before the deadline of the projects.

It is the reason why TOTAL has given his continuous support to the authors of this book for developing and promoting the molecular simulation techniques.

This remarkable piece of work is an important milestone in this process. It shows how mature this technique is. Of course there is still room for many improvements in the field of numerical techniques and in intermolecular potential description, but it is now well established that the molecular simulation can provide reliable data on many systems of interest in the Oil and Gas Industry. This is the end of the controversy about its usefulness for industrial applications.

One of the most important breakthrough is the transferability of the force fields used by the authors in the different application examples. It is possible to build up the improvements obtained on new chemical compounds since the description of the molecule is exactly at the right level of accuracy. This is particularly important for petroleum fluids which are mixtures of a huge number of isomers. The application of molecular simulation to the industrial separation processes by adsorption is also a promising perspective, as existing thermodynamic models are often limited in predicting the adsorption equilibria in zeolites.

We are entering a cycle of data production and this is a very good piece of news for our industry.

François Montel
Thermodynamics Expert, TOTAL

Table of Contents

Acknowledgements

It is a pleasure for us to acknowledge the support of Daniel Decroocq, former Director of Science at the Institut Français du Pétrole (IFP) who initiated the idea of writing the present book, and of Jacqueline Lecourtier, who continued his action. We are also deeply indebted with professor Alain Fuchs, head of the Laboratory of Chemical Physics (LCP) at the Université de Paris Sud, for his invaluable scientific advice and for his commitment in allocating appropriate resources to applied research in his group. Jean-Jacques Lacour, Director in charge of knowledge diffusion in IFP, and Jacques Jarrin, head of the Applied Chemistry and Physical Chemistry division of IFP, are gratefully acknowledged for supporting the writing of this book.

We are indebted with many colleagues for various types of contributions. Professor Keith Gubbins, from the university of Southern Carolina, provided us key helpful advice as he was visiting professor in LCP. Francois Montel is thanked for his interest in many studies shown in this book and for the support he obtained from his company, Total. Allan Mackie, university professor in Tarragona, has developed Monte Carlo programs while he was post-doc in IFP and subsequently contributed to intermolecular potential development. He has also kindly accepted to read and comment a preliminary version of the book. We are particularly thankful toward Jean-Pierre Bellat, Guy Weber and their colleagues of the University of Bourgogne for their measurements of adsorption isotherms. Similarly, the group of professor Jacques Jose in Lyon university performed vapour pressure measurements that were very useful to us. Vincent Gerbaud, from the CNRS-INP laboratory of Chemical Engineering in Toulouse, kindly provided us with the result of his group on nitriles. Véronique Lachet, research chemist in IFP, has contributed to several investigations of phase equilibria involving heavy hydrocarbons or polymers, extending thus the findings she had accumulated during her PhD on the adsorption of aromatics in Faujasite zeolites. We have benefited also from the important work of Jacqueline Ridard and Bernard Lévy in LCP to use ab initio calculations in the development of intermolecular potentials. Stéphane Requena (IFP) and Jean-Marie Teuler (LCP) made us profit of their impressive culture in computer science for code development and hardware implementation. Among the numerous other colleagues that have helped us, it is a plesure to cite Patrick Boisserpe, Theodorus de Bruin, Marie-France Couret, Jean Farago, Diego Klahr, Patrice Malfreyt, Pascal Mougin, Pascal Pernot, Bernard Rousseau, Véronique Ruffier-Meray and Michel Thomas.

Numerous young researchers of LCP and IFP have contributed to the results shown, in the frame of a fruitful cooperation. Roland Pellenq and Brigitte Neubauer implemented new algorithms and tested various potentials during their stays as post-docs. During his very productive PhD, Jérome Delhommelle simulated phase equilibria involving alkanes and polar compounds. Séverine Buttefey and Christèle Beauvais, whose PhDs were devoted to the adsorption in zeolites, are also gratefully acknowldged. Emeric Bourasseau, PhD student, provided a key effort in developing algorithms and potentials for new hydrocarbon families. Marie Lagache-Le Naour, also PhD student, tackled with success the determination of thermodynamic derivative properties, among other contributions. Oliver Contreras-Camacho, first as PhD student in Tarragona and then as post-doc in IFP, has developed potentials for aromatic hydrocarbons and simulated the adsorption of complex mixtures with polar components. Spyros Kranias obtained high quality results on ketones and aldehydes during his post-doc sponsored by ATO, before investigating branched alkanes in IFP. Pierre Pascual, PhD student, studied the adsorption of alkanes and alkenes in zeolites with new potentials, opening the way to very efficient applications. Although we cannot detail their contributions, we are glad to complement this list with other students and young researchers who provided significant achievements: Göktug Ahunbay, Eric Dehaudt, Grégoire Demoulin, Nicolas Desbiens, Aurélien Desloges, Xavier Guerrault, Mehalia Haboudou, Mohammed Hadjkali, Wang Ji, Hélène Kirsch, Vincent Poujol, Christian Tschirwitz, and Aurélie Wender.

1

Introduction

Understanding the properties of fluids and materials is often important in the gas and oil industry. In exploration and production, this mainly pertains to phase equilibria of crude oil and natural gases, but also to how these fluids interact with subsurface water, with reservoir rocks or with polymer materials used to coat pipes. In hydrocarbon processing, the design of efficient distillation- or adsorption-based separation processes also requires an intimate knowledge of phase equilibria when polar solvents such as methanol or amines may be involved, together with a good understanding of the physics of adsorption in microporous solids. As fluid and material properties cannot be experimentally measured across the entire, very broad range of diverse conditions encountered, the oil and gas industry makes extensive use of thermodynamic models. These include equations of state [Soave, 1972; Peng and Robinson, 1976, among others], activity coefficient models [Renon and Prausnitz, 1967; Fredenslund *et al.*, 1977; Huron and Vidal, 1979, among others] and corresponding state theories [Lee and Kesler, 1975, among others]. However, these classical models are sometimes inadequate when only a limited amount of experimental information is available to establish the key parametres. Certain examples of unsatisfactory predictions could be singled out, including the equilibrium properties of pure hydrocarbons that are not commercially available, and selective adsorption in microporous solids.

Over the last few decades, molecular simulation has emerged in the scientific community as a new corpus of theoretical methods which go far beyond the classical theories, and which may provide alternate ways of making predictions. Monte Carlo simulation is one such method which specifically addresses *equilibrium* properties (such as phase equilibria), as opposed to *dynamic* properties (such as viscosity). Before introducing Monte Carlo simulation itself, let us define more precisely molecular simulation.

Molecular simulation refers to computational methods in which molecular structure is explicitly taken into account. Depending on the type of problem addressed, molecular structure can be specified at very high resolution with every single electron modelled, or at coarser levels where the smallest particles represented individually are atoms or parts of molecules. Thus, the typical scale of molecular simulation ranges from less than 1 nanometre (1 nm $= 10^{-9}$ m) to a micrometre (1 μm $= 10^{-6}$ m) so molecular simulation is clearly part of the field of nanotechnology. Although this scale is very small compared with those accessible with most experimental techniques, it is often sufficiently representative to address chemical reactions, equilibrium properties and transport phenomena. This is not only true for systems exhibiting a well-defined structure at the nanometre level, such as microporous adsorbents or catalysts, but also for disordered systems like gases and liquids.

An important class of molecular simulation techniques are those of quantum chemistry which define molecular structure at the electronic level. The method depends on solving the basic equation of quantum mechanics – Schrödinger's equation – using numerical methods. By explicitly computing the distribution of the electrons, these techniques determine equilibrium conformations and the dynamics of molecular systems. However, quantum chemistry methods are limited to systems containing a few dozen atoms and to time frames of less than one nanosecond, even when several days of computer time are allocated for computation. Although many problems in chemistry can be tackled in this way, this scale is not suitable for addressing transport phenomena or thermodynamic properties.

This is why a second class of molecular simulation methods has been developed. These consider molecular structure at a coarser level, i.e. that of individual atoms or groups of atoms (Fig. 1.1). Among such methods, special mention can be made of two in particular, namely molecular dynamics, in which the equations of motion are solved to address dynamic properties, and Monte Carlo simulation which addresses equilibrium properties through the use of statistical methods. In essence, molecular dynamics is a versatile technique which has been very successful in analysing the dynamic properties of a large variety of systems, e.g. it is possible to make quantitative determinations of viscosity or diffusion coefficients in liquids using this technique. The particular advantage of Monte Carlo methods over those of molecular dynamics is that they are generally more efficient if relaxation times are high, i.e. for systems exhibiting low diffusion coefficients and high viscosity (polymers melts, dense liquids, etc.). They are also better at simulating fluid phase equilibria and adsorption in microporous solids. On the other hand, Monte Carlo methods do not provide any insight into the dynamics of molecular processes so although they can be used to determine thermodynamic properties in equilibrium conditions, they do not yield transport coefficients. Molecular dynamics and Monte Carlo simulations at the atomic level also require large amounts of computer time and this generally restricts their applicability to systems smaller than 5 to 10 nanometres containing just a few thousands or tens of thousands of atoms. When performed on systems containing a few hundreds of molecules, these techniques can be used to determine the average properties of fluids (e.g. density or viscosity) with reasonable statistical uncertainty.

However, some problems necessitate consideration of an even larger scale (e.g. the self-assembly of polymers in solution, protein folding); in this context, a third class of molecular simulation methods – dissipative particle dynamics – is now emerging. This is based on taking large portions of molecules as the individual sub-units. Although this kind of technique has not yet borne all its fruit, it is likely that abandoning the atomic level will make these methods more system-specific and less quantitative than either molecular dynamics or Monte Carlo methods.

The first Monte Carlo method was more or less contemporary with the first computers, dating back to the mid-twentieth century [Metropolis *et al.*, 1953]. Interestingly, because it involves the generation of random numbers, the method was named after the well-known casino town of Monte Carlo on the French Riviera. The method was soon improved and extended to cover flexible molecules at a simple level [Rosenbluth and Rosenbluth, 1955].

The method is closely linked to classical concepts of statistical thermodynamics. In the 1960's and 1970's, many theoretical concepts were developed to treat complex systems, as illustrated by the continued relevance of articles and textbooks written during this period [Widom, 1963; Münster, 1969; McQuarrie, 1976; among others]. However, applications were limited to small molecules for quite a long time and it was not until the 1980's that increased computer capacity made it possible to extend the scope of Monte Carlo methods significantly. This is probably why many important algorithms were only developed relatively recently, including the Gibbs ensemble technique [Panagiotopoulos, 1987] for the computation of fluid phase equilibria, as well as special algorithms to handle dense polar fluids like water [Cracknell, Nicholson *et al.*, 1990] or flexible molecules like alkanes and polymers [de Pablo *et al.*, 1992a; Dodd *et al.*, 1993; Smit *et al.*, 1995]. Meanwhile, Monte Carlo methods had been applied to the simulation of adsorption in microporous solids. In recent years, a great deal of research has focused on improving parametrisation for calculation of the equilibrium properties of fluids in chemical engineering [Martin and Siepmann, 1998; Nath *et al.*, 1998; Errington and Panagiotopoulos, 1999; Martin and Siepmann, 1999; Spyriouni *et al.*, 1999; Nath and de Pablo, 2000; Chen *et al.*, 2000; among others].

Considering how recently it was developed, it is not surprising that Monte Carlo simulation lags behind other molecular simulation methods in terms of industrial applications. In quantum chemistry, several methods are recognised standards at the academic level. They are used for a large variety of applications at the industrial level. In the oil and gas industry, quantum chemistry methods are particularly valuable when it comes to understanding the mechanisms of heterogeous catalysis. In molecular dynamics, specific parametrisations were developed to predict the properties of various liquids [Jorgensen *et al.*, 1983; Jorgensen and Madura, 1984] and these provided the basis of the molecular simulation methods used to investigate molecular interactions in biology. Molecular simulation thus emerged in the 1990's as a key tool to developing new drugs in the pharmaceutical industry, an application for which a precise account of molecular geometry is capital. In the oil and gas industry however, there has been no comparable development, although promising research work has been made on predicting transport properties. Monte Carlo simulation has been exploited less than molecular dynamics at the industrial level although a few industrial applications can be cited in the chemical industry, namely in gas separation (adsorption in microporous solids), special chemicals (prediction of phase diagrams for dangerous compounds or mixtures) and polymeric materials. In the oil and gas industry, the possible applications of Monte Carlo cover a large range of operations including reservoir engineering, gas processing, and separations. The major objective of this book is to review potential applications in this field, in order to promote the use of these techniques in the industry. Practical aspects of molecular simulation will be discussed in detail, including the parametrisation of intermolecular interactions as well as questions of software and hardware. The book contains three major chapters.

Basic aspects of Monte Carlo simulation methods are reviewed in Chapter 2. The purpose is not to go over details of the theory underlying molecular dynamics and Monte Carlo simulation for which there are already good textbooks on statistical mechanics [Münster, 1969; McQuarrie, 1976; Hansen and McDonald, 1986] and molecular simulation [Allen and Tildesley, 1987; Rowley, 1994; Frenkel and Smit, 1996]. Our aim in this section is rather to familiarise the reader with the principles of simulation so that he or she will be able to control the key points when the time comes to application. This will lead on to a review of several numerical methods that are specific to molecular simulation techniques. It is a general feature of computer-intensive methods that numerical aspects may influence the outcome of simulation and molecular simulation is no exception to this rule. In parallel, we will briefly outline the possible consequences of increasing computer capacity on the field of molecular simulation. We will also mention the major features of Gibbs software as developed jointly by IFP, CNRS and the Université de Paris Sud. This software has been used for most of the applications given in this book. The fact that these cover a broad range of diverse problems provides evidence that Gibbs software is now a useful, multipurpose tool. Its application is not limited to oil and gas technology and it can be used to address many problems encountered in the chemical industry as well as others.

The third chapter of this volume will be devoted to the simulation of fluid phase equilibria and thermodynamic properties. Practical examples will be presented, including the investigation of pure hydrocarbons and the study of multicomponent systems including toxic gases like H_2S at high pressure. In this field, equations of state (e.g. Soave and Peng-Robinson equations) and group contribution methods (e.g. UNIFAC) are often efficient. However, these classical methods are only reliable when a great deal of experimental data have been gathered on pure compounds and representative systems. As data acquisition is long and costly, there is a need for reasonably predictive methods when data are scarce. Why consider molecular simulation when it is so complex and requires so much computing time? Three main reasons can be emphasised. Firstly, in a geometrical context, the detailed structure of each molecule may be accounted for by molecular simulation, as well as its position with respect to neighbouring molecules. Secondly and more fundamentally, molecular simulation can separate the various contributions to overall energy (electrostatic, repulsion, dispersion, etc.) in the consistent theoretical framework of statistical mechanics; if we consider the example of butane (Fig. 1.1), the *trans* conformation is more abundant than the *gauche* because it is favoured from an energy standpoint, and any average properties computed by simulation will be to some extent influenced by this molecular conformation. Finally, a third reason for using molecular simulation is its ability to describe fluid structure. This is important in polar fluids and in near-critical conditions, as will be illustrated using examples. Apart from quantitative predictions, molecular simulation is also a good way of understanding what happens at the molecular level which is often the key to the design of better processes.

Another important application in the oil and gas industry relates to adsorption equilibria in the kind of microporous adsorbents (notably zeolites) that are used for industrial separations. These applications will be treated in Chapter 4. The need for new prediction methods is probably even greater for questions of adsorption than for fluid phase equilibria because classical adsorption theories are often found wanting when it comes to predicting selectivity patterns in industrially important adsorbents. Simulating adsorption takes advantage of the fact that many algorithms and parametres are the same as in fluid systems. Also, molecular simulation is an obvious technique in zeolites because the small size of the unit cell (typically just a few nanometres) means reasonable computing times when molecular structure is detailed at the atomic level. As we will see in Chapter 4, molecular simulation is not only able to predict adsorption behaviour in a given adsorbent, but also to predict how adsorption properties will

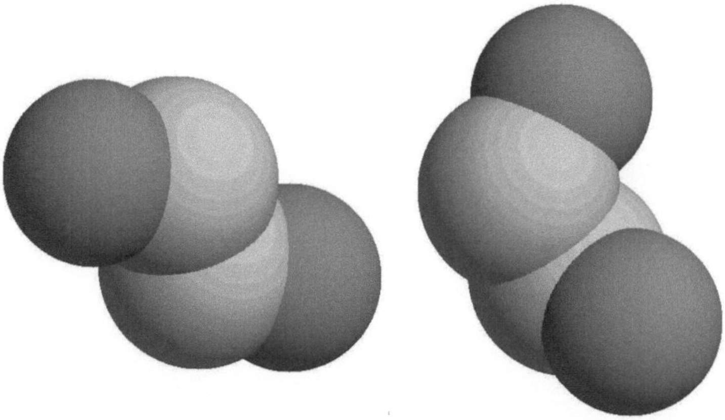

Figure 1.1 Representation of butane molecules with united atoms in two conformations. Each sphere represents either a methyl or a methylene group. The most stable conformation is the *trans* configuration (left) in which the four groups belong to the same plane. Another possible conformation is the *gauche* configuration (right) which is however less frequent because it corresponds to a higher potential energy.

change with small changes in zeolite structure. Therefore, molecular simulation may be considered as a way of understanding the basis of selectivity, and as a useful tool to guide the search for appropriate adsorbents.

2

Basics of Molecular Simulation

As outlined in the introduction, molecular simulation may be defined as a set of methods in which the individual position and conformation of every molecule in the system is explicitly accounted for. In contrast to quantum chemistry, the detailed electronic structure of each molecule is not considered in such simulations. Indeed, such detailed treatment is not relevant for thermodynamic properties, as the related molecular interactions do not involve major modifications of electronic structure. In compensation, molecular simulation can consider much larger systems, and this often makes it possible to derive macroscopic properties (i.e. that can be compared with measured quantities).

For systems at equilibrium, this connection between microscopic and macroscopic equilibrium systems is provided by the well-established framework of statistical thermodynamics. In this chapter, we will outline those basic concepts of statistical thermodynamics that are necessary for molecular simulation (Section 2.1). As the distribution of energy among molecules plays a central role in statistical mechanics, we will then review the way potential energy is modelled in molecular systems (Section 2.2). The next Section (2.3) introduces briefly the various types of Monte Carlo algorithms that will be exemplified in Chapters 3 and 4. The final Section (2.4) addresses more practical aspects of simulation, especially numerical methods, which should be known when undertaking actual simulations.

2.1 STATISTICAL THERMODYNAMICS

The link between microscopic and macroscopic properties is not straightforward, because microscopic systems behave differently.

Brownian motion is a well-known illustration of the difference between microscopic and macroscopic behaviour: particles of microscopic size make erratic movements (which can be seen in a light microscope) as a result of collisions with surrounding fluid molecules. For larger particles such as those we can see with our eyes this erratic motion is too small to be seen.

When dealing with microscopic systems, the consequences of such fluctuations may be counterintuitive. Let us consider the idealised case of a very small piston exerting pressure on a small volume of gas by way of example. If we were able to impose constant pressure in the gas through this very light piston, we would observe very rapid changes of volume (i.e. of piston position) with time (Fig. 2.1). This seems inconsistent with the well-known fact that pressure is fixed for a gas sample when you maintain its volume and temperature constant: this is the basis of equations of state, which express pressure as an analytical function of volume and temperature. The explanation is that *average* gas volume over a long enough period of time follows an equation of state, although *instantaneous* volume does not.

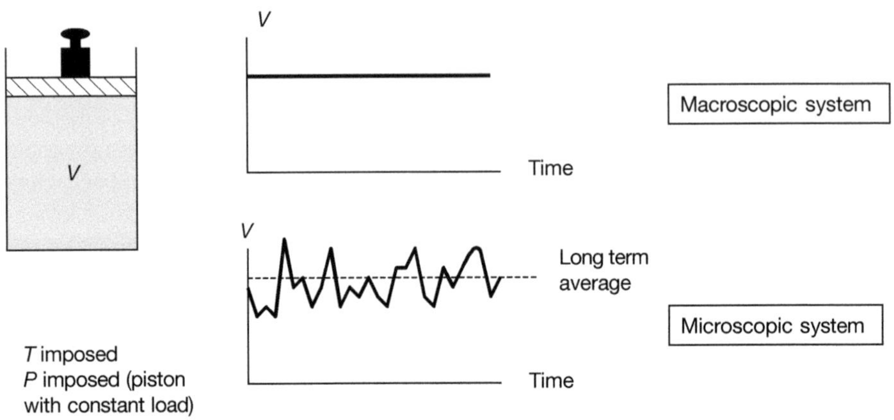

Figure 2.1 Behaviour of a gas sample under constant pressure at the macroscopic scale, where volume is constant, and at the microscopic scale, where volume fluctuations are significant.

This example illustrates the need to collect multiple snapshots of any microscopic system if a meaningful average property is to be ascertained. In statistical thermodynamics, the collection of snapshots which makes it possible to derive average properties is called a *statistical ensemble*. This is a key concept in molecular simulation.

2.1.1 Statistical Ensembles and Partition Functions

In exact terms, a statistical ensemble is a collection of various states of the system which differ *vis-à-vis* the positions and velocities of the component particles (Fig. 2.2). The space of all possible system states, which is of dimension $6N$ for N particles, is called the *phase space*. In order to obtain a statistical ensemble that is representative of a real system, it is essential that the occurrence of each state in the collection follows an appropriate probability distribution. As we will see later, there are several different types of statistical ensemble, depending on the type of system we are investigating and the conditions in which it is placed. First, we will start with the simplest case for an isothermal system, i.e. the canonical ensemble.

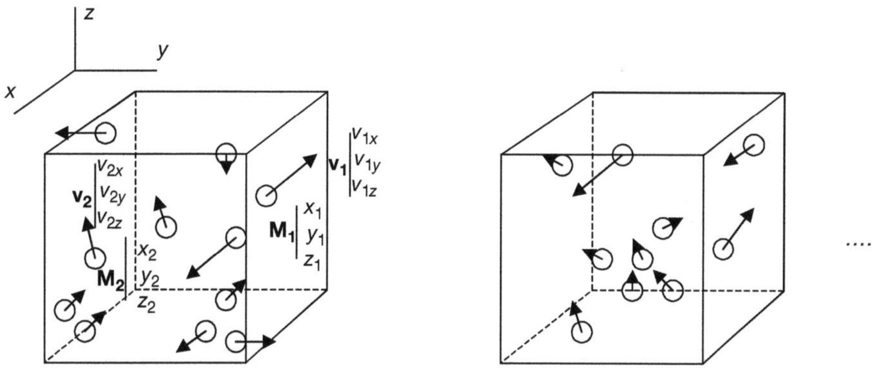

Figure 2.2 Schematic representation of a statistical ensemble in phase space, which may be seen as a collection of all possible configurations of the system when particle positions *and* velocities are varied.

A. Canonical or NVT Ensemble

If we wish to compute the properties of a system at imposed volume and temperature, we use a statistical ensemble which is called the *canonical ensemble* or *NVT* ensemble. The number of molecules N and the global volume V are identical for all system states belonging to the ensemble, but they differ in total energy which is a fluctuating variable in this ensemble (Fig. 2.3). When we say that the temperature T is imposed, it does not mean that every system state is at temperature T, but that the energies of the system states are displaying a specific distribution. It may be shown indeed that each state j of the canonical ensemble occurs with a probability proportional to $\exp(-E_j/kT)$ where k is the Boltzmann constant ($k = 1.381 \times 10^{-23}$ J.K^{-1}) and E_j is the total energy (kinetic + potential) of the system in state j.

The expression $\exp(-E_j/kT)$ is named the *Boltzmann factor* after the Austrian scientist who discovered it in the latter part of the nineteenth century. It expresses the concept that low energy states are favoured compared with high energy states. It also expresses the idea that increasing temperature broadens the energy distribution in the ensemble with the consequence that the average energy is increased. A general notation used in statistical mechanics is:

$$\beta = 1/kT \tag{2.1}$$

Thus, the probability of a given state j in the canonical ensemble is given by:

$$P_j = \frac{\exp(-\beta E_j)}{Q_{NVT}} \tag{2.2}$$

where

$$Q_{NVT} = \frac{1}{N!} \sum_{r_i} \sum_{p_i} \exp(-\beta E(r_i, p_i)) \tag{2.3}$$

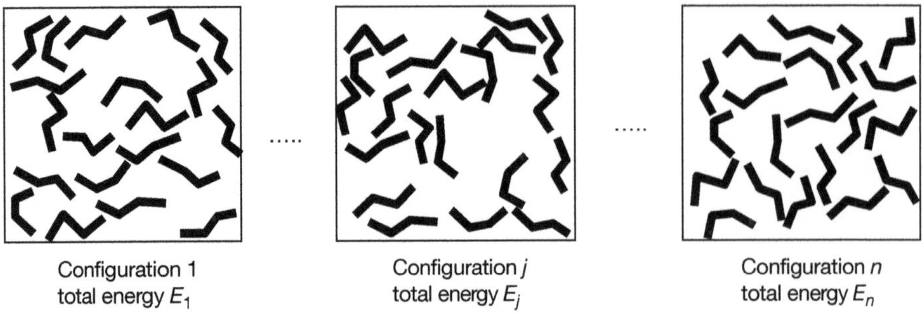

Figure 2.3 Schematic representation of the canonical statistical ensemble in the case of *n*-butane. The number of molecules, the volume and the temperature are imposed. The ensemble may be seen as a collection of system states differing by molecular positions, internal conformations and energy.

The expression Q_{NVT}, known as the *partition function*, is simply the sum of the Boltzmann factors for all possible different states in the phase space. In its expression, the factor $N! = 2 \times 3 \times 4 \times ... \times (N-1) \times N$ originates from the possible combinations of N identical particles which correspond to the same state.

As a result, all probabilities P_j sum to unity. Incidentally we may notice that these sums converge for a system of finite size because the number of states, although very large, is finite – this is a consequence of quantum mechanics. Consistently with quantum mechanics, the finite summation (2.3) may be transformed in a continuous integral:

$$Q_{NVT} = \frac{1}{h^{3N} N!} \int_{r_i} \int_{p_i} \exp(-\beta E(r_i, p_i)) \, dr_i \, dp_i \tag{2.4}$$

where the factor h^{3N} may be understood as the volume of an individual quantum state in the phase space (involving the Planck constant $h = 6.626\ 10^{-34}\ \text{J} \cdot \text{s}$).

Although it is generally not explicitly computed in molecular simulations, the partition function is a very important theoretical concept in statistical thermodynamics which provides the link with key thermodynamic functions. In the case of the canonical ensemble, the Helmholtz free energy A is expressed as:

$$A = -kT \ln Q_{NVT} \tag{2.5}$$

Using this relationship, numerous properties can be derived. The purpose of this chapter is not to develop such derivations, which are well explained in textbooks (see for instance [Mc Quarrie, 1976]). We will thus jump to the relationships that are directly applicable.

Let us now assume that we have a finite collection of system states which is representative of the *NVT* statistical ensemble, where the occurrence of each state is proportional to the Boltzmann factor, i.e. $\exp(-\beta E_j)$ (we will see in Section 2.3 how to build such a collection). Average properties can be obtained by simple arithmetic averaging over the n states composing the statistical ensemble. As is common in statistical mechanics, we will use the notation $\langle X \rangle$ for such averages:

$$\langle X \rangle = \frac{1}{n} \sum_{i=1}^{n} X_i \tag{2.6}$$

In this expression the summation runs over the n system replicas composing the statistical ensemble. For instance, the average energy is:

$$\langle E \rangle = \frac{1}{n} \sum_{i=1}^{n} E_i \tag{2.7}$$

Other statistical ensembles can be defined in the same way as the canonical ensemble – depending on the kind of problem being addressed – as summarised in Table 2.1. It may be observed from this table that, when an intensive variable is fixed, the associated extensive variable fluctuates:

- if temperature is fixed, energy fluctuates,
- if pressure is fixed, volume fluctuates,
- if chemical potential is fixed, the corresponding number of molecules fluctuates.

Table 2.1 Statistical ensembles.

Statistical ensemble	Imposed variables	Associated thermodynamic potential	Applications
Canonical ensemble	N, V, T	$A = E - TS$ (Helmholtz free energy)	Phase properties (P, H, C_v, μ...)
Grand Canonical ensemble	μ_i, V, T	PV (i.e. $E - TS - \Sigma\mu_i N_i$)	Adsorption isotherms, selectivities
Isothermal – isobaric ensemble	N, P, T	$G = H - TS$ (Gibbs free energy)	Phase properties (H, C_p, ρ, μ...)
Gibbs ensemble at imposed global volume (m phases)	$N = N_1 + \ldots N_m$, $V = V_1 + \ldots V_m$, T	$A = E - TS$ (Helmholtz free energy of the whole system)	Phase equilibrium of pure components and mixtures
Gibbs ensemble at imposed pressure (m phases)	$N = N_1 + \ldots . N_m$, P, T	$G = H - TS$ (Gibbs free energy of the whole system)	Phase equilibrium of mixtures

B. NPT Ensemble

If we want to simulate a system at imposed pressure and temperature, we use the *isothermal-isobaric* ensemble or *NPT* ensemble, where volume and energy are fluctuating variables (Fig. 2.4). The probability of any given state – with total energy E_j and volume V_j – is proportional to:

$$\exp(-\beta E_j - \beta P V_j) \tag{2.8}$$

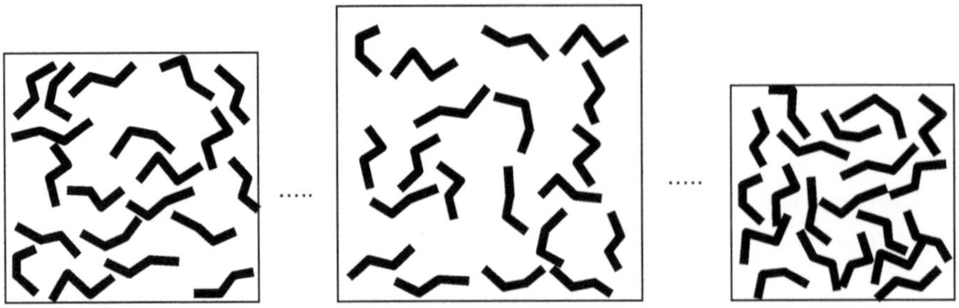

Figure 2.4 Schematic representation of the *NPT* statistical ensemble in which the number of molecules, the pressure and the temperature are imposed. The ensemble is composed of system states differing by molecular positions, internal conformations, velocities, energy and volume.

This ensemble may be used to compute the average energy using Eq. (2.6) and the average volume using:

$$\langle V \rangle = \frac{1}{n} \sum_{j=1}^{n} V_j \tag{2.9}$$

C. Grand Canonical or μVT Ensemble

If we want to simulate adsorption isotherms, either for pure compounds or multicomponent systems, we must be able to impose temperature and partial pressures to simulate a given point on the isotherm. For this purpose, the most convenient statistical ensemble is the *Grand Canonical ensemble,* also termed the μVT ensemble. In this ensemble, the chemical potential μ_i of every species i is specified as well as volume and temperature, while mole numbers fluctuate (Fig. 2.5). Indeed, specifying the chemical potential μ_i is equivalent to specifying the partial pressures prevailing in the gas phase in equilibrium with the adsorbent.

The chemical potential of each species appears explicitly in the probability of each system state in the Grand Canonical ensemble:

$$\exp(-\beta E_j + \beta N_{1j}\mu_1 + \beta N_{2j}\mu_2 ...) \tag{2.10}$$

In a grand canonical simulation, the main result is the average number of molecules $< N_{ij} >$ in the system for every species i. In the case of microporous adsorbents, this is the average number of adsorbed molecules. It is therefore easy to compare grand canonical simulation results with adsorption isotherms. It is important to make it clear that, for adsorption applications, the system considered comprises not only the adsorbed molecules but also the adsorbent. We will see in more detail how the Grand Canonical ensemble can be simulated

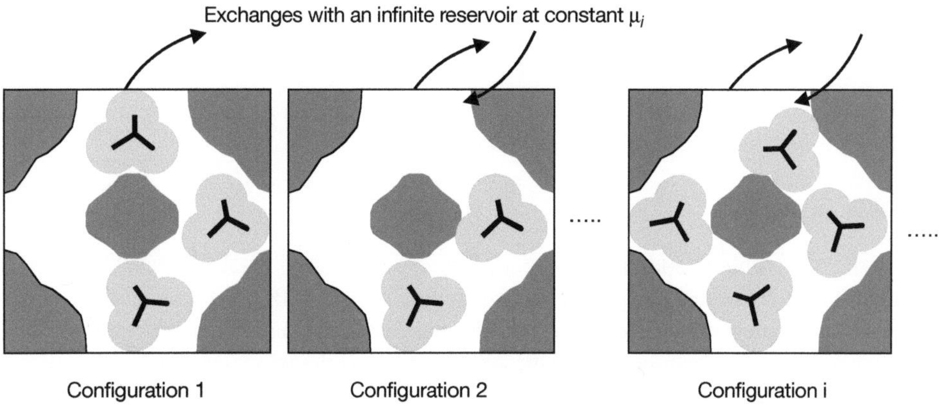

Figure 2.5 Schematic representation of the Grand Canonical ensemble in the case of isobutane molecules adsorbed in a microporous adsorbent. The imposed variables are the volume, the chemical potential of isobutane and the temperature. The system states composing the ensemble differ by the number of molecules, the molecular positions, the velocities and the energy.

by Monte Carlo techniques in Section 2.3.3 and how the adsorbent is taken into account in Section 2.4.2. Care must be also taken that the reference state for the chemical potential is not the same when defining the statistical ensemble with Eq. (2.10) as in classical thermodynamics. This point will be discussed in the context of the evaluation of chemical potential in Section 2.3.5.

D. Microcanonical or NVE Ensemble

The microcanonical or *NVE* ensemble, in which the number of molecules, global volume and internal energy are fixed, does not correspond to conditions often encountered in practice. However, this ensemble is important because it is naturally obtained by solving the equations of motion in molecular dynamics, as will be briefly discussed in Section 2.1.4. In this ensemble, every system state has the same probability. Temperature fluctuates, and the average temperature may be obtained by considering the average kinetic energy per degree of freedom of the system:

$$T = \frac{2\langle K \rangle}{k N_f} \tag{2.11}$$

where K is the kinetic energy of the system, expressed as a function of the masses m_i and the velocities v_i of its constituent particles:

$$K = \sum_{i=1}^{n} \frac{m_i v_i^2}{2} \tag{2.12}$$

and N_f is the total number of degrees of freedom, which can be obtained from

$$N_f = 3N - N_c \tag{2.13}$$

where N is the total number of atoms and N_c is the total number of independent constraints such as imposed bond lengths or imposed bond angles.

E. Gibbs ensemble

In theory, it is possible to address phase equilibrium, using either the *NVT* or *NPT* statistical ensemble. Then, phase separation would occur spontaneously within the system if this corresponds to maximum stability. However, it is obvious that in a simulation box with a few hundred molecules (such as can be dealt with within a reasonable computer time), a significant proportion of the molecules is close to the interface (Fig. 2.6). In these conditions, deriving bulk properties which are unaffected by interface considerations may be difficult. Direct simulation of the interface (e.g. [Goujon *et al.*, 2001]) is thus mostly used to investigate the properties of the interface itself, such as surface tension.

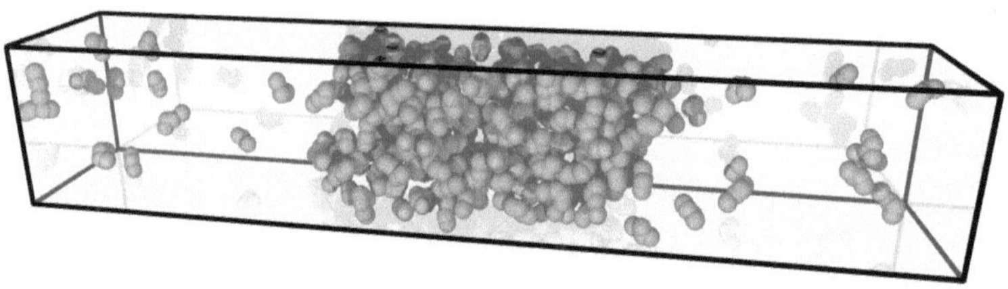

Figure 2.6 Direct simulation of a liquid-vapour interface in the case of *n*-pentane. Although the simulation box has been elongated, a significant fraction of molecules belongs to the interface (courtesy of P. Malfreyt).

A more efficient way of computing phase equilibria by molecular simulation uses the Gibbs ensemble [Panagiotopoulos, 1987], in which a separate simulation box is used for each phase without any explicit interfaces between fluid phases (Fig. 2.7). In this ensemble, both temperature and the total number of molecules are fixed, and we can impose either global volume (i.e. the sum of phase volumes) or pressure. Mole numbers therefore fluctuate in opposite directions in each phase. It may be demonstrated that chemical potentials are equal in both phases in this ensemble.

When pressure is imposed in the Gibbs ensemble, phase volumes fluctuate independently. This option can be used only when more than one component is considered. In fact, both phases would converge toward the same density in the case of a pure compound.

When global volume is imposed in the Gibbs ensemble, phase volumes fluctuate in opposite directions. This option is applicable either to single or multicomponent systems. In the case of a pure component, it allows determination of its equilibrium properties.

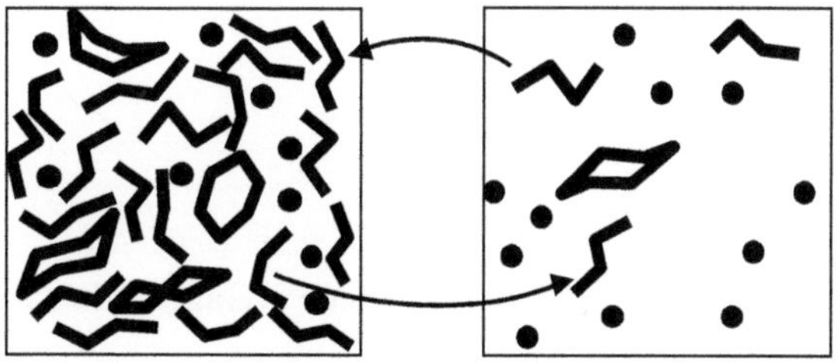

Figure 2.7 Liquid and vapour simulation boxes without interface used in Gibbs ensemble calculations. Molecules are transferred from the liquid to the vapour and *vice-versa*, so that the total number of molecules of each species remains constant.

The Gibbs ensemble is a particular case of the *NVT* or *NPT* statistical ensemble in which interface energy is not accounted for. This is consistent with the usual macroscopic definition of phase equilibrium which considers that interface energy does not have any influence, or that there is a planar interface (which is an equivalent assumption).

Phase volumes and energies may be derived by applying the usual averaging procedure defined by Eq. (2.6). Pressure may be computed for each phase as discussed in the next section. This may be used to check that mechanical equilibrium is reached, because pressure must be equal in both phases (to within the statistical uncertainties). When global volume is imposed, it provides the desired equilibrium pressure.

2.1.2 Determination of Average Properties

Average pressure is obtained by the following expression, where the subscript j refers to the jth configuration of the statistical ensemble:

$$\langle P \rangle = \frac{NkT}{V} + \langle W_j \rangle \tag{2.14}$$

where W_j is a term called the *Virial*, which accounts for intermolecular interactions through the forces $\vec{F}_k$ acting on molecule k:

$$\langle W_j \rangle = \frac{1}{3V} \sum_k \vec{r}_k \cdot \vec{F}_k \tag{2.15}$$

where $\vec{r}_k$ is the position of the molecular centre of mass.

The Virial may be equivalently expressed in terms of *atomic* instead of *molecular* forces and coordinates [Akkermans and Ciccotti, 2004]. If we recall that $k = R/N_a$, where $N_a = 6.022\ 10^{23}\ \text{mol}^{-1}$ is the Avogadro number, this expression reduces to the perfect gas law ($PV = RT$ for one mole) in the absence of intermolecular interactions. This is not surprising, since every fluid obeys the perfect gas law at very low densities, a state where intermolecular interactions (and thus intermolecular forces) tend to zero.

The average density (in molecules per unit volume) is derived using:

$$\rho = \frac{N}{\langle V \rangle} \tag{2.16}$$

It might also be written $\rho = \left\langle \frac{N}{V} \right\rangle$, i.e. the arithmetic average of fluctuating densities which is slightly different from Eq. (2.16) in the general case. The basic reason for preferring Eq. (2.16) is that the arithmetic average (2.6) applies only to *extensive* variables, i.e. those that vary linearly with the size of the system, such as energy, volume and number of molecules (this is indeed supported by the consideration of a super configuration constructed by gathering all the configurations of the statistical ensemble). Nevertheless both ways lead to similar averages when fluctuations are not larger than a few percentage points.

In statistical ensembles where temperature is imposed, deriving the average energy of the system through Eq. (2.7) is straightforward. However, it is worth commenting on the reference state for energy. Zero energy corresponds to zero kinetic energy and zero potential energy, i.e. the conditions of a dilute gas at 0 K. Once energy is known, enthalpy may be easily derived through $\langle H \rangle = \langle E \rangle + \langle P \rangle V$ in the case of the canonical ensemble or $\langle H \rangle = \langle E \rangle + P \langle V \rangle$ in the case of the *NPT* ensemble.

It is also theoretically possible to derive the chemical potential, as it is defined as a derivative of the Helmholtz energy:

$$\mu_i = \left(\frac{\partial A}{\partial N_i} \right)_{V,T,N_{j,j \neq i}} \tag{2.17}$$

which is connected to the partition function of the statistical ensemble through Eq. (2.5). However, it cannot be obtained simply as the average of a microscopic property. Its determination involves a specific procedure with test insertions of an additional molecule. For this reason, it will be discussed later together with Monte Carlo techniques (Section 2.3.5).

When comparing simulation results with macroscopic measurements for extensive properties, it is more convenient to express them on a molar basis. For instance, the molar volume v may be computed from a *NPT* simulation using:

$$v = \frac{N_a}{N} \langle V \rangle \tag{2.18}$$

The same conversion may be applied to other extensive variables like energy and enthalpy. Of course, intensive variables such as pressure do not need any correction factor to be compared with measurements, provided a consistent unit system is used.

2.1.3 Determination of Derivative Properties from Fluctuations

The information that can be derived from a statistical ensemble does not consist only in the average values of energy, volume, pressure, etc. The amplitude of the *fluctuations* around the average makes it possible to determine partial derivatives of the property considered. For instance, the standard deviation of energy in the canonical ensemble is linked with the partial derivative of energy versus temperature, i.e. the heat capacity C_v of the system:

$$C_v = \left(\frac{\partial E}{\partial T}\right)_V = \frac{1}{kT^2}\left(\langle E^2\rangle - \langle E\rangle^2\right) \tag{2.19}$$

In this expression, the right hand side represents energy fluctuations around the mean value which can be computed by molecular simulation. Here again, a conversion factor N_a/N must be applied to obtain molar heat capacity, and the reference state is the dilute gas at 0 K. Similarly, fluctuations in volume in the *NPT* ensemble may be used to obtain the compressibility coefficient:

$$\beta_T = -\frac{1}{V}\left(\frac{\partial V}{\partial P}\right)_T = \frac{1}{kT}\left(\langle V^2\rangle - \langle V\rangle^2\right) \tag{2.20}$$

In Section 2.4.8, we will discuss the application of such fluctuation formulae to the evaluation of other derivative properties like thermal expansivity and the Joule-Thomson coefficient.

2.1.4 Possible Ways of Simulating Ensembles

Conceptually, the simplest way to build a statistical ensemble is to solve the equations of motion for every atom in every molecule, accounting for intramolecular and intermolecular interactions inside the system: this way is known as *molecular dynamics*. An alternate way, known as Monte Carlo simulation, consists in using statistical methods to generate the collection of configurations with the desired probability distribution without solving the equations (Fig. 2.8). It has been shown that both methods yield identical ensemble averages (this important result is known as the ergodicity theorem). This book focuses on Monte Carlo simulation but molecular dynamics will be briefly discussed in order to make the reader aware of the principal differences between the two methods.

The Newton equations of motion, which have to be integrated in molecular dynamics, are expressed in the form:

$$\vec{F}_i = m_i \frac{d^2\vec{r}_i}{dt^2} \tag{2.21}$$

where

$\vec{r}_i$ is the position of an atom, i.e. a three dimensional vector (x_i, y_i, z_i),

$\vec{F}_i$ is the force acting on the atom from the other atoms of the system, which can be computed by deriving the potential energy U versus coordinates:

$$\vec{F}_i = \vec{\nabla}(U(\vec{r}_i)) \tag{2.22}$$

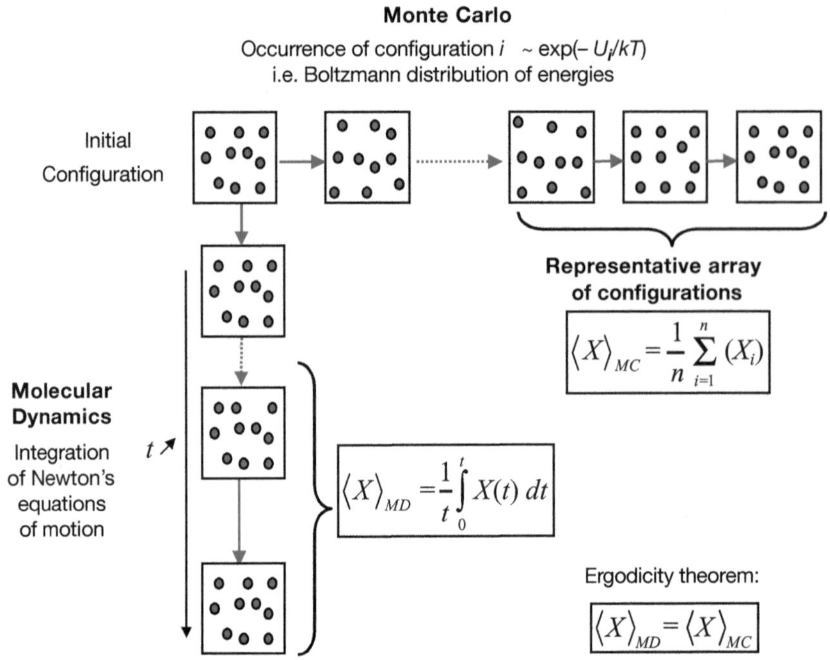

Figure 2.8 Comparison of molecular dynamics and Monte Carlo simulation in the perspective of simulating the canonical ensemble.

The solution of the equations of motion is such that total energy (i.e. the sum of kinetic and potential energy) is constant. Therefore, molecular dynamics simulates the microcanonical or *NVE* ensemble, unless specific steps are taken to force energy or volume to fluctuate.

In order to perform a molecular dynamics calculation at imposed temperature, the kinetic energy is changed by altering center of mass velocities at regular intervals: this is the purpose of thermostat techniques, which allow simulation in the *NVT* ensemble [Nosé, 1984; Hoover, 1985]. If a simulation is desired at imposed pressure, a specific technique has to be used to allow for volume variations ([Evans and Morris, 1983], among others). The aim of this type of technique (which are known as pressostat or barostat methods) is to obtain the correct probability distribution Eq. (2.8). It is the same as equilibrating the system with a pressure reservoir of infinite size. As a result, it is possible to simulate the *NVT* and *NPT* statistical ensembles with molecular dynamics.

It is uneasy to simulate open ensembles (Gibbs ensemble, Grand Canonical ensemble) with molecular dynamics because molecule insertions or deletions would cause excessive discontinuities in the integration of Newton's equations (at least in dense phases). As will be amply shown later, Monte Carlo is the ideal technique for such ensembles.

Molecular dynamics is most useful for simulating the dynamic behaviour of a system, particularly its transport properties. By transport properties, we mean here the coefficients which govern the rate of mass transfer (diffusion coefficients), heat transfer (thermal conductivity) or momentum transfer (viscosity). For instance, the self-diffusion coefficient of a component may be expressed from the mean squared displacement, according to:

$$D = \frac{\left\langle \left(r(t) - r(t_o) \right)^2 \right\rangle}{6(t - t_o)} \tag{2.23}$$

There are also ways to derive other transport coefficients (shear viscosity and thermal conductivity) using molecular dynamics. Some of these methods rely on simulations performed at equilibrium in the *NVE*, *NVT* or *NPT* ensemble [Dysthe *et al.*, 1999a, 1999b and 2000] but it is also possible to use non-equilibrium ensembles, in which perturbations or specific boundary conditions are imposed. As molecular dynamics uses the same type of potential energy models as the Monte Carlo techniques, it is clear that molecular simulation provides a consistent theoretical framework for transport and phase equilibrium properties. Although transport properties are not considered in this book, the perspective of related applications in this field is an additional reason for learning about molecular simulation.

2.2 POTENTIAL ENERGY OF MOLECULAR SYSTEMS

As we have seen in the previous section, energy is a central concept in statistical mechanics. The total energy E referred to in the previous section is the sum of the potential energy U and the kinetic energy K as defined by Eq. (2.12):

$$E = U + K \tag{2.24}$$

As will be seen in the next section, Monte Carlo simulation does not need to consider kinetic energy, the influence of which is accounted for by a suitable analytical integration. However, the potential energy of many system configurations needs to be evaluated to build a statistically representative ensemble. It is therefore essential that potential energy be accurately determined (in a reasonable computer time) in the course of a simulation. This is the role assigned to the *molecular potential*. More precisely, the role of the molecular potential is to provide a realistic prediction of the potential energy of the system as a function of atomic coordinates without having to solve the Schrödinger equation. Indeed, solving this equation accurately enough for a statistically representative simulation box containing several hundreds or thousands of atoms (such as are commonly addressed) is still impossible without excessive computer time.

2.2.1 Standard Decomposition of the Potential Energy

The potential energy U_{tot} of a group of molecules is classically subdivided according to two contributions:

$$U_{\text{tot}} = U_{\text{ext}} + U_{\text{int}} \tag{2.25}$$

where

U_{ext} is the external energy, also termed the intermolecular energy, i.e. the energy arising from the interaction between distinct molecules,

U_{int} is the intramolecular energy, which results from the interactions between the atoms belonging to the same molecule.

A. External Energy

Classically, the external (or intermolecular) energy is split up into a sum of four terms, as a result of a perturbation expansion of electronic charge distribution in the interacting molecules:

$$U_{\text{ext}} = U_{\text{el}} + U_{\text{pol}} + U_{\text{disp}} + U_{\text{rep}} \tag{2.26}$$

where

U_{el} is the electrostatic potential energy, which originates from the Coulombic forces between the permanent components of the electronic charge distribution,

U_{pol} is the polarisation energy, always attractive, which originates from the Coulombic interaction between 1° the charge distribution induced in a molecule by the electric field created by the permanent charge distribution of the surrounding molecules and 2° the permanent charge distribution of the surrounding molecules,

U_{disp} is the dispersion energy, always attractive, which is the Coulombic interaction between the fluctuating components of the charge distribution of the molecules,

U_{rep} is the repulsive energy, which prevents the molecules from overlapping significantly, as a result of the impossibility of electrons to occupy the same state.

B. Intramolecular Energy

The intramolecular energy U_{int} is expressed as the sum of the following contributions:

$$U_{\text{int}} = U_{\text{str}} + U_{\text{bend}} + U_{\text{tors}} + U_{\text{dn}} \tag{2.27}$$

where

U_{str} is the stretching energy, associated with the variations of bond length,

U_{bend} is the bending energy, arising from the variations of the angle formed by two successive chemical bonds,

U_{tors} is the torsion energy caused by the variations of the dihedral angles formed by four successive atoms in a chain,

U_{dn} is the distant neighbour energy resulting from the interaction between atoms separated by more than three chemical bonds.

In the following subsections, it will be seen how the various contributions to the external and internal energy are obtained in practice. Although dispersion energy is often the leading term among attractive energies, we will start with the electrostatic and polarisation energies, because this will help us introduce the basic notions underlying the theory of dispersion forces.

2.2.2 Electrostatic Energy

The permanent electronic charge distribution in a molecule is generally approximated by placing several partial charges (generally less than the charge of a single electron) at selected sites in the molecule. The location of these partial charges often coincides with atomic nuclei, but additional charges may be also placed at other sites to provide a better model. The amplitude and location of these charges may be determined by two ways. The first way – which is only valid for simple molecules – involves selecting charges on atomic nuclei so that they match the dipole or quadrupole moment. For instance, a molecule of CO_2 is represented by three partial charges (Fig. 2.9). As a result of symmetry in this example, the partial charges produce no dipole moment, but only a quadrupole moment agreeing with experimental measurements (the definitions of dipoles and quadrupoles will be reviewed below). The second way involves optimising the location and amplitude of partial charges in such a way that they produce an electrostatic field in agreement with quantum mechanical calculations at some distance from the molecule [see for instance Delhommelle *et al.*, 2000]. We will present this approach by way of examples in Section 3.

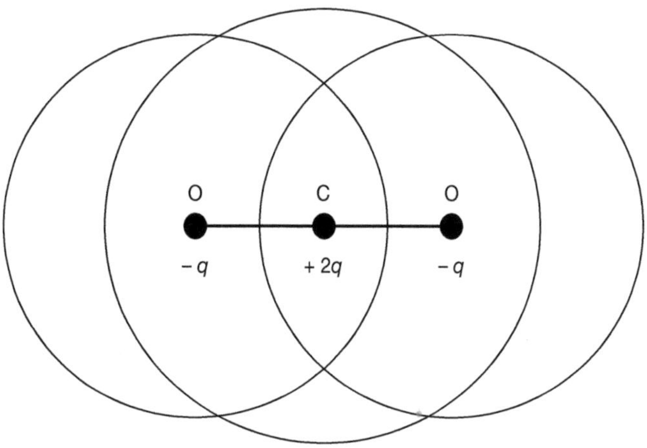

Figure 2.9 Representation of the carbon dioxide molecule in the model of Harris and Yung [1995], involving three electrostatic point charges located on the atom nuclei. The circles indicate the repulsion spheres of the Lennard-Jones potential.

On the basis of such discrete charge models, the electrostatic potential energy is computed from Coulomb's law:

$$U_{\mathrm{el}} = \sum_{\substack{i,j \\ i<j}} \frac{1}{4\pi\varepsilon_o} \frac{q_i q_j}{r_{ij}} \tag{2.28}$$

where the sum includes all possible pairs i, j of partial charges q_i, q_j, r_{ij} is the separation distance and $\varepsilon_o = 8.85419\ 10^{-12}\ C^2N^{-1}m^{-2}$.

If the related charges are of opposite signs, a negative (i.e. attractive) energy develops, while a positive (repulsive) energy corresponds to charges of the same sign. In molecular simulation, partial charges q_i are generally expressed in multiples of the electron charge ($1.6022\ 10^{-19}$ C). Unless ions are considered, the sum of partial charges belonging to a given molecule is zero.

For small molecules which are considered to be rigid, the summation (2.28) does not include pairs of charges associated with the same individual molecule, because this would have no physical meaning. However, for large flexible molecules containing more than one polar group (e.g. water-soluble polymers), the electrostatic forces between remote groups on a single molecule will need to be considered.

A characteristic feature of electrostatic energy is its long range of interaction, as it decreases relatively slowly with separation distance compared with repulsion and dispersion forces. As will be seen in Section 2.4.1, this requires specific numerical methods on boxes of finite dimensions, because the range of interaction is commonly greater than the box size.

A. Dipole Moment

With polar molecules (i.e. those with a significant electric dipole moment), electrostatic interactions make a major contribution to energy. The electric dipole moment of a molecule bearing opposite charges q and $-q$ separated by a distance d is defined as:

$$\mu_D = q\,d \qquad (2.29)$$

The classical unit for dipole moment is the Debye ($D = 3.334\ 10^{-30}$ C·m). A dipole moment can be also defined for molecules presenting a more complex charge distribution [Rivail, 1994]. In practical terms, when molecules possessing a dipole moment are placed in an electric field, they tend to align in a certain direction as if they were a small dipole made of two opposite charges separated by a fixed distance.

As seen in Table 2.2, noble gases (He, Ar) and saturated hydrocarbons (methane, ethane, ...) show zero or small dipole moments. In this case, the electrostatic energy is sufficiently small that no partial charges need to be considered in these molecules. As a consequence of Coulomb's law, the electrostatic energy between a non-polar molecule and any other molecule (polar or non-polar) is also considered to be zero when this approximation is made.

For highly polar molecules (i.e. those displaying a high dipole moment, such as water or alcohols), the electrostatic energy is a major component which must be carefully accounted for. Here, we will not further discuss when electrostatic energy has to be accounted for, as this topic will be addressed in more detail in Section 2.4.3.

B. Higher Order Moments

Some molecules whose symmetry precludes their having a dipole moment show evidence of a strongly heterogeneous intramolecular charge distribution. This is typical of molecules displaying point symmetry, such as CO_2, N_2, benzene, or paraxylene. In this case, electric moments of higher order are defined to describe their behaviour. More precisely, the quadrupole moment of a system of three aligned charges $(-q, 2q, +q)$ separated by a distance d is given by:

$$Q = -\frac{1}{2} q d^2 \tag{2.30}$$

For instance, CO_2 electrostatics may be represented by three partial charges as shown in Figure 2.9, where $Q = -14.3\ 10^{-40}\ \text{C}\cdot\text{m}^2$ [Graham *et al.*, 1998]. Similarly to the dipole moment, the quadrupole moment can be defined for any molecule presenting more complex charge distribution [Rivail, 1994].

Table 2.2 Dipole moments and polarizability of selected molecules.

	Molecule	Dipole moment (D)	Polarizability (Å^3)
Normal and branched alkanes	Methane	0	2.59
	Ethane	0	4.45
	Propane	0.084	6.29
	n-butane	0	8.22
	Isobutane	0.132	8.14
Olefins	Ehylene	0	4.25
	Propene	0.366	6.26
	1-butene (*cis* and skew)	0.36-0.44	7.97-8.52
	Isobutene	0.503	8.29
Aromatics	Benzene	0	10.3
	Toluene	0.375	12.28
	o-xylene	0.64	14.9
	p-xylene	0	14.9
Inorganic gases	He	0	0.205
	Ar	0	1.64
	N_2	0	1.74
	CO_2	0	2.91
	H_2S	0.97	3.78
Water and alcohols	H_2O	1.85	1.45
	Methanol	1.70	3.23
	Ethanol	1.69	5.11
	Propan-1-ol	1.56	6.74
	Butan-1-ol	1.66	8.88

$1\text{D} = 3.334\ 10^{-30}\ \text{C}\cdot\text{m}$

Source: Lide, D. R., Ed. (1992). *Handbook of Chemistry and Physics*. Boca Raton, CRC Press.

2.2.3 Polarisation Energy

If a molecule is placed within the electrostatic field E created by other molecules in the system, its internal charge distribution will change so that an additional – or induced – dipole moment μ_p is created:

$$\mu_p = \alpha_p E \tag{2.31}$$

where α_p is the polarisability of the molecule, and E the electric field. As a general rule, polarisability increases with the diameter of the electronic cloud, i.e. with molecular size. Within a first order approximation, the electric field may be obtained by applying Eq. (2.28):

$$E \approx \sum_j \frac{1}{4\pi\varepsilon_o} \frac{q_j}{r_j^2} \tag{2.32}$$

where the summation covers all the system charges apart from those of the molecule being considered, r_j being the distance between the polarised molecule and charge q_j.

The interaction energy of the induced dipole moment with the electrostatic field provides the polarisation energy:

$$U_{\text{pol}} = -\frac{1}{2}\alpha_P E^2 \tag{2.33}$$

Potential energy calculations may be time-consuming when polarisability effects are great because the approximation of Eq. (2.32) is no longer valid, and the electric field determination must take into account the influence of induced dipoles (the correction is often called back polarisation).

As a general rule, the polarisation energy is significant only when polar molecules or solids are present in the system because otherwise the electric field is not strong enough. In this case, non-polar molecules may also contribute to polarisation energy because their polarisability is significant. Indeed, there is no direct relationship between polarity and polarisability (Table 2.2). For instance, the polarisability of methane (2.59 $Å^3$) is greater than the polarisability of water (1.45 $Å^3$), although methane is non-polar while water has a large dipole moment.

The polarisability of many small molecules can be simply taken from experimental measurements. For complex molecules, it is found that polarisability may be often determined from group contributions, as illustrated by the linearity of the plots shown in Figure 2.10.

2.2.4 Dispersion and Repulsive Energy

Dispersion energy is often the main source of intermolecular attractive forces which explain the cohesion of liquids. It is also often a significant part of the adsorption energy in microporous solids. Repulsion prevents molecules from overlapping as they would do in liquids in the presence of attractive forces alone.

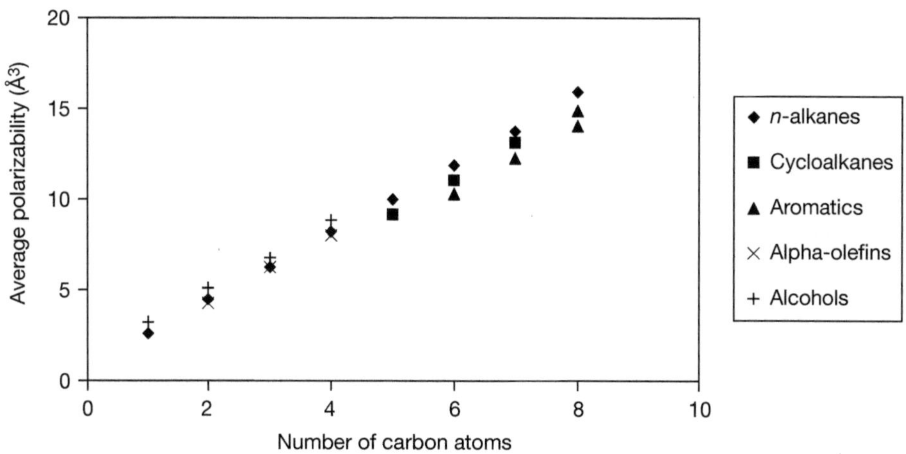

Figure 2.10 Dipole polarisability of hydrocarbons and alcohols versus carbon number.

A. Source of Dispersion Energy

A complete treatment of dispersion energy includes interactions between fluctuating dipoles, fluctuating quadrupoles, and higher moments of the electronic charge distribution. Also, dispersion energy does not include only pairwise interactions (i.e. involving two nuclei) but also three-body interactions, i.e. between three nuclei.

In these contributions, the leading term (which corresponds to the interaction between fluctuating dipoles) decreases with the sixth power of the distance separating the related nuclei:

$$U_{\text{disp}} = -\frac{3}{2}\left(\frac{E_i E_j}{E_i + E_j}\right)\frac{\alpha_i \alpha_j}{\left(4\pi\varepsilon_o\right)^2}\frac{1}{r_{ij}^6} \tag{2.34}$$

where α_i and α_j are the polarisabilities of the atoms, E_i and E_j the energies of first electronic transition in both atoms, and r_{ij} the separation distance.

When fluctuating quadrupoles and higher moments are included, higher order terms (e.g. $1/r_{ij}^8$, $1/r_{ij}^{10}$, etc.) are figured into the expression (2.34) but they are often neglected. Three-body contributions amount to 5-10% of the interaction in liquids [Axilrod and Teller, 1943], but they are generally not explicitly taken into account. These neglected terms are thus implicitly taken into consideration through the empirical calibration of effective dispersion parameters, using the general dependence of Eq. (2.34) with separation distance. Several examples of such calibrations will be presented in applications in this book.

An interesting question is whether the dispersion parameters could be derived by quantum mechanical calculations instead of being calibrated empirically. This is indeed possible for simple molecules with very few electrons and simple structure like H_2 or noble gases. It is, however, not yet possible for more complex molecules because it requires extremely accurate quantum calculations.

B. Source of Repulsive Energy

Repulsive energy results from the overlap between electronic charge distributions, which is limited by a basic principle of quantum chemistry, the Fermi exclusion principle. However, there is no theoretical expression comparable to Eq. (2.34) for repulsive energy. As a result, numerous empirical potentials have been proposed for repulsion, such as exponential $\exp(-b r_{ij})$ or power laws $1/r_{ij}^n$ with $n \geq 9$.

C. Lennard-Jones Potential

The most widespread expression of dispersion and repulsive energy is the Lennard-Jones potential, in which the repulsive energy varies with the 12th power and dispersion with the 6th power of separation distance. The Lennard-Jones interaction energy between two atoms (or more generally, between two force centres) is written as:

$$U_{LJ} = U_{rep} + U_{disp} = 4\varepsilon\left(\left(\frac{\sigma}{r}\right)^{12} - \left(\frac{\sigma}{r}\right)^{6}\right) \qquad (2.35)$$

where σ is the separation distance for which attractive and repulsive terms are exactly opposite making the Lennard-Jones interaction zero, and ε is the depth of the minimum interaction energy which corresponds to a separation distance $r_{min} = 2^{1/6}\sigma \approx 1.1225\sigma$ (Fig. 2.11). The attraction parameter ε is often given as ε/k, which has the dimension of temperature.

If we first consider the case of systems involving identical atoms, the parameters σ and ε may be considered as characteristic of the atom (or molecule) being considered. They are calibrated against experimental data such as critical temperature and density. The Lennard-Jones potential appears to account fairly well for the behaviour of simple fluids (Ar, Kr, ...) and molecular non-polar fluids with approximate spherical symmetry (methane, isopentane., etc.).

In the case of a system composed of different types of atom, the total Lennard-Jones energy is:

$$U_{LJ} = \sum_{\substack{i,j \\ i<j}} 4\varepsilon_{ij}\left(\left(\frac{\sigma_{ij}}{r_{ij}}\right)^{12} - \left(\frac{\sigma_{ij}}{r_{ij}}\right)^{6}\right) \qquad (2.36)$$

where the summation includes all pairs of atoms belonging to different molecules.

D. Combining Rules

In expression (2.36), the parameters σ_{ij} and ε_{ij} depend on the pair of atoms considered. Although they may be calibrated separately in some instances, they are generally derived

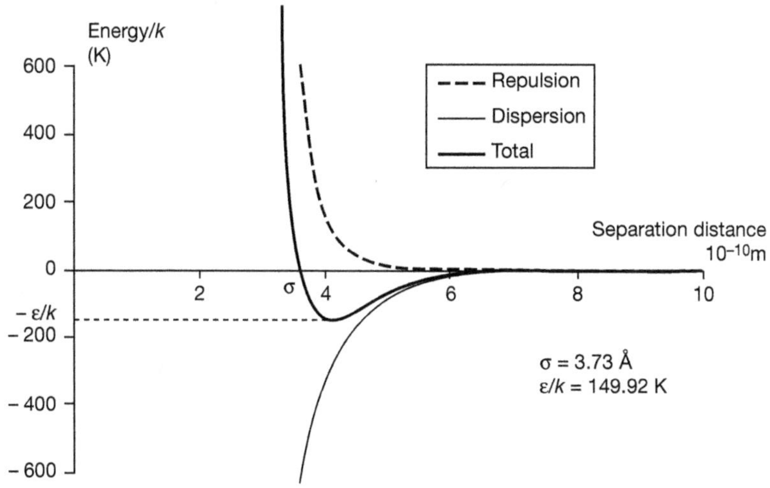

Figure 2.11 Lennard-Jones potential of methane. The sum of the long range negative dispersion term and the short range positive repulsion term results in a minimum around 4 Angstroems (1 Å = 10^{-10} m) which is thus the privileged distance between neighbour molecules in dense phases.

from pure component parameters ε_i and σ_i by appropriate combining rules. The most widely used combining rule is probably the Lorentz-Berthelot one:

$$\sigma_{ij} = \frac{\sigma_i + \sigma_j}{2} \tag{2.37}$$

$$\varepsilon_{ij} = \sqrt{\varepsilon_i \, \varepsilon_j} \tag{2.38}$$

While Eq. (2.37) has some reasonable basis because it corresponds to the limiting case of non-interpenetrating hard spheres (Fig. 2.12), the geometric mean of Eq. (2.38) is less intuitively clear. In fact, it has been shown that the form of the dispersion energy (Eq. 2.34) supports a geometric mean on the product $\varepsilon\sigma^6$ rather than on ε:

$$\varepsilon_{ij} \, \sigma_{ij}^6 = \sqrt{\varepsilon_i \, \sigma_i^6 \, \varepsilon_j \, \sigma_j^6} \tag{2.39}$$

This equation has been used in the combining rules of Kong [1973] and Waldmann-Hagler [cited by Delhommelle and Millié, 2001]. Compared with the ε obtained from the Lorentz-Berthelot combining rules, it results in lower ε for atoms with a substantial size difference but does not change significantly for atoms of equivalent size. Different combining rules for σ have been selected by Kong:

$$\varepsilon_{ij}\sigma_{ij}^{12} = \frac{\varepsilon_i\sigma_i^{12}}{2^{13}}\left[1+\left(\frac{\varepsilon_j\sigma_j^{12}}{\varepsilon_i\sigma_i^{12}}\right)^{\frac{1}{13}}\right]^{13} \tag{2.40}$$

and by Waldmann-Hagler:

$$\sigma_{ij} = \left(\frac{\sigma_i^6 + \sigma_j^6}{2} \right)^{\frac{1}{6}} \tag{2.41}$$

but they result generally in diameters close to Eq. (2.37).

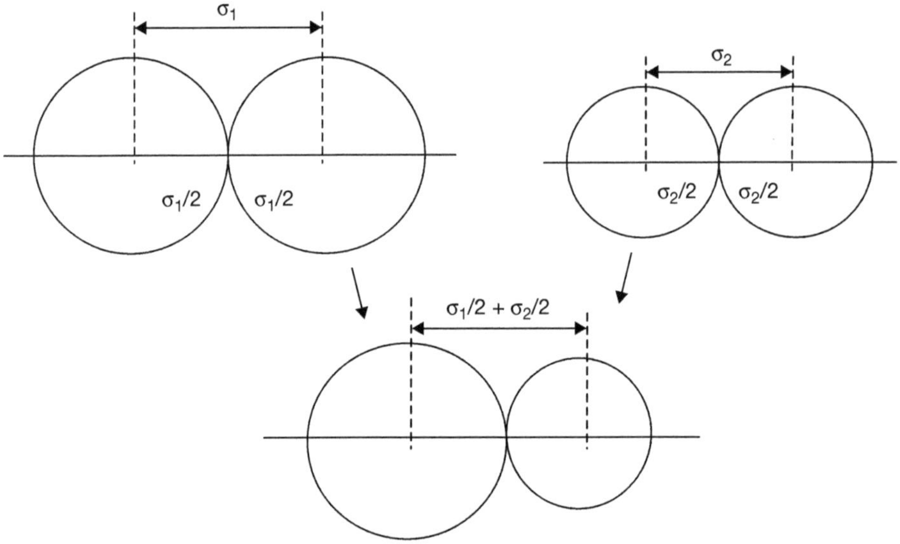

Figure 2.12 Justification of the arithmetic mean to obtain the interaction diameter σ between unlike groups in the case of hard spheres.

When the diameters of the groups are significantly different, the energy minimum is a few percentage points shallower with Eq. (2.39) than with Eq. (2.38), which results in a better representation of phase equilibria in mixtures of noble gases [Delhommelle and Millié, 2001]. Additional support for Eq. (2.39) based on volumetric properties will be presented in Chapter 3.

It is worth noting that in contrast to the equations of state, the properties of a pure component may depend on the combining rule if it is built with several different groups.

E. Range of Interaction of Intermolecular Forces

An important difference between repulsive/dispersion forces and electrostatic forces is their range of interaction, i.e. the separation distance at which they produce significant effects. As illustrated by Figure (2.13) which corresponds to the approximate case of two H_2S molecules with aligned dipole moments, the following ranges are observed:

- repulsion dominates at small separation distances, of the order of σ (here 3.73 Å) and below (see also Fig. 2.11)

– dispersion energy dominates at intermediate distances (here 4 to 6 Å), around the minimum of the potential
– the electrostatic dipole-dipole energy, which is attractive in this particular configuration, dominates at greater separation distances (here above 6 Å).

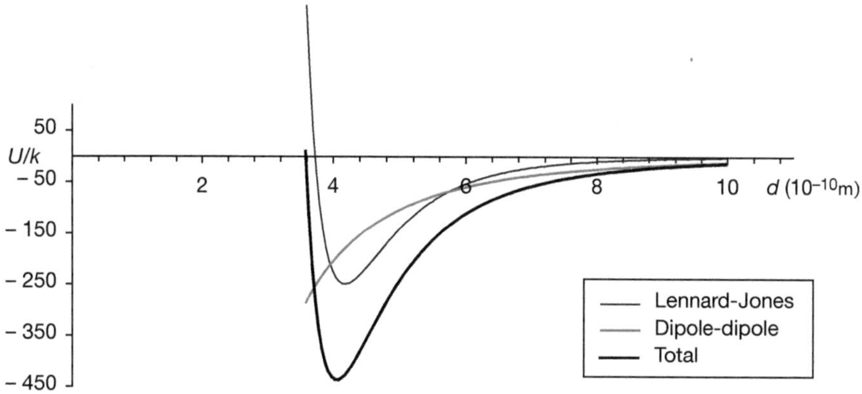

Figure 2.13 Comparison of the Lennard-Jones energy and the electrostatic dipole-dipole energy for a pair of H_2S molecules, assuming the dipoles aligned in the minimum energy configuration [Lennard-Jones parameters are taken from Kristof and Liszi, 1997].

Although the relative intensity of repulsion, dispersion and electrostatic energies depends greatly on the molecules under consideration and their orientations, the above ranking of the ranges of interaction is rather general. This may be understood by considering the dependence of the various types of interaction on separation distance. As a consequence of Coulomb's law (Eq. 2.28), it may be demonstrated that the electrostatic energy between permanent moments of the charge distribution varies according to $1/r$ for charge-charge interactions, according to $1/r^2$ for charge-dipole interactions, according to $1/r^3$ for dipole-dipole and charge-quadrupole interactions and according to $1/r^4$ for quadrupole-quadrupole interactions. The broader range of electrostatic interactions is due to their exponent, which is lower than the exponents of the dispersion energy (which varies according to $1/r^6$) and of the repulsive energy (which varies with $1/r^{12}$ according to Eq. 2.35).

F. Alternative Potentials for Repulsion and Dispersion

As there is no theoretical justification for the repulsive part of the Lennard-Jones potential, alternative choices exist. For instance, the Compass force field [Sun, 1998] makes use of a repulsion term in $1/r^9$.

Among the other types of potentials, we will just mention a few important ones:
– the square well potential, in which the molecules cannot come closer to one another than the distance σ, so that the molecules behave as attractive hard spheres. Although it looks like a very crude model, it is able to reproduce many features of liquids. It is often used in theoretical investigations but less in applied studies.

– the exp-6 potential, which is based on an exponential repulsion term:

$$U_{\text{rep}} = A\exp(-br) \tag{2.42}$$

As there are three adjustable parameters in this potential (including attraction), it is more flexible and may therefore provide therefore a more accurate description of fluid behaviour [Errington and Panagiotopoulos, 1998 and 1999a]. An additional combining rule is then needed because of the extra parameter compared with the Lennard-Jones potential.

The dispersion energy is sometimes represented by adding some higher order terms to the leading term in $1/r^6$. For instance, the adsorbate-zeolite potential used by Lachet *et al.* [1998 and 1999] contains additional terms in $1/r^8$ and $1/r^{10}$.

2.2.5 Internal Energy

Internal conformation changes cannot be neglected in flexible molecules like alkanes or polymers. In order to consider such changes in molecular simulation, we need to consider the internal potential energy of the molecule. Conceptually, the simplest way to do this would be to sum the values for dispersion and repulsive energy between the force centres belonging to the same molecule. However, it has been shown that this procedure inadequately reproduces major features like average bond angles and equilibrium molecular conformations. This is why the classical dispersion-repulsion potential between neighbouring atoms is replaced by more appropriate terms, namely stretching, bending and torsion interactions. These successive terms correspond to increases in the number of atoms involved, i.e. two in the case of stretching, three in bending and four in torsion.

A. Stretching Energy

The stretching energy U_{str} is the potential energy associated with the variation of bond length l around its mean value l_o. Therefore it involves two neighbouring atoms. Generally it is treated like a spring, i.e. characterised by a harmonic potential:

$$U_{\text{str}} = \frac{1}{2} k_{\text{str}} \left(l - l_o\right)^2 \tag{2.43}$$

where k_{str} is the stiffness of the bond.

At normal temperatures (say up to 700 K), this term is often neglected because the related vibrations are of small amplitude and simulations are sufficiently representative if bond length is set to the mean value l_o. A consequence of this approximation is that the contribution of bond stretching to internal energy and heat capacity is neglected.

B. Bending Energy

The bending energy U_{bend} is the potential energy associated with the angle θ between two successive bonds. As a consequence, it involves three atoms. It is generally treated by means of a harmonic potential of the same type as bond stretching:

$$\frac{U_{\text{bend}}}{k} = \frac{1}{2} k_{\text{bend}} \left(\theta - \theta_o\right)^2 \tag{2.44}$$

where θ_o is the mean value of the angle, k_{bend} a rigidity constant and k the Boltzmann constant. Both parameters are obtained from molecular structure and infrared spectroscopy [Ryckaert *et al.*, 1989]. In the case of saturated hydrocarbons, the average angle θ_o between successive carbon-carbon bonds is close to 110°, as a result of the sp^3 tetrahedral symmetry.

In some cases, the variations of θ are sufficiently small that they do not influence significantly simulation results. Then, it may be decided to impose constant bond angles, and to neglect the associated potential energy. This is for instance the case of CO_2, which may be assumed to be linear (in this example $\theta_o = 180°$).

An alternative formula for bending energy is [Toxvaerd, 1990]:

$$\frac{U_{\text{bend}}}{k} = \frac{1}{2} k'_{\text{bend}} (\cos\theta - \cos\theta_o)^2 \tag{2.45}$$

This expression may save computer time without introducing major changes compared to Eq. (2.44). However, care must be taken that first order equivalence requires a different constant k_{bend} (see Appendix 1).

C. Torsion Energy

Torsion energy U_{tors} is related to the dihedral angle φ (Fig. 2.14), which is defined from the coordinates of four successive atoms.

$$U_{\text{tors}}/k = \sum_{n=0}^{p} a'_j \cos(n\varphi) \tag{2.46}$$

The series comprises three or four terms so that the torsion potential exhibits several minima between 0 and 360° [Jorgensen and Madura, 1984]. A formally equivalent form of Eq. (2.46), which has been used by Toxvaerd [1997] for instance, is:

$$U_{\text{tors}}/k = \sum_{n=0}^{p} a_n (\cos\chi)^n \tag{2.47}$$

where the torsion angle χ is defined differently from the dihedral angle φ, differing by 180° (Fig. 2.14).

In the example of linear alkanes, the minimal torsion energy occurs at a torsion angle φ of 180° (i.e. a *trans* conformation) while two secondary minima occur at 60° and – 60° (i.e. *cis* conformations), as shown in Figure 2.15. As a result of the Boltzmann distribution of energies, the *trans* configurations are thus favoured over the *cis* conformations in linear alkanes (an example is given in Section 3.1).

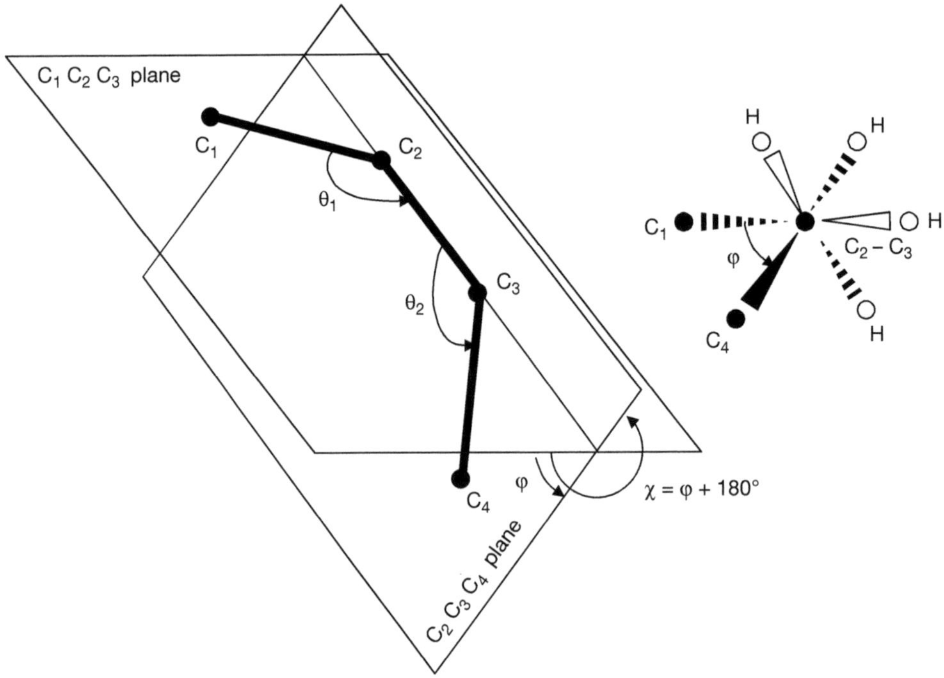

Figure 2.14 Definition of the bending angles θ and the dihedral torsion angle φ in the case of the *n*-butane molecule in perspective (left) and in Newmann projection (right). Note the alternative way of defining the torsion angle as $\chi = \varphi + 180°$.

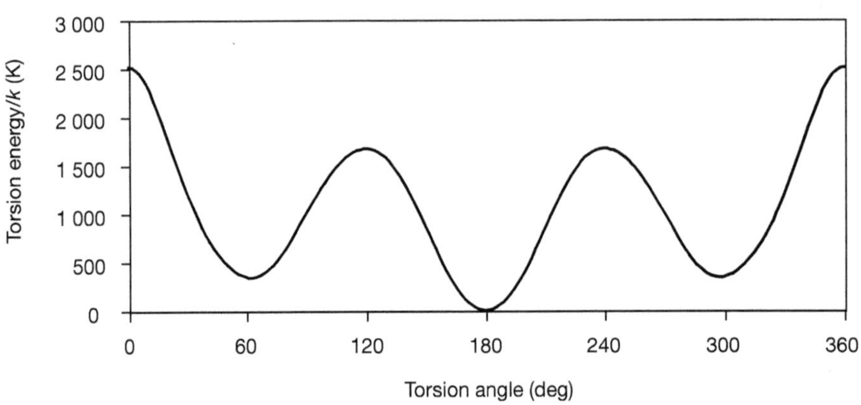

Figure 2.15 Torsional potential energy of the CH_x–CH_2–CH_2–CH_y sequence in *n*-alkanes computed after the parametrisation of Toxvaerd [1997]. The lower minimum ($\varphi = 180$) corresponds to the more stable conformation, while the two secondary minima ($\varphi = 60°$ and $\varphi = 300°$) correspond to *cis* conformations.

D. Distant Neighbour Internal Energy

The distant neighbour intramolecular interaction U_{dn} corresponds to the usual pair interactions between atoms that are not interacting through either stretching, bending or torsion, i.e. interactions between atoms separated by more than three bonds.

Distant neighbour energy may include dispersion, repulsion, electrostatic and polarisation components, exactly in the same way as is the case between separated molecules. For slightly polar molecules like the alkanes, the distant neighbour interaction is generally reduced to the Lennard-Jones energy, according to Eq. (2.36). The summation then covers the pairs of atoms belonging a single molecule but separated by more than three bonds. In long flexible polar molecules, internal electrostatic forces must be considered as well. Distant neighbour internal energy is essential to avoid overlap between distant parts of the molecule, and plays thus an important role in long chain molecules and polymers.

2.2.6 All Atoms vs United Atoms

When representing polyatomic molecules, two approaches may be used to represent the dispersion and repulsive forces:

- a separate force centre is assigned to each atom, located on its nucleus. Such methods are referred to as "All Atoms" or AA.
- a force centre is assigned to a group of atoms such as CH, CH_2 or CH_3. This technique, referred to as "United Atoms" can be divided into classical United Atoms (UA) and Anisotropic United Atoms (AUA) depending on the position of the force centres.

A. All Atoms

The advantage of All Atoms models is that they give a good account of molecular geometry and structure. The counterpart is that they require a great deal of computer time because of the large number of force centres, bending angles and torsion angles involved. Also, the parameters assigned to a given atom depend on its environment (for instance, the properties assigned to hydrogen atoms bonded to oxygen atoms will differ from those assigned to hydrogen atoms bonded to nitrogen atoms).

B. United Atoms

United Atoms methods generally neglect the hydrogen atoms while other atoms such as carbon, oxygen and sulphur are represented by separate force centres. The separation distances of Eqs. (2.35) and (2.36) are then counted between the nuclei of these major atoms, as if hydrogen atoms did not exist [Smit *et al.*, 1995; Nath *et al.*, 1998 and 2000; Martin and Siepmann, 1998 and 1999]. The influence of hydrogen atoms is considered through the parameterisation of potential parameters. For instance, the Lennard-Jones diameter σ assigned to CH_2 or CH_3 United Atoms is generally around 3.9 Å, i.e. significantly larger than the diameter of carbon in All Atoms methods, which is usually 3.5 Å [Jorgensen *et al.*, 1996].

United Atoms methods are often used for hydrocarbons because, with only one-third or one-quarter the number of force centres, they need less computer time. As computer time varies roughly with the square of the number of force centres, United Atoms methods take about one-tenth the amount of computer time as All Atoms methods. For fluid phase equilibria and adsorption applications, United Atoms methods are often used with success, especially for complex molecules which would be difficult to compute with All Atoms models. However, the related simplification of molecular structure may be limiting for some applications, as will be seen in Section 2.4.3.

C. *Anisotropic United Atoms*

In order to take better stock of the influence of hydrogen atoms in United Atoms potentials, it has been proposed shifting the force centre to an intermediate position between the major atom and the related hydrogen atoms [Toxvaerd, 1990 and 1997]. In this approach, referred to as "Anisotropic United Atoms" or AUA, the CH_2 force centre is thus located on the external bissector of the C–C–C angle and the CH_3 force centre is located on the C–C axis (Fig. 2.16). The distance from the major atom to the force centre is a parameter specific to the group under consideration. The positions of the atomic nuclei are constructed exactly in the same way as with classical United Atoms models, with the same bond lengths, and bending and torsion angles. The separation distances in Eqs. (2.35) and (2.36) are then counted between these force centres. As AUA provides a good compromise between simplicity and physical relevance, several applications of this method will be presented in Section 2.3. Related potential parameters are given in Appendix 1 and implementation details are given in Appendix 2.

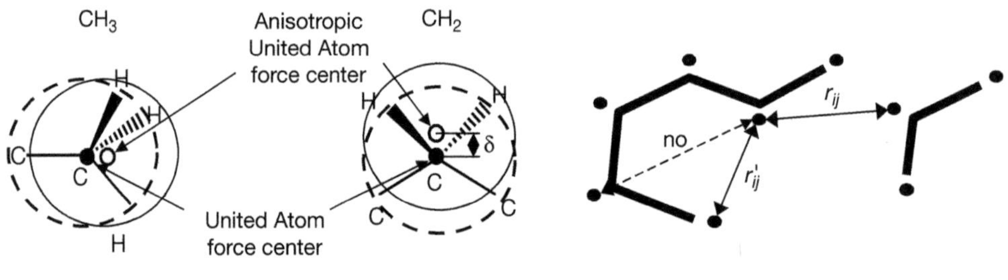

Figure 2.16 Location of the force centre in the Anisotropic United Atom model, compared with classical United Atoms.

2.3 MONTE CARLO SIMULATION PRINCIPLES

2.3.1 Basic Principle

As mentioned above, in the course of simulation, Monte Carlo methods consider only positions, not velocities. This does not mean that velocities and related kinetic energies are neglected: in fact, their contribution to the partition function is determined analytically, so that only particle positions and potential energy are under consideration when building the statistical ensemble. We briefly outline this derivation in the case of a canonical *NVT* ensemble of small rigid molecules.

A. Partition Function in the Configuration Space

The canonical partition function (Eq. 2.4) may be also expressed as:

$$Q_{NVT} = \frac{1}{h^{3N} N!} \int_{r_i} \exp(-\beta U(r_i)) dr_i \int_{p_i} \exp(-\beta K(p_i)) dp_i \tag{2.48}$$

where K is the kinetic energy and U the potential energy of the system. This factorization is possible because U depends only on positions r_i and K depends only on momenta p_i, so that the summations over p_i and r_i may be carried out independently.

The integral over momenta may be carried out analytically, and this may be shown to result in:

$$Q_{NVT} = \frac{V^N}{N! \Lambda^{3N}} \int_{s_i} \exp(-\beta U(s_i)) ds_i \tag{2.49}$$

where the integration variables s_i are dimensionless positions, and Λ is a temperature-dependant factor arising from the integration over momenta, called the de Broglie wavelength. For monoatomic particles of mass m, the de Broglie wavelength expresses as

$$\Lambda = \frac{h}{\sqrt{2\pi mkT}} \tag{2.50}$$

For polyatomic molecules, the de Broglie wavelength includes contributions from the rotation and internal degrees of freedom of the molecule. As the knowledge of Λ is not needed for most Monte Carlo algorithms, we refer the reader to McQuarrie [1976] where its expression can be found for monoatomic particles as well as for simple molecules like diatomic species.

In the case of complex molecules involving internal constraints (such as imposed bond lengths) and flexibility (either bending or torsional), Eq. (2.49) is no more exact. Indeed, the separation of the integrals over momenta and positions is no longer possible because of the constraints. As proposed by Go and Sheraga [1976], the Eq. (2.49) still holds if the amplitude of bending and torsional movements is sufficiently small. However, it may be argued that this equation introduces unjustified approximations for molecules which can assume different conformations, such as long chain alkanes [see the appendix of Ungerer *et al.*, 2001]. Nevertheless, we will consider Eq. (2.49) as a starting point, as has been done in most Monte Carlo studies to date. Then, the potential energy U appearing in this equation must be understood as the sum of external and intramolecular potential energies.

Eq. (2.49) means that in the configuration space (i.e. the space of positions, of dimension $3N$), the probability of occurrence of a given configuration (i.e. the Boltzmann factor) is:

$$\rho_{NVT} = \frac{1}{Q_{NVT}} \left(\frac{V^N}{N! \Lambda^{3N}} \exp(-\beta U) \right) \tag{2.51}$$

It may be pointed out that the probability density of the *NVT* ensemble in the configuration space (2.51) is not the same as in phase space (Eq. 2.2), as a further degree of volume and temperature dependence is introduced in factor V^N/Λ^{3N}.

As a result of Eqs. (2.49) and (2.51), it is sufficient to consider system *configurations* (characterised by position only) instead of system *states* in phase space (position coupled with momentum) to build the statistical ensemble.

For this purpose, the Monte Carlo technique makes use of a statistical method called a Markov chain: instead of solving the equations of motion in an iterative fashion as in molecular dynamics, this method transforms the system from one configuration to another according to a given transition probability, which is chosen to obtain the desired probability density of the statistical ensemble. After a sufficient number of iterations, the system has visited a representative subset of the statistical ensemble, and the collection of visited system configurations may serve to compute average properties. For this purpose, configuration changes are generated by various types of Monte Carlo moves (translation, rotation, etc.) which will be detailed in Sections 2.3.2 to 2.3.4.

B. Markov Chain

A Markov chain is characterised by the probability that a system in configuration a is transformed to configuration b. If this transition probability is noted π_{ab}, the condition for the Markov chain to converge toward the probability density ρ is given by the following stationary condition, given in matrix notation:

$$\pi\rho = \rho \tag{2.52}$$

where the dimension of the square matrix π and of the vector ρ is the number of all possible configurations (a huge number but nevertheless a finite one as a consequence of quantum mechanics principles).

In molecular simulation, we know the probability density of each configuration (for instance the Boltzmann Eq. 2.2 in the canonical ensemble), and we must determine the elements of the transition matrix in such a way that Eq. (2.52) is respected. For this purpose, it is sufficient to use the following equation, known as the *microscopic reversibility* condition, for every pair of configurations a and b:

$$\rho_a \pi_{ab} = \rho_b \pi_{ba} \tag{2.53}$$

In this equation, the left hand side is the flow of configurations a transformed into configurations b, while the right hand side corresponds to the reverse flow from b to a. In practice, it is thus sufficient to define π_{ab} and π_{ba} in such a way that both flows are equal.

C. The Metropolis Algorithm

The Metropolis algorithm [Metropolis *et al.*, 1953] is a way to generate a Markov chain where every iteration comprises two steps: in the first step, a new configuration is generated by a random change (for instance a random rotation or a random translation); in a second step, the new configuration is accepted or rejected according to a criterion designed to generate the desired probability distribution. The probability that the new configuration will be accepted is given by:

$$P_{acc}(\text{old} \rightarrow \text{new}) = \min\left(1, \frac{\rho_{new}}{\rho_{old}}\right) \quad (2.54)$$

where ρ stands for the probability density of the configuration in the statistical ensemble under consideration, and the acceptance probability P_{acc} is related to the transition probability of the Markov chain through:

$$\pi_{ab} = \frac{1}{\Omega} P_{acc}(a \rightarrow b) \quad (2.55)$$

where Ω stands for the number of accessible configurations.

It may be easily verified that the application of the Metropolis criterion (2.54) to the reverse flow (from the "new" to the "old" configuration) respects the microscopic reversibility condition stated by Eq. (2.53).

It is worth noticing that the Metropolis algorithm does not need to known the number of accessible configurations Ω to compute the acceptance probability. Also, the de Broglie wavelength Λ issued from the integration of the kinetic part of the energy is generally cancelling out because only the ratio of the probability densities appears in Eq. (2.54). As a consequence, the expression of the probability density is often rather simple. For instance, in the case of the *NVT* ensemble, it expresses as:

$$P_{acc}(\text{old} \rightarrow \text{new}) = \min\left(1, \exp\left(-\beta(U_{new} - U_{old})\right)\right) \quad (2.56)$$

This criterion is exploited in the following way:

- if $U_{new} < U_{old}$, i.e. $\exp\left(-\beta(U_{new} - U_{old})\right) > 1$, the new configuration is accepted, i.e. it is added to the ensemble;
- if $U_{new} > U_{old}$, a random number q is selected between 0 and 1, and the *new* configuration is accepted if $\exp\left(-\beta(U_{new} - U_{old})\right) > q$. Otherwise, the *old* configuration is added to the ensemble.

Monte Carlo methods make extensive use of random number generators to generate the new configurations and apply the acceptance criterion.

Once a sufficient number of configurations has been generated by the above procedure, they form a representative subset of the statistical ensemble, i.e. every accepted configuration appears proportionally to its Boltzmann factor. Then, standard averaging formulas such as Eqs. (2.6) or (2.7) can be used to derive macroscopic properties such as volume, potential energy, pressure, etc.

2.3.2 Standard Monte Carlo Moves Involving a Single Box

A. *Translation*

The first Monte Carlo move that we will consider is *translation* of an individual molecule, as sketched in Figure 2.17a-b. It consists of the following steps:

1. a translation vector with random components (t_x, t_y, t_z) is defined, and a molecule is selected randomly,
2. the translation vector (t_x, t_y, t_z) is applied to every atom of the molecule to obtain a new test configuration,
3. the potential energy U_{new} of the new test configuration is determined,
4. the acceptance criterion (2.56) is applied: if accepted, the new test configuration becomes the current configuration and all variables (energy, etc.) are updated; if rejected, the old configuration remains the current configuration.

This move does not change the internal conformation of the molecule. In practice, the components of the translation vector are selected in a finite interval $\left[-t_{max}, t_{max}\right]$ which is smaller than the simulation box. This is indeed compatible with the statistical procedure and makes it possible to control the average acceptance rate of the translation moves (see Section 2.4).

Translations would only be enough to explore the entire configuration space in the very simple case of monoatomic molecules in the *NVT* ensemble. As soon as more complex molecules or other ensembles are under consideration, additional MC moves have to be introduced.

B. Rotation

A second type of move is *rotation* of an individual molecule through a random angle α in a randomly chosen direction (Fig. 2.17c). In this move the molecule is considered as a rigid body, i.e. internal bond distances, bending angles and torsion angles are preserved. As with translations, the random variation α is restricted to an interval $\left[-\alpha_{max}, \alpha_{max}\right]$ of controlled amplitude in order to provide a good acceptance rate. Coupled with translations, rotations allow complete sampling of the configuration space at constant volume, provided the molecules under consideration are not subject to any internal deformations. In this move, the acceptance criterion (2.56) is applied without any change.

C. Volume Changes

When volume fluctuates, such as in the *NPT* ensemble, a specific MC move is used for *volume changes*, in which the simulation box is expanded or shrunk by an amount ΔV selected at random in an interval $\left[-\Delta V_{max}, \Delta V_{max}\right]$. In this move (Fig. 2.17d), the dimensionless positions of molecular centres of mass remain unchanged, as does the internal conformation of every molecule (this is required to respect constraints such as imposed bond lengths and angles). Therefore every molecule is translated, but the translation vector varies from one molecule to the other. The acceptance criterion incorporates the imposed pressure *P* according to:

$$P_{acc}(\text{old} \rightarrow \text{new}) = \min\left(1, \left(\frac{V+\Delta V}{V}\right)^N \exp\left(-\beta\left(U^{new} - U^{old} + P\Delta V\right)\right)\right) \quad (2.57)$$

where *N* is the total number of molecules in the simulation box.

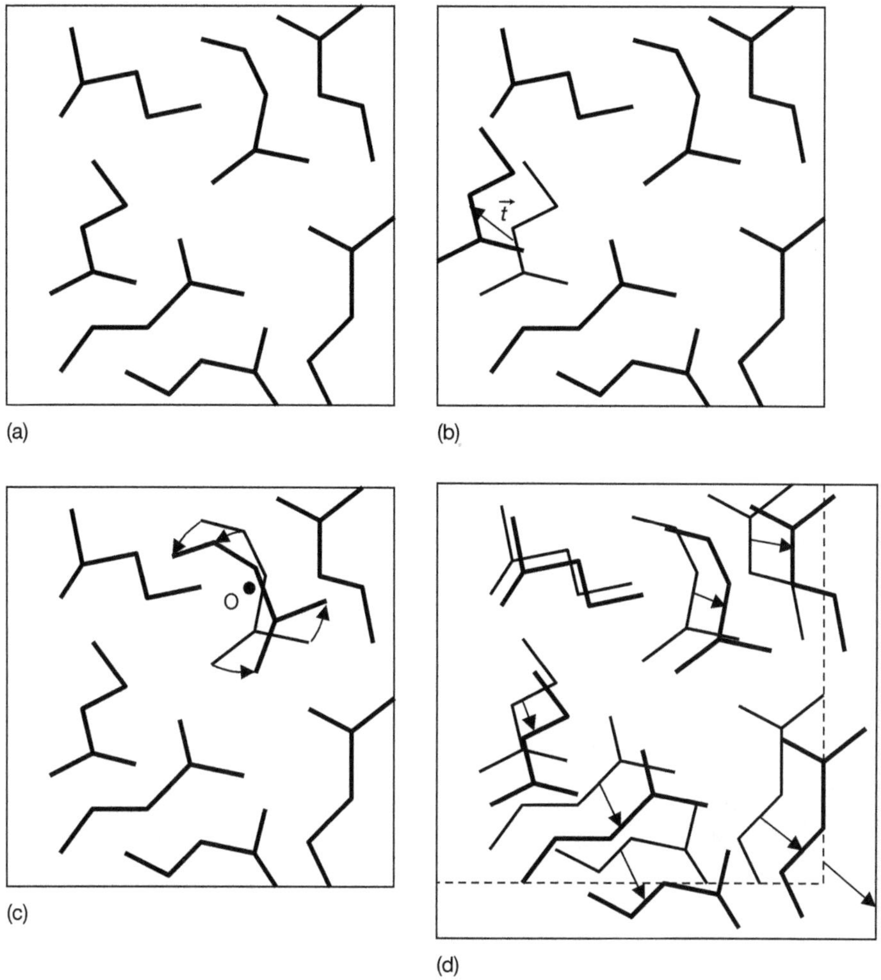

Figure 2.17 Elementary Monte Carlo moves of general purpose: (a) initial configuration (b) translation of a single molecule (c) rigid body rotation of a single molecule (d) volume change.

D. Flip Moves

In the case of flexible molecules like the alkanes, it is no longer enough to translate and to rotate the molecules but, in addition, their internal conformation has to be changed. For this purpose, the simplest move is the so-called *flip* move in which a randomly chosen atom of a chain is rotated around the axis formed by its two immediate neighbours (Fig. 2.18a), thereby preserving bond lengths. As with molecular rotations, the amplitude is selected within a finite interval [Bourasseau *et al.*, 2002b] and the acceptance probability is given by Eq. (2.56). This move is particularly useful when it comes to relaxing the inner part of a long chain molecule and flexible cyclic structures such as the cycloalkanes. Its limitation is that it cannot be applied to the terminal atoms of the chain nor to secondary carbons in branched molecules. Thus flip moves have to be complemented by other internal MC moves. These may be either similar internal rotations involving several atoms, as discussed by Dodd *et al.* [1993], or pivot or configurational bias regrowth (2.3.6). Specific aspects of the flip move with Anisotropic United Atoms potential are discussed in Appendix 2.

E. Reptation

Another very efficient MC move to simulate long chain molecules is *reptation* (Fig. 2.18b). This involves suppressing a segment of one or more atoms at one end of a randomly selected molecule, and then adding an identical segment at the other end in a random position (taking care of possible constraints such as fixed bond lengths or fixed bond angles). Then, the acceptance criterion of Eq. (2.56) is applied to select or reject the move. In effect, the reverse of a flip move, reptation is applicable only to linear molecules. For long chain molecules such as polymers, it is particularly efficient [Leblanc *et al.*, 2003]. Most often, the move is performed with a configurational bias as outlined in Section 2.3.6 and Appendix 2.

In order to ensure that microscopic reversibility is respected, it is essential that both forward and backward reptations are attempted, with equal frequency.

F. Pivot

When a molecule is composed of several more or less rigid parts (such as rings or branches) connected by a flexible chain, flip and reptation moves do not suffice to explore all possible internal configurations. Rotating part of the molecule around one of the atoms in a random rotation – the pivot move (Fig. 2.18c) – then makes more adequate sampling possible. The acceptance probability (2.56) is applied unchanged. This move is used for instance in the case of isoalkanes with multiple branches [Bourasseau, 2003].

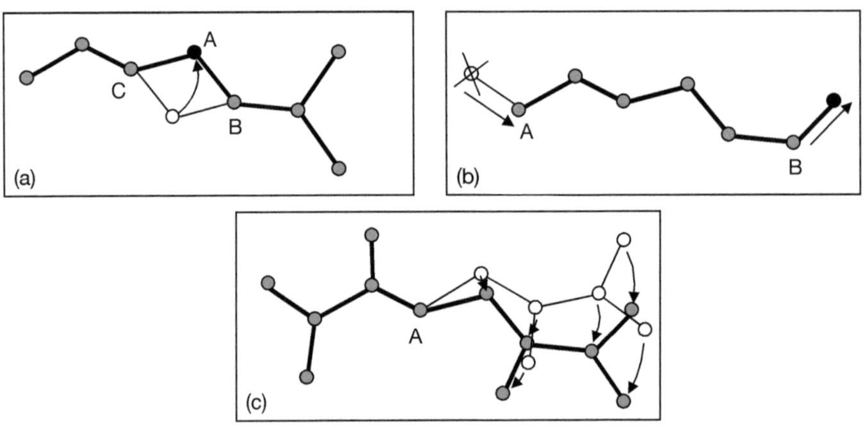

Figure 2.18 Elementary Monte Carlo moves contributing to the relaxation of flexible molecules: (a) flip, i.e. rotation of a single atom A around the axis B-C of its nearest neighbours (b) reptation (c) pivot, i.e. rotation of a part of a molecule around atom A.

G. Displacement, Regrowth

In a move that we will refer to here as *displacement*, a molecule is deleted at its original place in the simulation box and inserted again at a randomly selected position in a randomly selected orientation and conformation. Another move, called *partial regrowth* applies specifically to flexible molecules. This involves cutting one end off the molecule and allowing it to regrow at a randomly selected position. However these moves are generally used together with statistical bias, so they will be discussed in Section 2.3.6 and Appendix 2.

2.3.3 Insertion and Destruction Moves in the Grand Canonical ensemble

The fluctuation of mole numbers is the characteristic feature of grand canonical Monte Carlo simulation (GCMC). It is performed by means of two particular moves (Fig. 2.19):

- the *insertion* of a new molecule. In this move, the type i of molecule to be inserted is selected first at random. Then, a new molecule of type i is tentatively inserted in a randomly selected location in the box. If the molecule is not spherical, it must also be inserted in a random orientation. If it is flexible, a test insertion must be made in a random internal conformation, unless a configurational bias is used (see Section 2.3.6). The insertion is accepted if the following Metropolis acceptance criterion is satisfied:

$$P_{\text{acc}}(\text{insertion}) = \min\left(1, \frac{V}{(N_i+1)\Lambda_i^3}\exp\left(-\beta(U^{\text{new}}-U^{\text{old}})+\beta\mu_i\right)\right) \tag{2.58}$$

  where μ_i is the imposed chemical potential of molecular type i, N_i is the current number of molecules of type i in the simulation box, and the other symbols have the same meaning as in Eqs. (2.56) and (2.57).

- the deletion of an existing molecule of the simulation box. After selection of the molecular type i, a molecule is randomly chosen among those of type i in the current simulation box, and the following acceptance criterion is applied:

$$P_{\text{acc}}(\text{deletion}) = \min\left(1, \frac{N_i\Lambda_i^3}{V}\exp\left(-\beta(U^{\text{new}}-U^{\text{old}})-\beta\mu_i\right)\right) \tag{2.59}$$

For a given molecular type, it is important that an equal number of insertion attempts and deletion attempts are made, otherwise the desired probability distribution is not satisfied.

A problem arising from Eqs. (2.58) and (2.59) is that they correspond to an unphysical reference state for the chemical potential, as will be discussed in Section 2.3.5. If we introduce the chemical potential $\overline{\mu}_i = \mu_i - \mu_{io}$ (where μ_{io} is the chemical potential of a perfect gas of pure compound i under a reference pressure P_o at temperature T), the acceptance criteria can be expressed as:

$$P_{\text{acc}}(\text{insertion}) = \min\left(1, \frac{V P_o}{(N_i+1)kT}\exp\left(-\beta\,\Delta U_{\text{ext}}+\beta\overline{\mu}_i\right)\right) \tag{2.60}$$

$$P_{\text{acc}}(\text{deletion}) = \min\left(1, \frac{N_i\, kT}{V\, P_o} \exp\left(-\beta \Delta U_{\text{ext}} - \beta \overline{\mu}_i\right)\right) \tag{2.61}$$

where $\Delta U_{\text{ext}} = U_{\text{ext}}^{\text{new}} - U_{\text{ext}}^{\text{old}}$ is the change in *external* potential energy.

In addition to the more physical reference state, the expressions (2.60) and (2.61) present the advantage of avoiding the need to estimate the de Broglie wavelength Λ_i.

When density is low or moderate, the acceptance probability of GCMC moves is generally good. When density is high, or when large molecules are being dealt with, it is dramatically low if the moves are done exactly as outlined above (this applies also to adsorption in microporous adsorbents, which is a common application of GCMC). This is due to the difficulty of making insertions in dense phases because the random test position usually overlaps with existing molecules in the simulation box, causing a high positive energy variation ΔU_{ext} in Eq. (2.60) and a correspondingly low acceptance probability.

We will see in Section 2.3.6 and Appendix 2 how statistical bias may be used to increase the efficiency of GCMC moves.

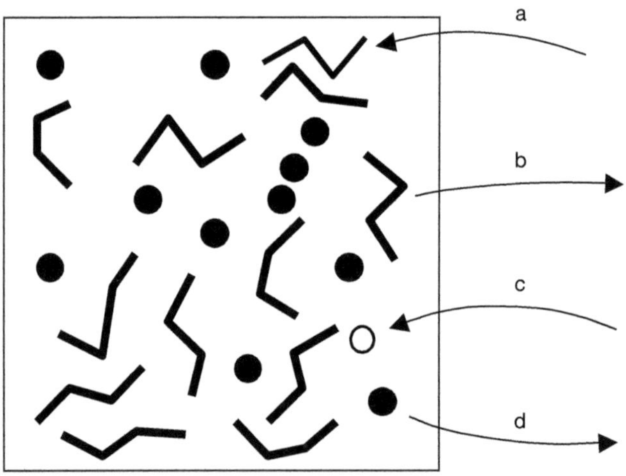

Figure 2.19 Elementary Monte Carlo moves specific of the Grand Canonical ensemble: (a, c) insertions of a new molecule (b, d) deletion of molecules.

2.3.4 Moves Specific to the Gibbs ensemble

As defined in Section 2.1, the Gibbs ensemble Monte Carlo (GEMC) involves two simulation boxes without any interface. In addition to the single box moves discussed in Section 2.3.2, specific moves are used to simulate this ensemble: transfers of molecules between the boxes, which aim at imposing equal chemical potentials in both phases, and coupled volume changes which provide mechanical equilibrium.

A. Transfers

The transfer move involves deleting a randomly chosen molecule in one phase and randomly inserting another one of the same type in the other phase (Fig. 2.20). It may be viewed as the coupling of a grand canonical destruction in one phase and insertion in the other phase. The acceptance probability for the transfer of a molecule of type i from box A to box B is:

$$p_{\mathrm{acc}}(\text{transfer}) = \min\left(1, \frac{N_i^A}{V^A}\frac{V^B}{N_i^B+1}\exp\left(-\beta\Delta U^A - \beta\Delta U^B\right)\right) \tag{2.62}$$

where ΔU^A and ΔU^B are the variations of potential energy in both phases, N_i^A and N_i^B being the number of moles of type i in each phase before the transfer.

Transfer moves are essential in the Gibbs ensemble because they make it possible to satisfy the condition of average equal chemical potentials in both phases ($\langle\mu_i^A\rangle = \langle\mu_i^B\rangle$).

For the same reason as with the GCMC moves presented in the previous section, GEMC transfers have a low acceptance probability when one or both phases are particularly dense, or when large molecules are involved. In such cases, statistical bias methods are used to improve acceptance rates, as will be discussed in Section 2.3.6.

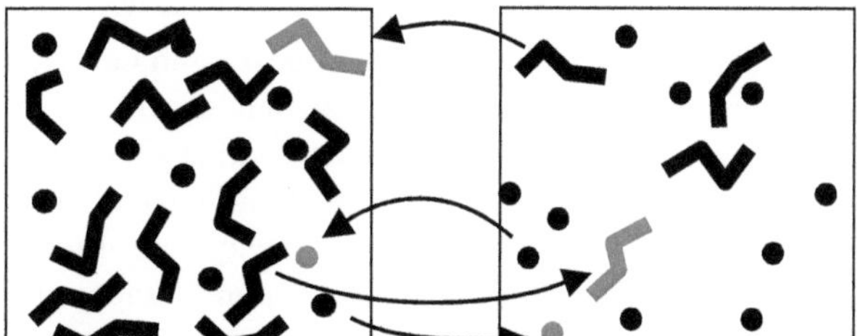

Figure 2.20 Monte Carlo moves of molecule transfers from one phase to the other, characteristic of the Gibbs ensemble.

B. Volume Changes

In the Gibbs ensemble at imposed pressure, volume change moves are applied independently in every simulation box, and the acceptance probability is the same as Eq. (2.57). This ensures that the simulation is performed at the requested pressure.

In the Gibbs ensemble at imposed global volume, inverse volume changes $+\Delta V$ and $-\Delta V$ are applied to simulation boxes A and B. This move is also essential because it provides the mechanical equilibrium condition that average pressures must be equal in both phases $\langle P^A\rangle = \langle P^B\rangle$. The acceptance probability is:

$$P_{\mathrm{acc}} = \min\left(1, \left(\frac{V^A+\Delta V}{V^A}\right)^{N^A}\left(\frac{V^B-\Delta V}{V^B}\right)^{N^B}\exp\left(-\beta\Delta U^A - \beta\Delta U^B\right)\right) \tag{2.63}$$

2.3.5 Evaluation of the Chemical Potential

In contrast to volume, pressure and energy, the chemical potential cannot be obtained by simple averaging in the *NVT* or *NPT* ensembles. If it has to be evaluated, the only way is to perform a large number of test insertions in the system (Widom tests). These tests are exactly like the GCMC insertions of Figure 2.19, except that they are not effective, i.e. the list of molecules of the box is not updated. In the *NVT* ensemble, the chemical potential is thus obtained by the following relationship:

$$\mu_i = -kT \ln\left\langle \frac{V}{(N_i+1)\Lambda_i^3} \exp\left(-\beta \Delta U^+\right)\right\rangle \tag{2.64}$$

where ΔU^+ is the variation of total potential energy of the system when a test molecule is inserted in it, and the brackets refer to an average over all test insertions in the *NVT* ensemble. A slightly different relationship applies to the *NPT* ensemble. When this method is applied to a liquid, most test insertions result in extensive overlapping between the inserted molecule and existing molecules, so that very high values of ΔU^+ are found. The test insertions falling in "holes" between the liquid molecules correspond to low values of ΔU^+, so that they contribute much more to the average than those resulting in overlapping.

Care must be taken that the chemical potential μ_i appearing in Eq. (2.64) and in the Boltzmann factor (2.10) are not expressed in either the same units or with the same reference state as in classical thermodynamics. The difference in units is rather simple, since μ_i is expressed per molecule and not per mole of substance, and both definitions differ thus by a factor N_a, the Avogadro number. However, the problem of the reference state is more complex. As it is important for simulating adsorption phenomena in the Grand Canonical ensemble, we will treat it in detail.

The implicit reference state in Eq. (2.64) may be readily obtained by searching the conditions for which the chemical potential is zero. In the limit of large mole numbers, this corresponds to $\Delta U^+ = 0$ with a standard density $N_i / V = \Lambda_i^{-3}$. The condition $\Delta U^+ = 0$ corresponds to a perfect gas (null intermolecular energy, so $\Delta U_{\text{ext}}^+ = 0$) with zero intramolecular energy ($\Delta U_{\text{int}}^+ = 0$), i.e. an unphysical state.

By comparison, the classical thermodynamic reference state is a perfect gas of pure component *i* under a reference pressure P_o and temperature *T*, which possesses a distribution of intramolecular energies but no intermolecular energy ($\Delta U_{\text{ext}}^+ = 0$ and $\Delta U_{\text{int}}^+ \neq 0$). Applying Eq. (2.64) to a perfect gas in this reference state, it can be shown that the corresponding chemical potential is:

$$\mu_{io} = -kT \ln\left\langle \frac{kT}{P_o \Lambda_i^3} \exp\left(-\beta \Delta U_{\text{int}}^+\right)\right\rangle \tag{2.65}$$

Then expression (2.64) may be rewritten in the general case:

$$\mu_i = \mu_{io} + \overline{\mu}_i \tag{2.66}$$

where

$$\bar{\mu}_i = -kT \ln\left\langle \frac{P_o V}{kT(N_i+1)} \exp\left(-\beta \Delta U^{+}_{\text{ext}}\right)\right\rangle \tag{2.67}$$

is the chemical potential expressed with the classical reference state. It may be noticed that the de Broglie wavelength Λ_i cancels out in this expression.

The Widom test is known to converge very slowly in dense liquids or, more generally, in condensed phases [Kofke and Cummings, 1997]. In the same way as Monte Carlo moves, statistical bias techniques may be used to improve convergence in such cases, as we will see in Section 2.3.6 and in Appendix 2.

2.3.6 Statistical Bias Monte Carlo Moves

If the molecule under consideration is a large species, there is little chance of finding a "hole" corresponding to its particular shape in a condensed phase when attempting to insert it in a random fashion. As mentioned above, this makes the acceptance ratio of GCMC and GEMC moves very low. In order to increase it, non-random MC moves are made so that favourable positions are preferentially sampled. This means that the sampling is biased, and the expressions of acceptance ratios for insertions and deletions (2.60 and 2.61) have to be corrected for the bias. Here, we will restrict the presentation to configurational bias and reservoir bias, but numerous other statistical bias moves have been developed for the insertion or deletion of large molecules in dense phases [Kaminsky, 1994; Spyriouni *et al.*, 1998; Boulougouris *et al.*, 1999 and 2001].

A. *Configurational Bias*

Configurational bias Monte Carlo, also referred to as CBMC, addresses the case of long linear or branched molecules that can adopt numerous conformations. This method [de Pablo *et al.*, 1992; Smit *et al.*, 1995] takes advantage of the flexibility of the molecule to grow it step by step, testing several possible random locations r_k, $k = 1 \ldots k_{\text{max}}$ for the next atom (Fig. 2.21). The final position of the new atom is selected from among the tested locations with a probability:

$$p(r_i) = \frac{\exp\left(-\beta\, u(r_i)\right)}{\sum_{k=1}^{k\,\max} \exp\left(-\beta\, u(r_k)\right)} \tag{2.68}$$

where $u(r_k)$ is the increment of potential energy associated with the addition of a new atom in position r_k. Once a position r_i is selected, the same process is applied to the next atom, and so on until the end of the chain is reached. As $u(r_k)$ includes internal potential energy (bending, torsion, etc.) as well as external energy, the molecule is thus reconstructed in a non-random way. Once the whole molecule has been regrown, the move is accepted according to a modified acceptance probability, so that its geometry is statistically representative at the temperature under consideration.

CBMC can be used with the following Monte Carlo moves:

- insertions and deletions in grand canonical simulations (Section 2.3.3)
- transfers in the Gibbs ensemble (see Section 2.3.4)
- Widom tests (Section 2.3.5)
- partial regrowth, reptation and displacements (Section 2.3.2)

The acceptance probability of these CBMC moves is provided in Appendix 2. The number of locations tested for each atom, k_{max}, does not need to be the same for all atoms of the molecule, e.g. a greater number of test locations is often used for the first atom (see also Section 2.4.5 for the selection of appropriate values).

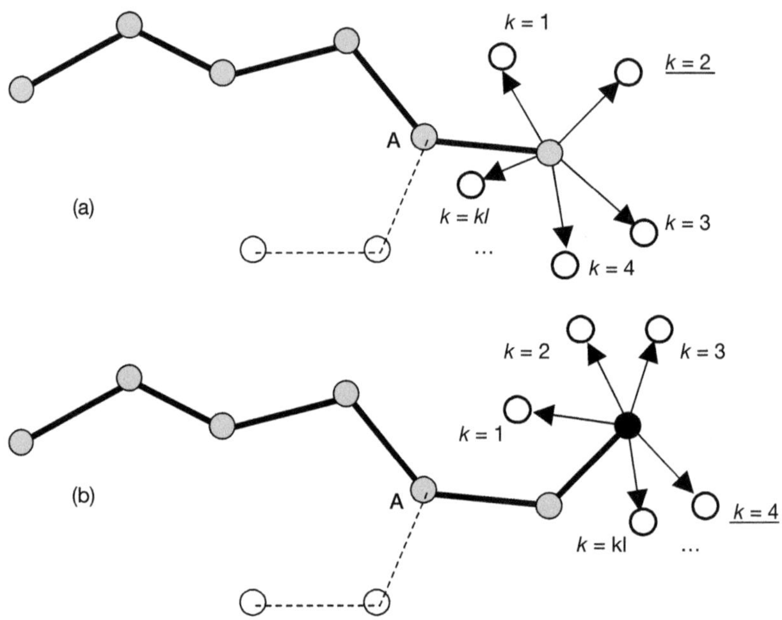

Figure 2.21 Schematic example of configurational bias applied to the regrowth of a segment of two atoms from an existing chain of six atoms (a) test of k_{max} positions to place the first new atom and selection of its position (here $k = 2$), (b) test of k_{max} positions for the second new atom and selection of its position (here $k = 4$). The dotted part of the chain is its previous conformation.

B. Reservoir Bias

Configurational bias is applicable to flexible cyclic molecules, [Neubauer *et al.*, 1999a] but the constraint of closing the ring often makes for a poor average acceptance rate. A satisfactory way of solving this problem is to use a *canonical reservoir* of molecular conformations for the ring, i.e. a collection of molecular conformations in which the Boltzmann distribution of internal energies is respected (Fig. 2.22). This means that the probability of occurrence of a given conformation in the reservoir is proportional to $\exp(-\beta U_{\text{int}})$. In practice, this reservoir is created by performing repeated internal moves such as the flip described in Section 2.3.2.

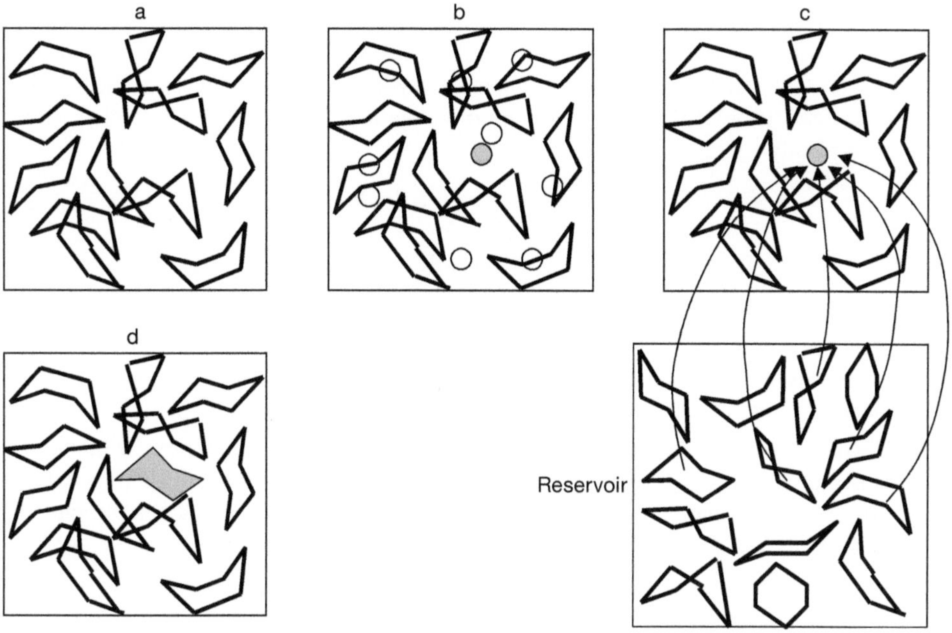

Figure 2.22 Schematic example of reservoir bias algorithm applied to the case of insertion of a new cyclohexane molecule in a box (a) initial configuration (b) test insertion of Lennard-Jones particles in several locations and selection of a favourable location (c) test of several molecular conformations taken from a canonical reservoir for insertion at the selected location (d) configuration after successful insertion.

In the case of grand canonical insertions, a molecular conformation is selected at random in the reservoir and is tentatively inserted [Errington and Panagiotopoulos, 1999]. The reservoir bias may also be used to improve the efficiency of CBMC algorithms with branched molecules by picking bending angles from a suitable reservoir [Macedonia and Maginn, 1999] instead of generating them repeatedly during the regrowth process (Branch point sampling).

In the case of Gibbs ensemble transfers, reservoir bias has been exploited with an additional preliminary biasing step to find suitable "holes" in the liquid [Bourasseau *et al.*, 2002a]. This involves the following stages:

- in the first step, several random locations for the centre of mass r_k are tested with a very simple potential (single Lennard Jones Atom). One of these is selected on the basis of a similar criterion as CBMC moves, i.e. with a probability

$$p(r_i) = \frac{\exp(-\beta u_{LJ}(r_i))}{\sum_{k=1}^{k\max} \exp(-\beta u_{LJ}(r_k))} \tag{2.69}$$

where u_{LJ} is the interaction energy of the Lennard Jones force centre with the system.

– in the second step, several molecular conformations c_k are randomly picked in the reservoir and tentatively inserted in the system with the centre of mass at the location identified in the first step, in a random orientation. One of these is selected with a probability

$$p(c_i) = \frac{\exp(-\beta U_{ext}(c_i))}{\sum_{k=1}^{k\max} \exp(-\beta U_{ext}(c_k))} \tag{2.70}$$

– the move is accepted or rejected with an acceptance criterion which corrects for the bias introduced by the first two steps (see Appendix 2 for the corresponding acceptance probabilities).

The reservoir bias move affords significant saving in computer time for cyclic molecules in condensed phases. It is also very efficient for small flexible molecules with a limited number of different conformations, such as propane or butane. The concept may also applied to GCMC insertions and deletions.

In the case of rigid molecules, the reservoir of conformations is useless but the two-step procedure outlined above provides a very significant improvement in GCMC and GEMC compared with the unbiased moves discussed in Sections 2.3.3 and 2.3.4.

2.3.7 Determination of Bubble Points and Dew Points

The Gibbs ensemble method (Sections 2.1.1 and 2.3.4) only makes it possible to compute phase equilibria for systems in which the *total* number of molecules is specified for each molecular species. In industrial applications, it is also often necessary to predict the conditions when a liquid of given composition (e.g. crude oil or a stream in a process plant) separates into two phases. This equilibrium condition – known as the bubble point – corresponds to a system in which the *liquid* phase composition is imposed. Similarly, a dew point corresponds to a phase equilibrium in which the composition of the *vapour* phase is imposed. These problems cannot be addressed directly with the Gibbs ensemble method, and specific methods must be used for this purpose in the general context of a multicomponent system [Vrabec and Fischer, 1995; Escobedo, 1998 and 1999; Ungerer *et al.*, 1999 and 2001].

Let us consider for instance the problem of computing a bubble point at imposed temperature (Fig. 2.23). The natural way to impose the composition of the liquid phase is to fix the number of moles of every species in this phase. Consequently, the associated intensive variables (i.e. the chemical potential of every species of the liquid phase) must fluctuate. On the other hand, the number of moles in the vapour phase fluctuates (it cannot be fixed, because it is a desired result in the problem!) and thus the associated intensive variable – the chemical potential in the vapour phase – is fixed. As a result, chemical potentials cannot be equal at all times: they coincide only in the thermodynamic limit, i.e. for a very long simulation. This is the essential difference between a pseudo-ensemble and a true statistical ensemble.

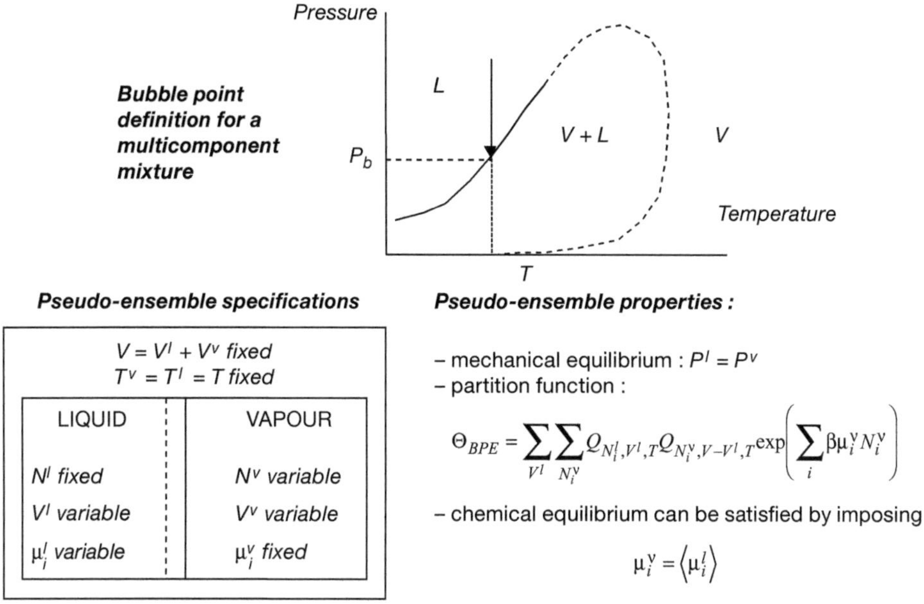

Figure 2.23 Specifications of the Bubble point pseudo-ensemble.

In order to simulate bubble points, existing methods are based on chemical potential estimations in the liquid phase by Widom test insertions (which may be using CBMC or reservoir bias if needed). These chemical potentials are then imposed on the vapour phase which is treated in a way similar to the Grand Canonical ensemble (Fig. 2.24). This process must be repeated a sufficient number of times so that pressure equilibrium is also reached. In the examples of bubble point calculations that will be shown in Section 2.3, pressure equilibrium is achieved by coupling the simulation boxes: they are subject to the constraint that their global volume is constant, i.e. inverse volume changes occur in the two phases [Ungerer *et al.*, 1999]. In order to perform insertions and deletions in the vapour phase, a procedure similar to the Gibbs ensemble transfer move is implemented. The same acceptance probabilities are applied (Eq. 2.62) but the liquid phase is updated when the move is accepted, the vapour phase being unchanged. The procedure is also applicable with configurational bias for flexible molecules [Ungerer *et al.*, 2001].

It is also possible to consider uncoupled volume changes, which require a specific iterative algorithm [Escobedo, 1998]. This method presents the advantage of being applicable to dew points, for which convergence is more difficult to achieve [Escobedo, 2000].

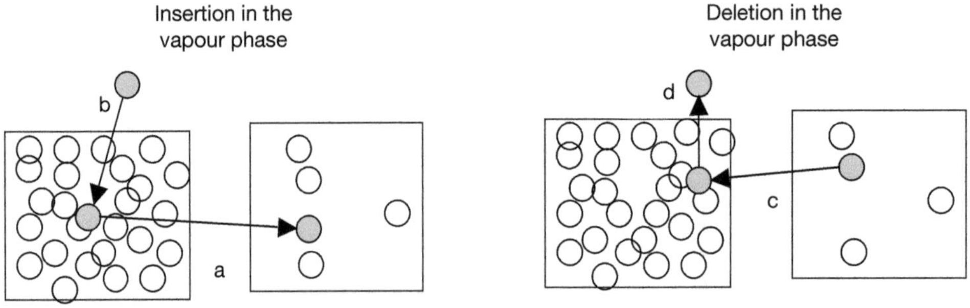

Figure 2.24 Monte Carlo moves used for the direct simulation of bubble points (a) if a molecule transfer from the liquid to the vapour is accepted according to the Gibbs ensemble prescription, the leaving molecule is replaced (b) if a transfer is accepted from the vapour to the liquid, the molecule is not inserted in the liquid.

2.3.8 Thermodynamic Integration

As mentioned in Section 2.3.4, the acceptance probability of Gibbs ensemble transfer moves decreases drastically with temperature in dense liquids. With pure compounds of moderate complexity, it is therefore difficult to obtain a precise result for the most sensitive simulation, i.e. vapour pressure significantly below the normal boiling point. Thermodynamic integration is a very efficient way to extend vapour pressure calculations to lower temperatures: this involves exploiting the Clapeyron equation:

$$\frac{dP_{\text{sat}}}{dT} = \frac{\Delta_{\text{vap}} H}{T\,\Delta_{\text{vap}} V} \tag{2.71}$$

or, equivalently:

$$\frac{d\left(\ln P_{\text{sat}}\right)}{d\left(1/T\right)} = -\frac{T\,\Delta_{\text{vap}} H}{P_{\text{sat}}\,\Delta_{\text{vap}} V} \tag{2.72}$$

where P_{sat} is vapour pressure, $\Delta_{\text{vap}} H$ is the change in enthalpy and $\Delta_{\text{vap}} V$ is the change in volume accompanying vaporization in saturated conditions. The principle of thermodynamic integration involves estimating first P_{sat} at temperatures $T^{(0)}$ and $T^{(1)}$ where Gibbs ensemble calculations are tractable, and using then monophasic simulations to evaluate the right hand side of Eq. (2.72) at lower temperatures [Kofke, 1993]. A suitable numerical scheme is then used to integrate the Clapeyron equation and yield P_{sat} at lower temperatures. Rather than high-order numerical schemes which provide a misleading impression of accuracy, a second-order numerical scheme with regularly spaced values of $1/T$ is sufficient to perform this integration [Ungerer *et al.*, 2000]:

$$\frac{\ln P_{\text{sat}}^{(n+1)} - \ln P_{\text{sat}}^{(n-1)}}{1/T^{(n+1)} - 1/T^{(n-1)}} = -\frac{T^{(n)}\,\Delta_{\text{vap}} H^{(n)}}{P_{\text{sat}}^{(n)}\,\Delta_{\text{vap}} V^{(n)}} \tag{2.73}$$

where the superscript n is the iteration counter. Selecting regularly spaced $1/T$ values means that the difference $1/T^{(n)} - 1/T^{(n-1)}$ is the same for every n. A more practical equation is obtained by expressing the desired unknown $P_{\text{sat}}^{(n+1)}$:

$$\ln P_{\text{sat}}^{(n+1)} = \ln P_{\text{sat}}^{(n-1)} - \frac{T^{(n)} \Delta_{\text{vap}} H^{(n)}}{P_{\text{sat}}^{(n)} \Delta_{\text{vap}} V^{(n)}} \left(1/T^{(n+1)} - 1/T^{(n-1)}\right) \tag{2.74}$$

In this equation, the right hand side may be evaluated from monophasic *NPT* simulations of the liquid and vapour phases conducted at temperatures $T^{(n-1)}$ and $T^{(n)}$ and pressures $P_{\text{sat}}^{(n-1)}$ and $P_{\text{sat}}^{(n)}$. Simulations of the vapour and of the liquid at a given temperature differ only by their initialization, and they make possible straightforward calculation of the molar property changes $\Delta_{\text{vap}} H$ and $\Delta_{\text{vap}} V$. It is thus possible to evaluate the saturation pressure $P_{\text{sat}}^{(n+1)}$ at which the next two monophasic *NPT* simulations at temperature $T^{(n+1)}$ must be performed. This process must be started from two well-converged Gibbs ensemble calculations yielding $P_{\text{sat}}^{(0)}$ and $P_{\text{sat}}^{(1)}$, but it can be repeated as many times as necessary to extrapolate estimated saturation pressure values to lower temperatures.

In practice, it is often sufficient to use an approximate form of the Clapeyron equation, which is generally valid at temperatures below the normal boiling point:

$$\frac{d(\ln P_{\text{sat}})}{d(1/T)} \approx -\frac{\Delta_{\text{vap}} h}{R} \approx -\frac{-u_{\text{ext}}^{\text{liq}} + RT}{R} \tag{2.75}$$

where $u_{\text{liq}}^{\text{ext}} = \frac{N_a}{N} U_{\text{ext}}^{\text{liq}}$ is the molar external potential energy of the liquid phase, and $\Delta_{\text{vap}} h$ is the *molar* vaporization enthalpy. Among others, the approximations involved in Eq. (2.75) are that the vapour obeys the perfect gas law and that the internal potential energy is the same in the liquid and vapour phases. Eq. (2.74) simplifies then to:

$$\ln P_{\text{sat}}^{(n+1)} = \ln P_{\text{sat}}^{(n-1)} - \frac{-u_{\text{ext}}^{\text{liq}(n)} + RT}{R} \left(1/T^{(n+1)} - 1/T^{(n-1)}\right) \tag{2.76}$$

The advantage of Eq. (2.76) over Eq. (2.74) is mainly that it requires a simulation of the liquid phase only at every temperature $T^{(n)}$. Normally this simulation should be performed at pressure $P_{\text{sat}}^{(n)}$ which may be estimated by extrapolating from the high temperature points in the Clapeyron diagram.

The process of thermodynamic integration may appear complex but it is simply equivalent to exploiting the known slope in a Clapeyron plot (i.e. the vaporization enthalpy) to place the successive points (Fig. 2.25). It may even be further simplified if we note that below the normal boiling point, the potential external energy of a liquid depends very little on the exact value of pressure between 0 and 1 bar, so that preliminary rough estimates of $P_{\text{sat}}^{(n)}$ are sufficient to conduct the monophasic liquid *NPT* simulations. Then the properties of the liquid phase may be obtained by setting the pressure to zero. Rigorously speaking, the liquid is metastable at zero pressure but there is no significant difference with the actual pressure as a result of the usual statistical uncertainties. This may be used to process the *NPT* simulations in a parallel rather than sequential scheme, thereby affording considerable time-saving on parallel computers or networks.

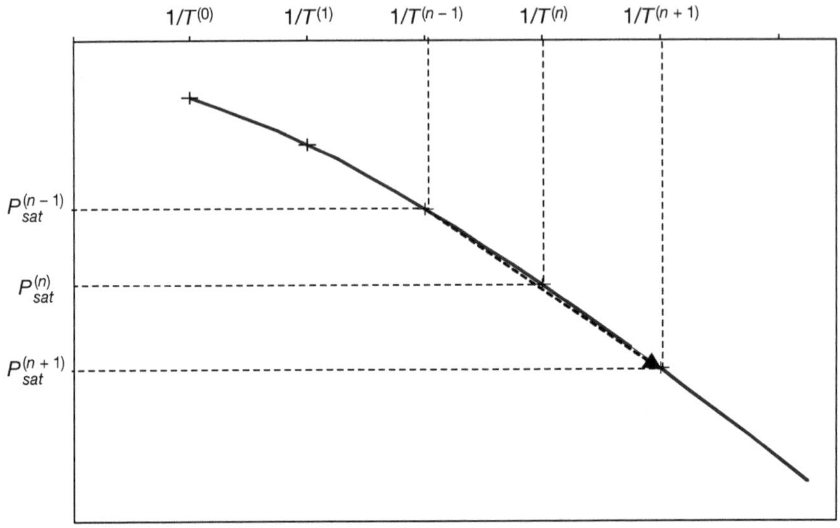

Figure 2.25 Principle of the iterative determination of vapour pressures by thermodynamic integration of the Clapeyron equation.

Thermodynamic integration may also be applied to multicomponent systems in which case it is known as Gibbs-Duhem integration. The integration process is more complex, since it occurs in a multidimensional space. The reader is referred to related articles [Mehta and Kofke, 1994; Escobedo and de Pablo, 1997; Brennan and Madden, 2003] for additional information.

2.3.9 Parallel Tempering

Parallel tempering involves conducting several Monte Carlo simulations of the same system at different temperatures simultaneously. At regular intervals, attempts are made to exchange configurations between two temperatures. In the *NVT* ensemble, the acceptance probability is given by:

$$p(\text{exchange}) = \min\left(1, \exp\left((\beta_i - \beta_j)(U_i - U_j)\right)\right) \tag{2.77}$$

where U_i and U_j are the potential energies of the current configurations of the simulations conducted at $\beta_i = 1/kT_i$ and $\beta_j = 1/kT_j$.

As a result, the simulation of a given temperature is subject to an important configurational change every time such an exchange is accepted (Fig. 2.26). The different temperatures are selected in such a way that the energy histograms overlap sufficiently, otherwise the acceptance ratio of the exchanges of configurations between two temperatures is excessively low.

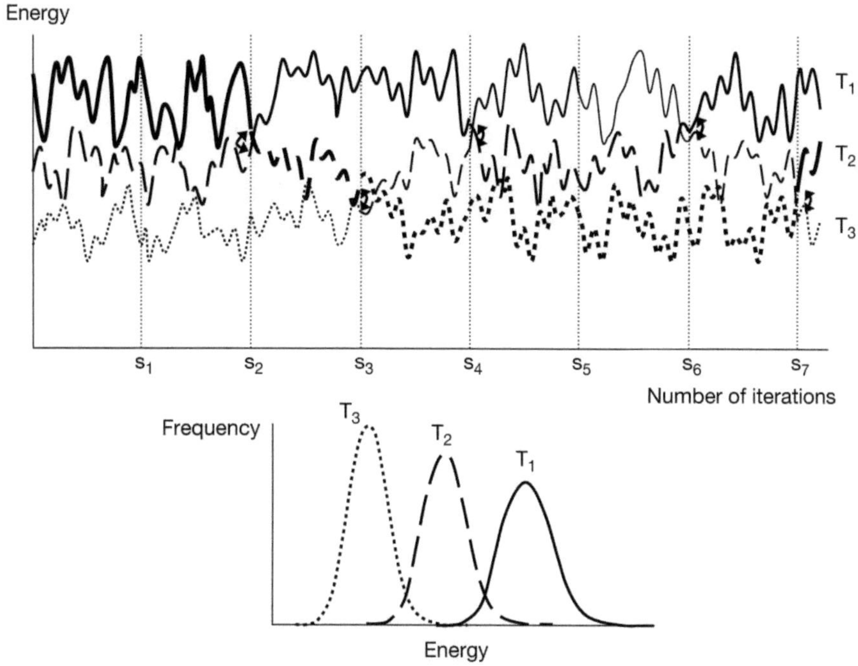

Figure 2.26 Principle of the parallel tempering method in the case of three temperatures T_1 (full lined), T_2 (dashed line) and T_3 (dotted line). Three configurations are used (thick, medium and thin lines). As a result of temperature exchange between configurations, a given configuration experiences several temperatures and a given temperature experiences several configurations successively.

Parallel tempering is useful for exploring configuration space in an efficient manner, while avoiding the low temperature simulations to be trapped in a local potential energy minimum. Also, parallel tempering makes optimal use of parallel computers and it is compatible with statistical bias techniques. It is not only suitable for simulations in the *NVT* or *NPT* ensembles, but also in the Grand canonical ensemble [Yan and de Pablo, 1999] as well as for pure components and multicomponent systems.

As a consequence, parallel tempering is a very attractive option for many systems. It is particularly useful in problems where high energy barriers separate visited configurations. Typical examples are polymer materials or cations in zeolites (among others). Quite often, the range of temperatures used in such parallel tempering applications may seem strange, because the upper temperatures are very high. High temperatures are indeed needed to produce more rapid configurational changes through the usual Monte Carlo moves, even though they are unrealistic for the process under consideration. This provides the low temperature simulations with a larger array of possible configurations through exchange moves, i.e. a more efficient exploration of configuration space.

2.4 PRACTICAL IMPLEMENTATION

This section is devoted specially to the numerous special tricks that are used in addition to statistical theory to make simulations work in practice. These tricks range from applied mathematical techniques to the fruit of practical experience in selecting parameters. Although the subject has been treated to some extent in textbooks [Allen and Tildesley, 1987; Frenkel and Smit, 1996], it is probably the most difficult obstacle for beginners, so it will be covered in detail. In all fields of simulation which make an extensive use of computer memory and speed, a basic knowledge of the numerical methods involved is of great help, if not a prerequisite. This holds for reservoir engineering, seismic processing and fluid mechanics. And molecular simulation is by no means an exception to this rule.

2.4.1 What is Exactly a Simulation Box?

A. Periodic Boundary Conditions

If we consider an isolated simulation box comprising 1 000 molecules ordered on a cubic grid, the number of molecules in contact with the outside is 488, so that almost 50% of the molecules belong to the boundaries. If interactions were computed only within such an isolated box, the simulation would be highly subject to boundary effects and would fail to be adequately representative of a bulk or homogeneous system: the fact that molecules located at the boundary have fewer neighbours would lead to reduced attraction energy *vis-à-vis* their surroundings and this would affect the results. Using larger simulation boxes would reduce the fraction of molecules located at the boundaries of the box but computer time (which rises according to the square of the number of molecules) seriously restricts the practicability of this option.

The use of *periodic boundary conditions* is the classical way to solve the problem. This technique involves repeating identical replicas of the simulation box in all space directions (Fig. 2.27). In consequence, the molecules located close to the boundaries of the simulations box have the molecules of the opposite boundary as immediate neighbours and there is no more boundary effect. This method does not entail storing more atomic positions in the computer memory than in the case of an isolated simulation box because the positions in the replicas can be obtained from those in the original box by elementary translations. As illustrated in Figure 2.27, there are two ways to define the simulation box, depending on whether the molecules crossing the boundary are cut or left intact, but they are strictly equivalent.

The distances separating an atom from other atoms are computed using a modified procedure known as the *minimum image convention*. This procedure is equivalent to computing the separation distances in a new box centred around the selected atom (Fig. 2.28). Alternatively, periodic

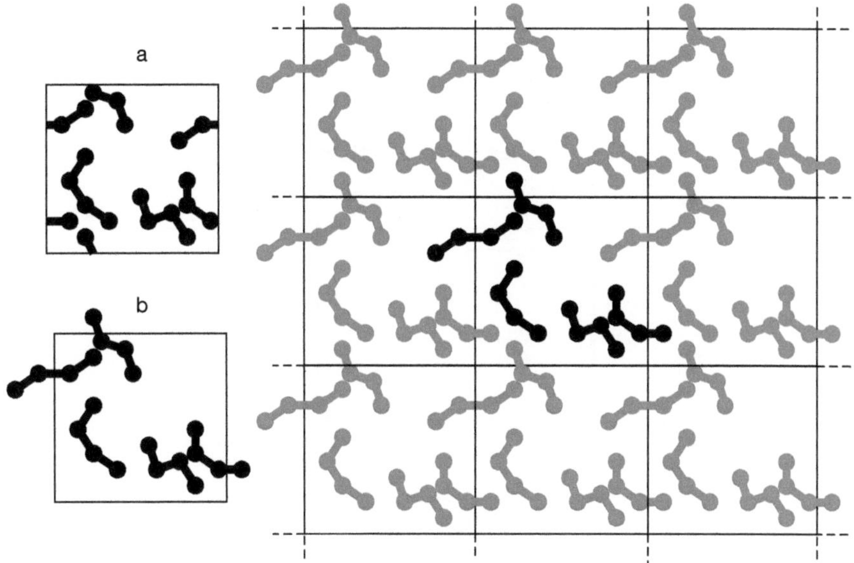

Figure 2.27 Periodic boundary conditions used to avoid the influence of system boundaries on simulation results. The simulated system is equivalent to an infinite crystal whose unit cell is the simulation box. (a) periodic boundary condition applied to atomic coordinates, cutting the molecules at box boundaries (b) periodic boundary condition applied to centre of mass coordinates, leaving molecules intact when crossing box boundaries.

boundary conditions may be viewed as assimilating the system to a crystal in which the elementary cell is the simulation box. Thus it may account for the amorphous character of a fluid at small scale. It is also an obvious way of treating crystalline adsorbents as will be seen in Section 2.4.2.

B. Cut-off of Intermolecular Interactions

As discussed in Section 2.2, intermolecular interactions – especially dispersion – repulsion forces – decrease with separation distance, and thus, to avoid time-consuming computations, it would be tempting to try to ignore the potential energy associated with separation distances greater than some threshold value. This is in effect a common procedure, and the threshold is known as the *cut-off distance*. From the example of Figure 2.13, we would say that a typical value for the cut-off distance would be10 Å.

However, the situation is not so simple because the number of interacting neighbours increases with separation distance, and this may compensate for the decrease of potential energy *per neighbour* with separation distance. This is because the average number of neighbours located at some separation distance r increases as r^2 in an homogeneous system. In consequence, the interaction energy of a given atom with *all* its neighbours decreases more slowly than suggested by the potential energy curve of a single pair as shown in Figure 2.13. If these effects are accounted for, it is found that the validity of imposing a cut-off distance depends on the range of the interaction considered.

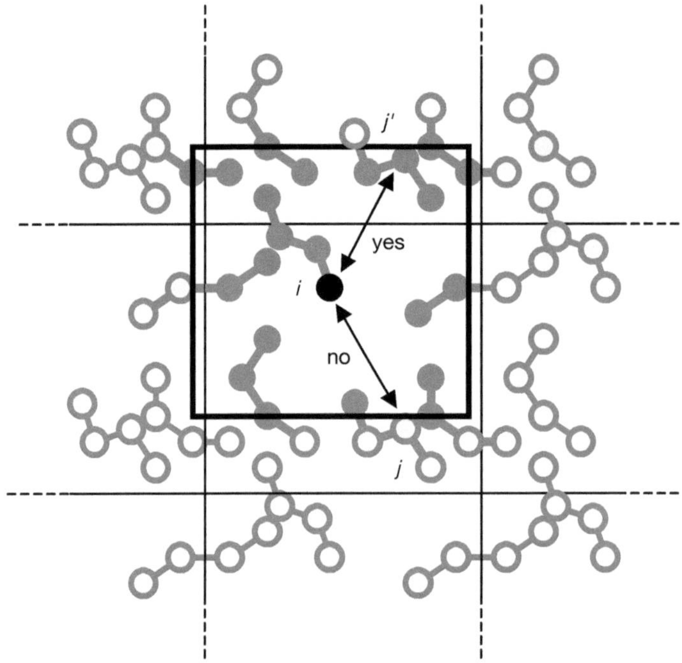

Figure 2.28 Minimum image convention used to compute molecular interactions of a force centre *i* together with periodic boundary conditions. Only the interactions with the closest image of other particles are counted. This is equivalent to restrict interactions to a box centred on force centre *i* (thick line).

Repulsive interactions have a very short range and they can be truncated beyond a cut-off of 8-10 Å without affecting results. Generally, no attempt is made to implement an approximate evaluation of the ignored repulsive interactions beyond the cut-off point.

Dispersion interactions decrease with $1/r^6$ so they are not compensated by the increasing number of neighbours. With typical cut-off values like 10 Å, the ignored part of the dispersion energy amounts to a few percentage points which is sufficient to cause significant distortion of the final results. In consequence, the ignored energy is estimated with the system considered as, being homogeneous, a procedure often referred to as long range correction [Allen and Tildesley, 1987]. The resulting working equation for the Lennard-Jones energy is thus given by:

$$U_{\mathrm{LJ}} = \sum_{\substack{i,j \\ i<j \\ r_{ij}<r_c}} 4\varepsilon \left(\left(\frac{\varepsilon}{r_{ij}} \right)^{12} - \left(\frac{\sigma}{r_{ij}} \right)^{6} \right) - \frac{8\pi\rho\varepsilon\sigma^6}{3r_c^3} \tag{2.78}$$

where r_c is the cut-off distance and r_{ij} is the separation distance between Lennard-Jones force centers i and j computed with the minimum image convention, and ρ is the average number density of force centres in the simulation box. A consequence of using the minimum image convention is that the cut-off cannot be greater than half the box size (it can be shown that this would lead to an improper account of all interactions). This condition must be respected at all stages of the simulation.

Quite often, the cut-off distance r_c is set to exactly half the box length. In the *NPT* or Gibbs ensembles, the cut-off distance is then matched to differences in box size. This presents the advantage of counting as many interactions as possible in an explicit fashion. The counterpart is that computing time may be longer in large systems, and small systematic differences may occur if the same system is treated with different box sizes.

Electrostatic interactions – which have a longer range of interaction – cannot always be treated by truncating any interactions beyond some threshold distance. This is particularly true for interactions between point charges (e.g. cations in a zeolite) because the energy *per pair of charges* only decreases according to $1/r$. In such a case, the energy does not converge when the cut-off distance is increased so that the Ewald summation method presented below has to be applied. In the case of strongly polar molecules like water, the electrostatic energy *per pair of molecules* decreases with the dipole-dipole interaction, i.e. according to $1/r^3$ if the molecules are taken in a fixed orientation. The energy is then converging slowly when the cut-off distance is increased. Truncation should be then replaced with more rigorous integration methods like reaction field [Neumann, 1985; Hansen and McDonald, 1986] or Ewald summation. Finally, a cut-off can be applied for moderately polar and quadrupolar molecules without introducing too much distortion because, in this case, among the attractive interactions, electrostatic energy plays a relatively small role compared with that of dispersion. When a cut-off is applied to electrostatic energy in polar molecules, care is taken that point charges belonging to the same molecule are treated in the same way (i.e. either considered or ignored); if this is not done, some molecules near the cut-off will bear net charges, thus changing significantly the interaction.

C. Ewald Summation

Ewald summation is a precise integration method used to compute the electrostatic energy with periodic boundary conditions. Basically, the problem stems from the fact that the electrostatic energy of an elementary charge with another charge is infinite – either positively or negatively so – when periodic boundary conditions are applied. Yet, the electrostatic energy of the system is finite if the net charge of the elementary simulation box is zero: this is because infinite terms of opposed signs compensate in the total sum (for those readers familiar with mathematics, it is a conditionally converging sum). Ewald summation has been treated in the literature [Allen and Tildesley, 1987, Nymand and Linse, 2000], so here we will just provide a simplified account to explain the importance of the numerical parameters.

Ewald summation involves adding a Gaussian distribution of screening charges $q_i^{\text{gauss}}(r) = -q_i \left(\frac{\alpha}{\sqrt{\pi}}\right)^3 \exp\left(-\alpha^2 \left|r - r_i\right|^2\right)$ around every point charge q_i located at position r_i so that screened charges $q_i^{\text{sc}}(r) = q_i + q_i^{\text{gauss}}(r)$ can be defined (Fig. 2.29). The whole charge of the Gaussian distribution is exactly opposite to the related point charge. The interaction energy U_j of any point charge q_j with the other charges may be computed from the electrostatic potential $\Phi(r)$ through $U_j = q_j\Phi(r_j)$. Using the general principle of superimposing states in electrostatics, the total electrostatic potential $\Phi(r_j)$ may be written as the sum of two terms:

- the contribution $\Phi'(r_j)$ due to screened charges $q_i^{sc}(r_j)$
- the contribution $\Phi''(r_j)$ due to Gaussian distributions $-q_i^{gauss}(r_j)$ opposite in sign to the screening charges,
- a correction factor for the screening charges of the point charge q_j itself.

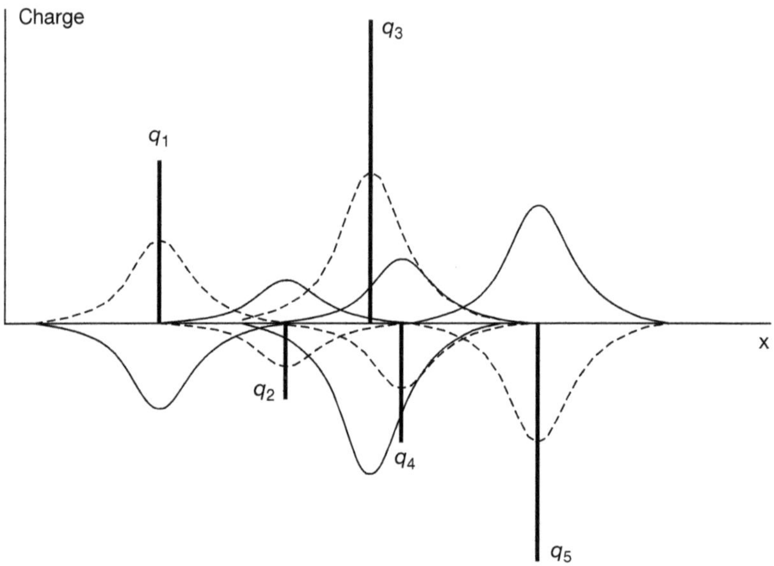

Figure 2.29 Electrostatic charge decomposition used in the Ewald summation method. Every point charge q_i (vertical bars) is associated with a screening gaussian distribution of opposite charge $q_i^{gauss}(r)$ (full line). The distribution of point charges is expressed as the sum of screened charges $(q_i + q_i^{gauss}(r))$ which converges fast and the opposite distributions ($-q_i^{gauss}(r)$, dotted lines) which is integrated by Fourier analysis.

The first term $\Phi'(r)$ is obtained by summing the contributions of all screened charges within a spherical cut-off (generally taken as half the box length). This contribution is indeed converging faster because the screened charges $q_i^{sc}(r_j)$ are neutrally charged entities. It can be computed rather simply from standard library functions, making the computation of

$\Phi'(r)$ somewhat similar to Lennard-Jones interactions. The second term $\Phi''(r_j)$ is obtained after careful derivation using Fourier transformations. Its final expression involves a summation of terms in Fourier space, i.e. a summation over vectors $\vec{k} = \left(\frac{2\pi}{L_x}k_x, \frac{2\pi}{L_y}k_y, \frac{2\pi}{L_z}k_z\right)$.

$$\Phi''(r_j) = \sum_{k_x=-k_{\text{ew max}}}^{k_{\text{ew max}}} \sum_{k_y=-k_{\text{ew max}}}^{k_{\text{ew max}}} \sum_{k_y=-k_{\text{ew max}}}^{k_{\text{ew max}}} \phi''(k_x, k_y, k_z) \tag{2.79}$$

where ϕ'' includes contributions for all screening charges $-q_i^{\text{gauss}}(r)$, and $k_{\text{ew max}}$ is an integer defining a cutoff distance in Fourier space.

The delicate point in using Ewald summation is the selection of the parameter α, which defines the range of the screening Gaussian distributions, and the cutoff distance $k_{\text{ew max}}$ in Fourier space. The condition that the Gaussian distribution must vanish within the dimensions of the simulation box may be expressed approximately as $\alpha > 2\pi/L$ where L is the minimum box dimension (the parameter α has the dimension of an inverse length). The cutoff $k_{\text{ew max}}$ must be sufficient to provide convergence of $\Phi''(r_j)$. It is observed that screening with very narrow Gaussian distributions (i.e. with high values of α) requires higher values of $k_{\text{ew max}}$, so that it is important to avoid excessive values of α to keep computing time within reasonable limit. Ewald summation is responsible for significant increases in computing time, mainly due to the triple summation involved in Eq. (2.79), the magnitude of which varies like $k_{\text{ew max}}^3$.

For instance, in the case of liquid water, when a small simulation box is used (20-25 Å), the optimal α is approximately 0.3 Å^{-1}, and $k_{\text{ew max}} = 6-7$ proves sufficient to make U'' converge. Slightly different values are obtained in zeolites. Nevertheless, it is recommended that the convergence of Ewald summation be tested with various $k_{\text{ew max}}$ values on a single configuration every time a new system is considered, as this does not require much computer time.

2.4.2 Modelling Microporous Adsorbents

Adsorption in a microporous solid is generally simulated in the Grand Canonical ensemble, as introduced in Section 2.3.3. It is not fundamentally different from the grand canonical simulation of fluid phases, but the specific features of adsorbents entail both simplifications and additional constraints.

The major simplification is the assumption that the solid has a negligible vapour pressure. This means that the atoms of the solid are not subject to insertion and destruction moves, and they are not treated in the same way as molecular species. This simplification is rather universal.

A second simplification, known as the Kiselev approximation, is that the solid is unchanged by adsorption of guest molecules [Fuchs and Cheetham, 2001]. This is a frequent approximation, because adsorption is often defined by reference to the lack of significant structural change in the adsorbent. In this case, the time taken to compute the interaction energy between guest molecules and the solid may be considerably shortened. Prior to starting the grand canonical simulation, the interaction energy between the solid and a single center of force is computed for various positions of the center on a three dimensional grid and stored (Fig. 2.30). During grand canonical simulation, interaction energies with the solid at any test position are obtained by interpolating in the precomputed energy grid. This takes much less computer time than computing hundreds or thousands of elementary interactions. The interaction energy between the atoms of the solid would play no role in the simulation, so it is not accounted for. On the other hand, the interaction energy between guest molecules must be accounted for. Although the related attraction energy is generally low, its contribution is essential to prevent the molecules from overlapping. If electrostatic interactions must be accounted for, it is also possible to store the electrostatic energy of a unit charge with the solid computed by Ewald summation on a grid.

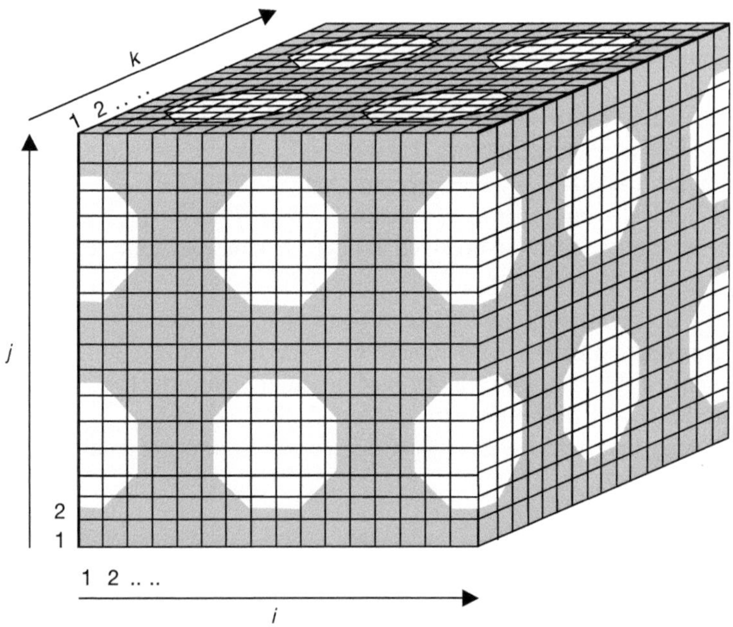

Figure 2.30 Schematic example of a precomputed energy grid used to simulate adsorption in the Grand Canonical ensemble. The interaction energy of an individual force centre with the solid is stored for every possible location *i, j, k* of the force centre on the grid. Energy is set to an arbitrary high value where the force centre overlaps with the atoms of the solid (shaded zones).

However, it is not always possible to consider the solid as being unchanged since significant displacements of atoms may occur. These may be of two types:

– vibrations of the solid structure which are known to influence transport properties but are generally considered to have a negligible influence on equilibrium properties,

- displacements of non-bonded atoms of the solid structure which may influence adsorption if their amplitude is significant. This occurs particularly with charge-compensating cations in zeolites when polar molecules are adsorbed.

When cation displacements cannot be ignored, it is still possible to treat the framework atoms as inert and the cations as mobile guests. This will be discussed in more detail in Section 2.4.

From a simulation point of view, there are mainly two types of microporous adsorbents: amorphous solids, that comprise mainly activated carbons, and crystalline solids, which comprise zeolites.

In the case of amorphous adsorbents, the problem of generating a representative solid structure is a significant problem. The Reverse Monte Carlo method has been deviced for this purpose, using informations from X-ray scattering and adsorption isotherms of small molecules [Pikunic *et al.*, 2001].

In the case of zeolites, it is of course required that periodic boundary conditions produce a solid structure compatible with the crystal structure of the zeolite. This requires that the simulation box is a multiple of the elementary crystal cell. When the crystal structure is cubic, a cubic simulation box is used in the same way as for fluid phases. When the structure is orthorhombic, i.e. the elementary cell has 90° angles but different cell lengths in x, y and z directions, the simulation box must have similarly different box lengths in x, y and z. Other structures (monoclinic, triclinic...) must be treated with non-orthogonal simulation boxes and specific boundary conditions [Allen and Tildesley, 1987].

2.4.3 What Type of Potential to Use?

Despite the unavoidable uncertainties attached to the statistical nature of Monte Carlo simulations, the factor which limits the accuracy of predictions is most often the availability of a suitable intermolecular potential. Besides, there are often several possible intermolecular potentials for a given application, with various levels of approximation as discussed in Section 2.2. In consequence, choosing a potential for a new application is, in practice, a delicate task.

A. Rigid vs Flexible

In many cases, the molecules can be assumed to be rigid which means that their intramolecular interaction energy U_{int} is considered constant. This is the case with noble gases (helium, neon, argon, krypton) – which is not surprising because they are monoatomic. It is also a common practice in the case of polyatomic molecules where internal deformations are small such as methane, ethane, H_2S, CO_2, ethylene, nitrogen, benzene, toluene, xylenes, etc. In polyatomic molecules, it is obvious that intramolecular energy cannot be ignored when estimating ideal gas properties like heat capacity (see Section 2.3.8 about the decomposition of heat capacity in ideal and residual contributions).

B. All Atoms or United Atoms?

As they are more representative of molecular structure, All Atoms potentials might be considered more accurate than United Atoms potentials. Related parameters are also available for most functional groups. However, several All Atoms potentials are rather poor in predicting equilibrium properties [Chen *et al.*, 1998]. The most likely reason for this is that their parameters are fitted only on the cohesive energy and liquid density in ambient conditions [Jorgensen *et al.*, 1993; Jorgensen and Madura, 1994; Faller *et al.*, 1999]. Some All Atoms potentials have been developed to provide better predictions of fluid phase equilibria [Chen and Siepmann, 1999] but they are still limited to linear alkanes. Although fluid-solid equilibria shall not be discussed in this book, it is useful to know that All Atoms potentials seem to be required in order to get a reasonable solid structure for linear alkanes [Polson and Frenkel, 1998a and 1998b].

United Atoms potentials have been calibrated on fluid phase equilibria for more chemical families (Table 2.3) but there are still numerous polar groups for which parameters have been calibrated on liquid properties only. Well-calibrated United Atoms or Anisotropic United Atoms potentials are to be preferred when available, because they require much less computing time and can thus be applied to larger systems.

In the specific case of adsorbate-zeolite interactions, some potentials use specifically calibrated Lennard-Jones coefficients σ_{ij} or ε_{ij} (i indexing the atoms of the zeolite and j the centers of force of the adsorbates). This may be done without explicitly stating the atomic parameters for the oxygen atoms of the zeolite [Krishna *et al.*, 1998; Vlugt *et al.*, 1999]. The combining rules discussed in Section 2.2.4. are then used for guest-guest, but not for guest-host interactions. Alternatively, effective Lennard-Jones parameters can be attributed to oxygen atoms to obtain guest-host parameters from routine combining rules [Pascual *et al.*, 2003].

C. When does Electrostatic Energy Need to be Considered?

The question of whether or not electrostatic interactions and polarisation are to be taken into account or ignored is also delicate when moderately polar molecules or functions are under consideration. As the number of point charges in a molecule is generally equal to or greater than the number of Lennard-Jones force centres, this question significantly affects computer time. We may consider three kinds of systems:

- low polarity systems, such as fluid alkanes,
- moderately polar systems, such as organic molecules with a moderate dipole moment (e.g. toluene, 0.375 D) or a quadrupole moment (CO_2),
- strongly polar systems characterized by a big dipole moment (e.g. water, 1.85 D; methanol, 1.70 D) or ions (e.g. zeolites with charge-compensating cations).

In the first case (low polarity), equilibrium properties are not very dependent on electrostatic interactions.

In the second case (moderate polarity), electrostatic energy is not required to produce an accurate representation of equilibrium properties. However, its influence is sufficient that a recalibration of the Lennard-Jones potential parameters is required when changing from an electrostatic model to a non-electrostatic model or *vice-versa*.

Table 2.3 Parameters of intermolecular potentials based on United Atoms (UA) and Anisotropic United Atoms (AUA) for selected groups (see also Appendix 1 for the AUA potential).

Family	Ref.	Type	Investigated molecules	Groups	σ (Å)	ε/k (K)	δ (Å)	α
n-alkanes	[1]	LJ-UA	Methane	CH_4	3.73	149.92	–	–
	[2]	Exp-6	Methane	CH_4	3.741	160.3	–	15
	[3] [a]	LJ-UA	C_5 to C_{48}	$-CH_3$	3.93	114	–	–
				$-CH_2$	3.93	47	–	–
	[4]	LJ-UA	C_2 to nC_{12}	$-CH_3$	3.75	98	–	–
				$-CH_2-$	3.95	46	–	–
	[5]	LJ-UA	C_2 to nC_{24}	$-CH_3$	3.91	45.8	–	–
				$-CH_2-$	3.91	104	–	–
	[6]	Exp-6	C_2 to nC_{78}	$-CH_3$	3.679	129.6	0.303 [e]	16
				$-CH_2-$	4.00	73.5	–	22
	[7]	LJ- AUA	C_2 to nC_{20}	$-CH_3$	3.607	120.15	0.216	–
				$-CH_2-$	3.461	86.29	0.384	–
Branched alkanes	[4]	LJ-UA	C_4 to C_8 isoalkanes	$-CH_3$	3.75	98	–	–
				>CH–	4.68	10	–	–
	[8] [b]	LJ-UA	C_4 to C_8 isoalkanes	$-CH_3$	3.79-3.90	70-102.6	–	–
				>CH–	3.85	39.7	–	–
	[9] [c]	LJ-AUA	C_4 and C_7 isoalkanes	>CH–	3.362	50.98	0.646	–
Cyclic alkanes	[10]	LJ-UA	cyC_5, cyC_6, cyC_8	$-CH_2-$	3.85	50.5	–	–
	[11] [c]	LJ-AUA	cyC_5, cyC_6, cyC_8	cyclic $-CH_2-$	3.461	90.09	0.336	–
	[12]	Exp-6	cyC_6	cyclic $-CH_2-$	3.91	77.4	-	20
Olefins	[13]	LJ-UA	1-butene, 1-hexene, 1-octene	$=CH_2$	3.905	89.93	–	–
				=CH–	3.915	81.69	–	–
	[14] [c]	LJ-AUA	C_2 to C_8 olefins and diolefins	$=CH_2$	3.48	111.1	0.295	–
				=CH–	3.32	90.6	0.414	–
				>C=	3.02	61.9	–	–
Aromatics	[12]	Exp-6	Benzene	arom. CH	3.71	74.06	–	20
	[15]	LJ-UA	Benzene	arom.CH	3.695	50.5	–	–
	[16, 17, 18] [c]	LJ-AUA	C_6 to C_{18} aromatics	arom. CH	3.246	89.41	0.407	–
				arom. C	3.246	37.72	0	–

[a] No long distance correction. [b] CH_3 parameters depend on group position. [c] see Section 3. [d] fixed bond angles. [e] use of a longer distance for the C–C bond involving CH_3, equivalent to use AUA for CH_3.

[1] Möller *et al.* [1992]. [2] Errington and Panagiotopoulos [1998]. [3] Smit *et al.* [1995]. [4] Martin and Siepmann [1998]. [5] Nath *et al.* [1998]. [6] Errington and Panagiotopoulos [1999]. [7] Ungerer *et al.* [2000]. [8] Nath and dePablo [2000]. [9] Bourasseau *et al.* [2002]. [10] Neubauer *et al.* [1999]. [11] Bourasseau *et al.* [2002]. [12] Errington and Panagiotopoulos [1999]. [13] Spyriouni *et al.* [1999]. [14] Bourasseau *et al.* [2003]. [15] Wick *et al.* [2000]. [16-17] Contreras-Camacho *et al.* [2004a and 2004b]. [18] Ahunbay *et al.* [2005].

In the third case (strong polarity), electrostatic energy cannot be ignored, and polarisability is also likely to have significant effects. Here again, recalibration of either Lennard-Jones parameters or point charges may be required when changing from a polarisable to a non-polarisable model.

When polar and non polar components are present in the same system, there is no single solution. If a few highly polar molecules are dispersed in a low-polarity fluid (such as CO_2 or water molecules in a natural gas), electrostatic issues can be ignored. If we are interested in the solubility of an alkane in an aqueous phase or in its adsorption in a zeolite, we should ideally account for its polarisation… but this is still seldom done in practice, due to the lack of relevant, well-tested models.

D. Validation

Whatever the type of potential selected, it is highly recommended to check that its implementation provides the same statistical averages as reported in the literature. Indeed, significant discrepancies can arise for various reasons. One reason is linked with the evaluation of the energy model, which is sensitive to box sizes, cut-off radius and the calculation method used for long-range electrostatic interactions. Another reason is statistical: the length of the simulation run, the frequencies of MC movements, and statistical bias.

Predictions should also be validated by comparison with reliable experimental information in representative conditions on pure compounds as well as mixtures. Although molecular simulation results are sometimes compared with equations of state calculations in the literature, the latter cannot be used as a reliable reference: we should not forget that equations of state are subject to fundamental drawbacks (such as incorrect near-critical behaviour, no detailed account of molecular structure, etc.) and it would be contradictory to validate molecular simulations on the basis of such results. There is however one substantial exception, namely the quasi-experimental equations of state that have been developed for numerous pure fluids like methane [Friend *et al.*, 1991], CO_2 [Angus *et al.*, 1973], H_2S [Goodwin, 1983] among many others. Such equations of state must be regarded as state-of-the-art interpolations between critically evaluated experimental data, which serve a very different purpose than engineering equations of state.

2.4.4 Optimisation of the Intermolecular Potential

The reader may be surprised that a book devoted to industrial applications contains a section focused on potential optimisation, because the development of intermolecular potentials is generally considered as a complex subject, to be tackled only by experienced researchers. The first basic reason for doing so is that recent methods make optimisation feasible with ordinary workstations. Another reason is that the natural tendency of engineers is to calibrate prediction methods on relevant data for the target system or process. It is thus likely that future applications of Monte Carlo simulation will reveal an increasing need for potential refinement tools in order to provide specific matches for a given property or particular P, T conditions. Rather than letting users work out their own trial-and-error optimisation systems, we prefer to provide them with procedures that have been well tested in research laboratories, at least for fluid phase equilibria.

Of course, the following developments are somewhat more specialised, and the first-time reader may skip to the next section without any problem. The main part of the following section is devoted to the optimisation of the dispersion-repulsion potential. A brief outline of how to determine electrostatic charges from quantum mechanical calculations is also provided in the end of this section, as this may be necessary for the simulation of polar molecules.

A. General Principle

The problem of optimising intermolecular potential parameters is a particular case of parameter identification with an *a priori* model (e.g. as discussed by Walter and Pronzato [1997]). The recommended approach for such problems involves the following steps:

1. definition of an error criterion that must be minimised, which ideally involves a large set of experimental data points to match,
2. determination of optimum parameters *and* their uncertainties, based on an extensive probing of parameter space,
3. validation of the model by comparison with a set of data that have not been used in the error criterion.

In such an approach, the systematic probing of parameter space in step 2 is intended to test the occurrence of several equivalent local minima, which is the sign of an ill-defined problem arising from over-abundant parameters with respect to the amount of independent experimental information. It must not be confused with local optimisation methods, such as gradient methods and the simplex method, which aim at finding the nearest local optimum, irrespective of the possible presence of better optima elsewhere.

This systematic approach is subject to two major difficulties in molecular simulation. The first is the cost of evaluating the error criterion, which involves as many Monte Carlo simulations as the number of reference state points considered. Thus it is difficult to consider a systematic probing of phase space, which involves hundreds or thousands of evaluations of the error criterion. The second difficulty is the statistical uncertainty associated with every molecular simulation result. As an illustration of this difficulty, the local gradient estimated from two points may indicate a wrong search direction (Fig. 2.31).

Nevertheless, some steps can be taken to make the optimisation procedure more efficient and its output more reliable. These consist mainly in defining a suitable error criterion, and then in using an efficient method to search for the parameters which minimise the error criterion. We also propose a series of systematic sensitivity tests that do not require additional molecular simulations and yet give a reasonable assessment of those parameters that can be optimised simultaneously.

B. Definition of the Error Criterion

An error criterion is most simply defined as the mean square relative deviation between experimental data and model results:

$$F(y) = \frac{1}{n}\sum_{i=1}^{n}\left(\frac{f_i^{\text{calc}}(y) - f_i^{\text{exp}}}{f_i^{\text{exp}}}\right)^2 \tag{2.80}$$

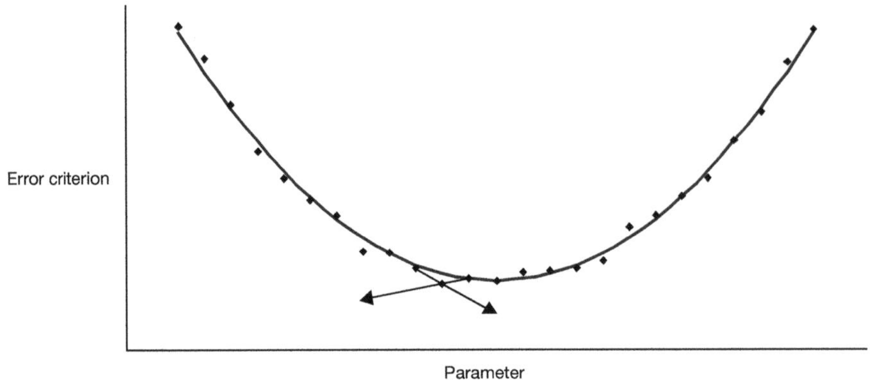

Figure 2.31 Problem raised by statistical uncertainties in the minimisation of the error criterion. The true value of the criterion is indicated by the full line, and evaluations of the criterion from simulations are shown as points. The gradient of the error criterion obtained from two simulations with close values of the optimised parameter may be significantly different from the true gradient.

where n is the number of reference data points f_i^{exp} and $f_i^{calc}(y)$ is the simulation result with potential parameters $y = (y_1, y_2, \ldots y_p)$.

However, this expression suffers from a fundamental drawback because the differences $(f_i^{calc} - f_i^{exp})$ may have strongly varying uncertainties. If we imagine for instance that one of the f_i^{exp} has a particularly high uncertainty, F will be excessively influenced by the error on this term and the parameters obtained by minimising F will be biased to match this false measurement.

A better alternative is the maximum likelihood criterion:

$$F(y) = \frac{1}{n}\sum_{i=1}^{n} \frac{\left(f_i^{calc}(y) - f_i^{exp}\right)^2}{s_i^2} \tag{2.81}$$

where s_i is the estimated statistical uncertainty on the difference $(f_i^{calc} - f_i^{exp})$, which may be evaluated from the uncertainties computed variable s_i^{calc} on computed results and s_i^{exp} on experimental measurements according to:

$$s_i^2 = {s_i^{exp}}^2 + {s_i^{calc}}^2 \tag{2.82}$$

When experimental uncertainty is small, as is often the case on well known compounds, s_i is approximately the uncertainty on the simulation result.

Apart from the application of a maximum likelihood criterion, an important step in making the optimisation process more efficient is to maximise the number of independent experimental data points f_i^{calc} that are considered in a given simulation run. In the simulation of liquid-vapour equilibria, the usual procedure of fitting the pure compound coexistence curve [Smit *et al.*, 1995; Nath *et al.*, 1998; Nath and de Pablo, 2000; Martin and Siepmann, 1998 and 1999] is equivalent to comparing two experimental data points, the vapour and liquid densities, with the results of a single Gibbs ensemble simulation. Taking vaporization enthalpy as a third data point is a way of adding independent information into the error criterion with no extra simulation cost [Ungerer *et al.*, 2000]. However, it would be useless to consider vapour pressure as a fourth variable in addition to vapour density, because both variables contain the same essential information. Vapour pressure is however preferable because it is measured directly more often than vapour density.

A final way of improving the error criterion is in the selection of state points and components. If we draw a straight line from a set of points, we know that there is much less uncertainty if the slope is determined on the basis of two points which are far apart rather than many points that are relatively close to each other. Similarly, potential parameters are more tightly controlled if two state points at very different temperatures are used rather than a large number of state points at relatively close temperatures. For instance, adding a monophasic state point at a low temperature is a way of tying down potential parameters more firmly than a large number of state points between the boiling point and the critical temperature.

The selection of reference components is also a delicate point. Selecting reference compounds of low molecular weight has the advantage that reliable experimental information is more frequently available and simulation statistics are better. However, extending the range of molecular weights is desirable because this extends the temperature range of reference equilibrium data. When several groups are optimised simultaneously, it is desirable to select components where one group is well represented and the others minon, so that a given compound brings information specific to a given group.

C. Searching for the Local Minimum

The minimum condition for the error criterion F defined in Eq. (2.81) is that the partial derivatives of F with respect to potential parameters y_j must be zero.

$$\frac{\partial F}{\partial y_j} = \frac{1}{n}\sum_{i=1}^{n} \frac{2\left(f_i^{\text{calc}} - f_i^{\text{exp}}\right)\dfrac{\partial f_i^{\text{calc}}}{\partial y_j}}{s_i^2} = 0 \quad \text{for } j = 1, \ldots, p \tag{2.83}$$

This expression may be parametrized by using the first-order Taylor expansion of f_i^{calc} around the starting point of the optimisation process, noted y^0:

$$f_i^{\text{calc}}(y^0 + \Delta y) = f_i^{\text{calc}}\left(y^0\right) + \sum_{k=1}^{p} \frac{\partial f_i^{\text{calc}}}{\partial y_k}(y^0)\,\Delta y_k \quad \text{for } i = 1, \ldots, n \tag{2.84}$$

where $\Delta y_k = y_k - y_k^0$

The minimum condition can then be expressed as:

$$\sum_{i=1}^{n} \frac{\left(f_i^{\text{calc}}\left(y^0\right) - f_i^{\text{exp}} + \displaystyle\sum_{k=1}^{p} \frac{\partial f_i^{\text{calc}}}{\partial y_k}\Delta y_k\right)\dfrac{\partial f_i^{\text{calc}}}{\partial y_j}}{s_i^2} = 0 \quad \text{for } j = 1, \ldots, p \tag{2.85}$$

The minimum of F is thus obtained from a linear system $A\Delta y = B$ with p unknowns Δy_k, $k = 1...p$ and p equations. If p is greater than n, the system is undetermined, as expected when attempting to determine more parameters than reference data points.

A key point in exploiting Eq. (2.85) is the evaluation of the partial derivatives $\frac{\partial f_i^{\text{calc}}}{\partial y_k}(y^0)$. This evaluation is based on the following fluctuation formula, which is true for any property $\langle X \rangle$ that can be obtained by an average in a statistical ensemble [Gray and Gubbins, 1984]:

$$\frac{\partial \langle X \rangle}{\partial y_k} = \left\langle \frac{\partial X}{\partial y_k} \right\rangle - \beta \left(\left\langle X \frac{\partial U}{\partial y_k} \right\rangle - \langle X \rangle \left\langle \frac{\partial U}{\partial y_k} \right\rangle \right) \tag{2.86}$$

where β is $1/kT$ and U is the potential energy, including internal contributions such as bending and torsion. A similar expression has been already used by Lyubartsev and Laaksonen [1995] and by Toth [2001] to optimize intermolecular potential parameters for the representation of liquid structure. Eq. (2.86) shows that the derivative of the average is not equal to the average of the instantaneous derivatives, but contains an additional term involving the derivative of potential energy $\partial U / \partial y_k$.

As an example, let us consider the liquid density obtained from a monophasic simulation in the *NPT* ensemble:

$$\rho = \frac{M_w N}{\langle V \rangle N_A} \tag{2.87}$$

where M_w is molecular weight, N_A is the Avogadro number and $\langle V \rangle$ the ensemble average of system volume with N molecules. Applying Eq. (2.86) to obtain the derivative of ρ versus the Lennard-Jones diameter σ yields:

$$\frac{\partial \rho}{\partial \sigma} = \frac{\beta M_w N}{\langle V \rangle^2 N_A} \left(\left\langle V \frac{\partial U}{\partial \sigma} \right\rangle - \langle V \rangle \left\langle \frac{\partial U}{\partial \sigma} \right\rangle \right) \tag{2.88}$$

Similar expressions can be obtained for other properties computed in the monophasic *NPT* ensemble or in the Gibbs ensemble. The evaluation of the instantaneous derivative $\frac{\partial U}{\partial y_k}$ in Eq. (2.86) is made by finite differences, i.e. by changing y_k by a small amount δ_k:

$$\frac{\partial U}{\partial y_k} \approx \frac{U(y_1, ..., y_k + \delta_k, ..., y_p) - U(y_1, ..., y_k, ..., y_p)}{\delta_k} \tag{2.89}$$

As it is sufficient to retain one configuration per 1 000 Monte Carlo iterations to evaluate the partial derivatives, the computational overhead is negligible, and the simulation progresses at the same speed as usual. However, convergence of partial derivatives requires longer simulation runs by a factor of two or so. This is not surprising, as the evaluation of thermodynamic derivative properties from statistical fluctuations requires longer runs as well [Lagache *et al.*, 2001]. In the cases investigated so far, the typical accuracy that can be reached on partial derivatives is 5-10%.

Once the partial derivatives $\frac{\partial f_i^{\text{calc}}}{\partial y_k}(y^0)$ are evaluated, the solution of Eq. (2.85) by standard methods of linear algebra provides the evaluated optimum y^1. At this stage, the error criterion (2.81) is determined again to check this new set of potential parameters. Ideally, the process of optimisation should be repeated to generate increasingly accurate estimates of the optimum y^2, y^3, etc. In fact, it appears sometimes useless because applications may be facing one of the two following cases:

case 1 – the maximum likelihood criterion $F(y^1)$ is significantly lower than 1, indicating that the deviation of the optimised model is lower than statistical uncertainty on average, and the simulation results $f_i^{\text{calc}}(y^1)$ agree reasonably with the estimates of Eq. (2.84).

case 2 – some of the optimised parameters y^1 have unrealistic values (such as negative energies for instance) and/or some simulation results $f_i^{\text{calc}}(y^1)$ do not agree at all with the estimates of the Taylor expansion (2.84).

In case 1, improving potential parameters without changing the error criterion is hopeless, and the optimisation is successful after one evaluation of the error criterion and related derivatives. Additional tests with new molecules are recommended to validate the optimised parameter set. Yet, it may happen that this validation is not fully successful, the most likely explanation being that some parameters are not well determined. This was the case of our first attempt to optimise the olefin potential, as we will discuss it in Section 3.1. In such circumstances, enlarging the experimental database and reducing parameter space are favourable ways toward significant improvements.

In case 2, the optimisation process has failed, and improvements are unlikely to be provided by additional iterations. This unfavourable case is found particularly when initialisation is poor, or when some parameters relate to "hidden" groups, i.e. atoms with more energetic substituents such as the carbon of CO_2 or the central carbon of neopentane. Here again, systematic sensitivity testing and reducing the parameter space may provide key improvements.

D. Sensitivity Tests and Parameter Space Reduction

Many tests do not require additional simulation runs, only additional solutions of the optimisation problem (2.85) which are obtained very fast on today's computers. This may serve either to correct for initial mistakes (such as optimising too many parameters at the same time) or to evaluate more thoroughly an apparently satisfactory optimum before a time-consuming validation campaign is launched.

Testing the sensitivity to small reference data changes or to simulation results may identify unstable optimisation problems linked with flat optima or bad initialisation, because the optimised parameters are often extremely sensitive in such cases. More generally, the dispersion of optimised parameters during repeated tests is a way to estimate their uncertainty.

The sensitivity to a reduction in the number of parameters can also be useful when it comes to analysing the stability of the optimisation process. This may be done in two ways:

– reduce the number of optimised groups from a given data set (a subset may be used as well). Obtaining small parameter changes with respect to the full optimisation is a good indication that a meaningful optimum has been found. This is illustrated by Table (2.4) in the case of the CH_2 and CH_3 groups of *n*-alkanes.
– reduce the number of parameters for poorly determined groups (generally hidden atoms less energetic than their neighbours). This is a good way to stabilise the optimisation when the variations of two parameters (such as ε and σ) compensate each other in some way.

However, care is warranted because these tests may be inappropriate if the starting point is too far from the optimum or if potential improvement requires coupled variations of several parameters.

Sensitivity tests to the initial parameters are also recommended if there is some doubt about the optimum, but it requires the error criterion to be computed again.

Table 2.4 Sensitivity of the optimization of the CH_2 and CH_3 groups of *n*-alkanes to the selection of optimized groups.

Parameter	Starting point [Toxvaerd, 1997]	Simultaneous optimization of CH_2 and CH_3 [Ungerer, Beauvais *et al.*, 2000]	Optimization of CH_2 alone	Optimization of CH_3 alone
ε/k (CH_2) (K)	79.87	86.29	87.03	
σ (CH_2) (10^{-10} m)	3.516	3.461	3.509	
δ_{AUA} (CH_2) (10^{-10} m)	0.40	0.384	0.419	
ε/k (CH_3) (K)	119.8	120.15		128.05
σ (CH_3) (10^{-10} m)	3.516	3.607		3.597
δ_{AUA} (CH_3) (10^{-10} m)	0.18	0.216		0.219

E. Optimisation of Electrostatic Charges from Quantum Mechanical Calculations

The determination of electrostatic charges has been investigated in detail for many chemical families in the literature [see for instance Jorgensen *et al.*, 1993; Jorgensen *et al.*, 1983; Jorgensen, 1986; Jorgensen, 1986]. This determination is based on quantum chemistry calculations on the isolated molecule. The charges are placed on the atomic nuclei, and their intensity is optimized to reproduce the best possible electrostatic field around the molecule. This optimisation is based on a least-squares procedure comparable to that discussed above in the context of dispersion and repulsion parameters.

Although many problems can be addressed with Jorgensen's work, there are still cases where optimisation of the partial electrostatic charges may be useful. This is particularly true for poorly known flexible molecules. In Section 3.4, a few such cases (e.g. organo-mercuric compounds, organic sulphides and thiols) will be reviewed. The brief outline provided here is therefore devoted specifically to the specific optimisation method applicable to flexible polar molecules, as developed by Delhommelle *et al.* [2000].

In the case of a flexible molecule, several different conformations may correspond to minimum energy, with each conformation having a different electronic structure. For instance, the molecule of butane displays three energy minima at torsion angles of 0, 120° and 240°. In order to sample the possible changes of electronic structure with conformational changes, the molecular geometry is optimised under constrained torsion angles. This geometrical optimisation is a classical feature offered by quantum chemistry packages. The quantum chemistry calculation provides also the electrostatic potential $V(P_k)$ on a grid of points P_k around the molecule for each optimised geometry.

The electrostatic potential created by a set of N partial electrostatic point charges q_i is given by the alternative expression of the Coulomb's law (2.28):

$$V_{fit}(P_k) = \sum_{i=1}^{N} \frac{q_i}{4\pi\varepsilon_o r_{ik}} \tag{2.90}$$

where r_{ik} is the distance between P_k and the charge q_i.

Then the optimisation consists simply in minimising the least-squares criterion:

$$F(q_1, q_2, \ldots q_N) = \sum_k \left(V(P_k) - V_{fit}(P_k)\right)^2 \tag{2.91}$$

The problem encountered with routine minimisation procedures provided in quantum chemistry packages is that several sets of electrostatic charges may account for the electrostatic potential generated by a given molecular conformation. In other words, the problem is underdetermined, and erratic variations of electrostatic charges are found with different molecular conformations. In order to solve this problem, it has been proposed that the space of optimised parameters q_i be restricted through a systematic procedure involving eigenvalue analysis of the linearised optimisation problem [Levy and Enescu, 1998]. Without entering into the mathematical details of this procedure, it can be said that the eigenvalue analysis identifies a subspace which concentrates most of the influence of the charges on the least-squares criterion. Another efficient reduction of the optimised space is to consider that the charges attributed to identical groups must be the same all over the molecule [Delhommelle, *et al.*, 2000]. When both ways of reducing the dimension of the space of optimised charges are implemented, it is found that the optimum electrostatic charges depend much less on the molecular conformation. This point will be illustrated in the examples of Section 3.3.

2.4.5 Selection of Numerical Parameters

A. Statistical Bias Parameters

As exposed in Section 2.3.6, configurational bias Monte Carlo (CBMC) is based on generating a set of test positions for the next atom when growing a molecule. Typical values for the number of tests positions are $k_{max} = 5-50$ for the first atom and $k_{max} = 5-10$ for the next ones, the larger values being used in dense liquids or industrial adsorbents. As a general rule, larger k_{max} will improve acceptance probabilities for CBMC (regrowth, GCMC insertions and destructions, GEMC transfers), but the average computer time per move will increase.

In reservoir bias, it is recommended that numerous positions (typically 5 to 50) are tested in the first step of the move, i.e. selection of a favourable place for insertion. Indeed, this first step involves single Lennard-Jones particle insertions, which are less time-consuming than whole molecule insertions. The diameter of the test Lennard-Jones particle must be such that it is included in the whole molecule, otherwise equilibrium may be more difficult to reach [Bourasseau *et al.*, 2002a]. For elongated or flat molecules, it should be thus taken as the diameter of a typical individual group rather than as the average molecular diameter. In the second step, the typical number of trial molecular orientations and conformations is between 5 and 20. Here again, increasing the number of tested positions increases the acceptance ratio.

Whether the statistical bias technique is configurational bias or reservoir bias, increasing the number of trial positions beyond some point is useless because the acceptance ratio of these moves is ultimately limited by the availability of suitable empty space in the liquid or adsorbent (Fig. 2.32). This is why we do not consider more than approximately 10 trial positions during the regrowth process in CBMC, or 20 trial conformations in the second step of the reservoir bias. However, we recommend that users carry out their own tests!

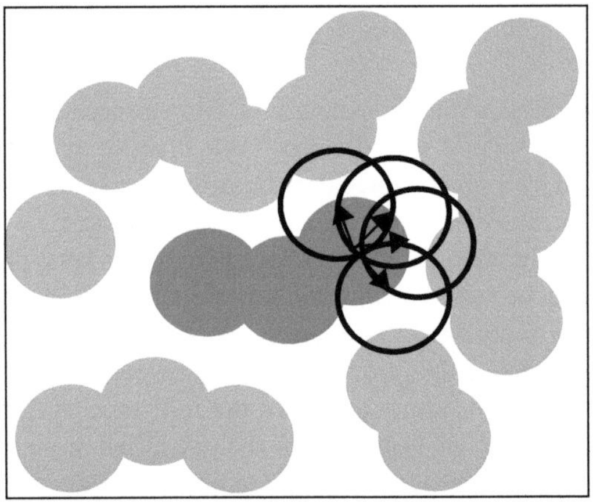

Figure 2.32 Selection of the adequate number of test positions in the configurational bias. Using several test positions increases the probability to find a position that does not overlap with neighbours, if there is any. Using several tens of test positions would not be much more successful because the efficiency would be limited by the availability of empty space in the system.

B. Frequency of the Various Monte Carlo Moves

Selecting the frequencies of the various Monte Carlo moves is a delicate problem. These parameters do not influence the final results but they control the efficiency of the exploration of configuration space and they must be adequately matched to the type of simulation considered. Although self-adapting methods have been proposed for this purpose in the specific case of polymers [Leblanc *et al.*, 2003], there is no well-established general rule for selecting optimal frequencies. Nevertheless, the following guidelines may be proposed as a starting point, keeping in mind that users are encouraged to devise more appropriate procedures for their own applications:

- volume changes: a frequency p/N may be used for a p-phase system with N molecules,
- transfers and insertions/deletions: the frequency of transfers in the Gibbs ensemble and insertions/deletions in the Grand Canonical ensemble is generally comprised between 30% (high temperature, i.e. low density liquids) and 80% (low temperature). This frequency should be calibrated in such a way that the total number of accepted transfer moves is at least one tenth of the accepted internal moves (translations, rotations, ...),
- internal moves (translation, rotation, partial regrowth, flip, reptation, pivot): the global frequency of these moves must add up to 1 with volume changes, transfers or insertions/destructions. Their distribution depends mainly on molecular type:
  - monoatomic molecules without electrostatic charges (methane, noble gases): translations only,
  - monoatomic molecules with electrostatic charges and rigid polyatomic molecules: translations and rotations with equal frequencies,
  - flexible molecules of moderate size (10 atoms or less): translation, rotation and regrowth with equal frequencies (include reptation as well in the case of linear molecules, and pivot in the case of highly branched molecules),
  - flexible molecules of large size (more than 10 atoms): translation, rotation, regrowth, flip with matched frequencies (fewer translations and rotations than flips and regrowth); include reptation as well in the case of linear molecules, and pivot in the case of highly branched molecules.

When large energy barriers are suspected inside the simulation box in the *NVT* or *NPT* monophasic ensembles, it may be useful to use displacement moves as well, since other internal moves are not efficient when it comes to jumping over such barriers.

2.4.6 Selection of System Size and Initial Conditions

For fluid systems, the total number N of molecules in Monte Carlo simulations is generally between 100 and 2000, the lower end of the range corresponding to large molecules which would require excessive computer time otherwise, the upper end to particularly difficult systems, e.g. near-critical conditions, multicomponent systems, and phase equilibria involving low solubilities. When simulating phase equilibria, it is generally recommended that a simulation box contain not less than 200 molecules for a liquid and 10 molecules for a vapour (to avoid significant finite size effects). The relationship between density ρ (kg/m^3) and box length L (m) is: $\frac{NM_w}{N_A L^3} = \rho$ where N_A is Avogadro number and M_w molecular weight (kg/mol). In consequence the typical size for a simulation box is 20 to 40 Å for a liquid phase and

30 to 200 Å for a vapour phase. The usual procedure is to start the simulation with molecules placed on a regular cubic lattice, or with a configuration coming from a previous simulation of the system.

In the case of simulations at imposed pressure – either monophasic or multiphasic – the initial box size should not influence the final result. However, starting from an excessively low density may cause the simulation to converge towards a vapour-like system while a liquid would be more stable (in the same way as with equations of state). Starting from a density higher than the expected liquid density is not recommended because it often results in molecular overlap, and this may cause numerical problems because of extremely high initial energies.

In the case of simulations at imposed global volume – generally in the Gibbs ensemble – global volume controls the respective proportions of the various phases. Care must be thus taken that the global volume and mole numbers for each species are high enough to provide representative average numbers in every phase (Fig. 2.33).

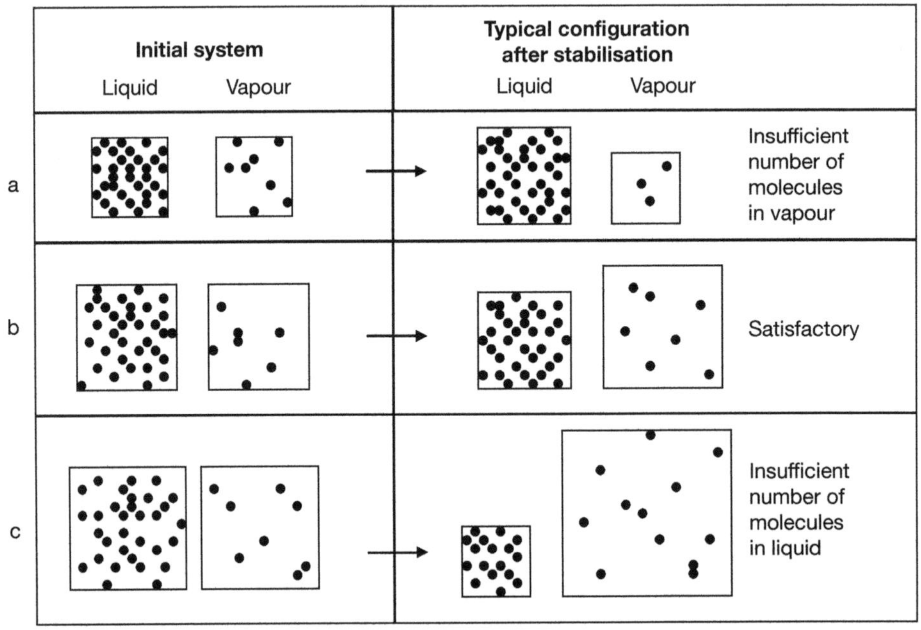

Figure 2.33 Influence of system size on a Gibbs ensemble simulation at constant global volume.

When simulating adsorption in crystalline microporous materials, the simulation box length must be a multiple of the elementary crystal cell in either direction. In order to treat representative systems, it is common to consider 500 atoms or more to define the adsorbent, so that at least 10-20 guest molecules are found on average. In the case of amorphous adsorbents much larger systems are needed [Pikunic *et al.*, 2001]. The system is generally initialised without any guest molecule, and progressive filling is provided by GCMC insertion moves.

2.4.7 Convergence and Statistical Uncertainties

A. Frequency of Storage

If variables like volume, energy, pressure are saved for all steps of the Markov chain, it is observed that they are correlate closely with one another, i.e. the variable at step n is much closer from its value at iteration $(n - 1)$ than it would be expected from the average amplitude of fluctuations. In consequence, it is not necessary to store the results for all iterations because most of the information stored would be redundant. The common practice is to store results only once over several hundreds of elementary steps. Also, this saves computer time because it would be very time-consuming to compute the pressure at each step since this involves a double sum over force centres and charges. This frequency of storage is typically every 500, 1 000, 2 000 steps, depending on the system being considered. Doing this is theoretically justified because a random subset of a statistical ensemble is still a statistical ensemble. Collecting configurations with a large frequency does not alter averages nor fluctuations. It is also possible to store intermediate averages for properties like volume or energy, i.e. an arithmetic average over the few hundred steps since the last stored value. This allows final averages to be computed without any loss in accuracy.

B. Equilibration

When a simulation is initialised as discussed in Section 2.4.6, the initial configuration is generally more ordered than those associated with the statistical ensemble. Also, its density and energy may be far outside the possible range of variation in the statistical ensemble considered. In consequence, a simulation always starts with an equilibration period during which molecular positions get more random and variables like density and energy progressively reach a regime of fluctuations around long-term averages (Fig. 2.34). The collection of the configurations of the statistical ensemble must start only once this regime is reached, otherwise the averages are influenced by the initial configuration. It is thus important to determine when the equilibration period is finished. As illustrated in Figure 2.34, this may be highly variable. Generally, equilibration is longer for complex molecules (especially flexible), for high densities, for low temperatures and for molecules displaying strong polar interactions like water and alcohols. In some cases like high molecular weight polymers, equilibration is very slow although specific statistical bias moves are used.

Unfortunately, there is no automatic reliable procedure to test whether or not equilibration has been reached. The minimum check is to verify that density and energy have reached equilibrium by plotting them as in Figure 2.34. In addition, decorrelation of molecular positions from their initial values can be tested by checking that centres of mass have been changed by as much as half the molecular length on average [Allen and Tildesley, 1987]. In the case of flexible molecules, it is wise to check that internal conformations have reached equilibrium, i.e. that the most frequent conformations are indeed the most stable according to the classic rules of organic chemistry [Vollhardt and Schorre, 1999]. This may be done by plotting configurations on the screen.

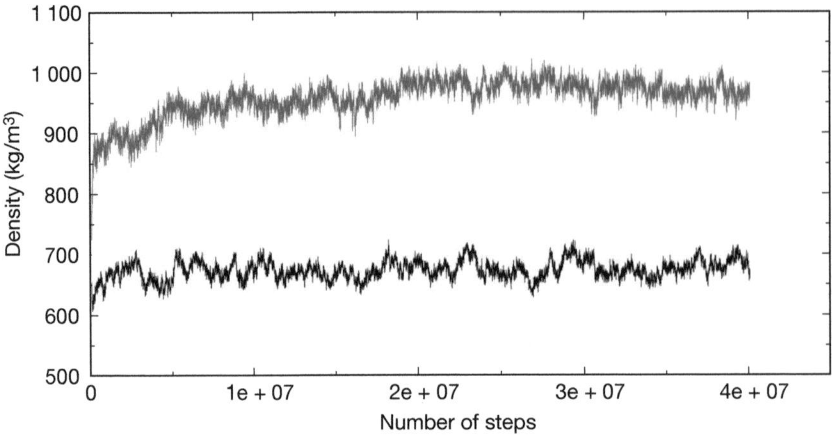

Figure 2.34 Example of equilibration of density in a difficult case, the liquid-liquid phase equilibrium calculation of the water-hydrogen sulphide system at 343 K, 7 MPa. The denser aqueous phase requires approximately 2 10^7 Monte Carlo steps to equilibrate, i.e. to reach a regime of fluctuations around 970 kg/m^3. The H_2S-rich liquid phase equilibrates in approximately 2 10^6 steps.

C. Length of the Averaging Period

Once the system has reached equilibrium, how long should the simulation continue collecting averages? The answer depends on the desired precision. Indeed, the statistical uncertainty on averages decreases with increasing number of elementary steps. Nevertheless, there are minimum requirements that have to be respected.

Referring to monophasic systems, Allen and Tildesley [1987] recommended collecting averages over at least 10 000 to 20 000 Monte Carlo cycles (for a phase with N molecules, a Monte Carlo cycle comprises one volume change move, N translations and N rotations). For complex systems made of large flexible molecules, mixtures or polar fluids, Monte Carlo cycles must contain also internal moves such as flips, and the averaging period should be longer because it takes more time to explore the configuration space.

In Gibbs ensemble simulations, an additional criterion is that the number of accepted transfer moves is 10 times larger than the number of molecules in the simulation boxes, so that every molecule has undergone several transfers on average. This criterion make simulations much longer in dense phases (typically at temperatures lower than the normal boiling point). Similarly, the number of accepted insertion or deletion moves in the Grand Canonical ensemble should be at least an order of magnitude larger than the average number of molecules.

D. Estimation of Statistical Uncertainties

If a stored array of results X_i such as volume or pressure behaves like a random statistical variable, the standard deviation σ on its average can be evaluated with the classical standard expression:

$$\sigma^2 = \frac{1}{n-1}\left(\left\langle X^2\right\rangle_{\text{run}} - \left\langle X\right\rangle_{\text{run}}^2\right) \tag{2.92}$$

Where the subscript *run* refers to an average over the whole interval. The standard deviation decreases thus like $\frac{1}{\sqrt{n-1}}$, i.e. approximately $\frac{1}{\sqrt{n}}$ for large values of the number of stored results n.

As discussed above, the results of a simulation are correlated. Storing results every 500 or 1 000 iterations is generally not sufficient to decorrelate them, so that Eq. (2.92) is not valid. The recommended method involves estimating the degree of correlation by performing averages on subsets, known as block averages [Allen and Tildesley, 1987]. If the file is decomposed in n_b blocks containing τ_b results each, it is thus possible to estimate a "statistical inefficiency":

$$\varsigma = \lim_{\tau_b \to \infty} \frac{\tau_b \sigma_b^2}{\sigma^2} \tag{2.93}$$

where σ^2 is the uncorrelated standard deviation computed according to (2.92), and $\sigma_b^2 = \frac{1}{n_b - 1}\sum_{b=1}^{n_b}\left(\left\langle X\right\rangle_b - \left\langle X\right\rangle_{\text{run}}\right)^2$ is the standard deviation observed among block averages of length τ_b. The standard deviation on the run average $\left\langle X\right\rangle_{\text{run}}$ is then:

$$\sigma_{\text{run}}^2 = \varsigma\sigma^2 \tag{2.94}$$

The standard deviation should not be confused with the uncertainty, which is generally meant as a confidence interval. If we want to be 99% sure that the true average is in the confidence interval, the uncertainty must be set to 2.5σ.

2.4.8 Calculation of Thermodynamic Properties

Several useful thermodynamic properties cannot be determined from simple averages in the same way as pressure, energy, or enthalpy can. This is particularly true for critical properties, derivative properties, excess volume and excess enthalpy. In the following paragraph, extensive properties for a given number of molecules will be noted with upper case symbols, while molar properties will be noted with lower case letters.

A. Critical Properties of Pure Compounds

Because of the increasing characteristic size of the density fluctuations, Gibbs ensemble simulations cannot be performed in the close vicinity of the critical point. Therefore, the determination of the critical temperature T_c and of the critical density ρ_c are generally based on the knowledge of several liquid-vapour coexistence points, assuming the following scaling law:

$$\rho^L - \rho^V = A\left(T - T_c\right)^\beta \tag{2.95}$$

where $\beta \approx 0.325$ is a characteristic universal exponent [see for instance Wilding, 1995; Barrat and Hansen, 2003].

It is also assumed that the densities of the coexisting liquid and vapour obey the so-called law of rectilinear diameters:

$$\frac{\rho^V + \rho^L}{2} - \rho_c = B(T - T_c) \tag{2.96}$$

Practically, T_c and ρ_c are regressed by a least square method minimizing the average deviation between the above equations and the simulated coexistence densities. Examples are given in Chapter 3.

Once these properties are obtained, the critical pressure may then be estimated by extrapolating the Clapeyron plot to $1/T_c$. However, the critical pressure obtained by this procedure is often subject to significant uncertainties.

B. Critical Properties of Binary Mixtures

When a binary mixture involves a liquid component and a supercritical component, the isothermal (P, x) diagram may present a liquid-vapour critical point which corresponds to the higher pressure of the two-phase envelope. For the same reason as pure compounds, Gibbs ensemble simulations cannot be performed close to the critical point. The critical pressure, critical density and critical compositions must be determined by some extrapolation, based on adequate scaling laws. The problem is more delicate with binary systems than with pure compounds, because it requires the simultaneous representation of two coexistence curves in the pressure-density (P, ρ) diagram and in the pressure-composition (P, x) diagram. While the pressure-density coexistence curve of binary mixtures is much alike the temperature-density curve of pure compounds, the pressure-composition coexistence curve may display a quite different shape, depending on the temperature considered. This is illustrated by the butane-carbon dioxide system, for which accurate data are available from Hsu *et al.* [1985]. At the temperature of 377.6 K, the (P, x) coexistence curve is rather flat near the critical point, while it is much narrower and sharp at 319.3 K (Fig. 2.35).

In order to match the pressure-density behaviour, the same scaling laws as pure compounds apply, except that temperature is replaced by pressure:

$$\rho^L - \rho^V = \gamma(P - P_c)^\beta \tag{2.97}$$

$$\frac{\rho^V + \rho^L}{2} - \rho_c = \lambda(P - P_c) \tag{2.98}$$

In the near-critical region of binary mixtures, the composition difference has been found to behave in the same way as density, i.e. the leading term is of the type $(P - P_c)^\beta$ [Rainwater and Williamson, 1986; Moldover and Rainwater, 1988]. In an investigation of the scaling behavior of Lennard-Jones fluid mixtures by histogram reweighting, Potoff and

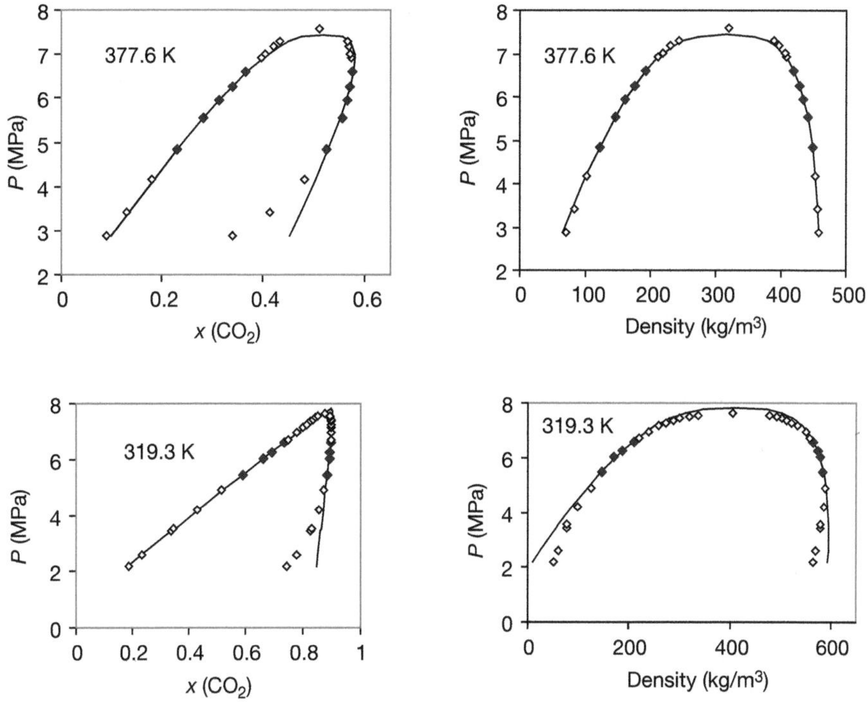

Figure 2.35 Liquid-vapour coexistence curves of the butane-carbon dioxide system at 377.6 K and 319.3 K in the pressure-composition diagram (left) and pressure-density diagram (right). Experimental data, indicated by symbols, are taken from Hsu *et al.* [1985]. The regressed scaling laws (2.101) and (2.102) are indicated by the full line. The reference data points for the regression are shown as full symbols.

Panagiotopoulos [1998] found that the same critical exponents apply to mixtures and pure fluids. Here it is proposed to represent the pressure-composition diagram by complementing the near-critical scaling law with a linear term:

$$x_v - x_l = \lambda_1(P_c - P) + \mu(P_c - P)^\beta \tag{2.99}$$

Compared with pure components, this may be seen as adding a higher order term in a Taylor-like expansion to match the observed behaviour at some distance from the critical point. In the close vicinity to the critical point, the linear term vanishes compared with the β-exponent term.

An equivalent relationship as Eq. (2.98) may be used for the mid-composition:

$$\frac{1}{2}(x_v + x_l) = \lambda_2(P_c - P) \tag{2.100}$$

The resulting expressions for the coexisting densities and compositions are then given by:

$$\rho_i = \rho_c + \varepsilon \frac{\gamma}{2}(P_c - P)^\beta + \lambda(P_c - P) \tag{2.101}$$

$$x_\varphi = x_c + \left(\lambda_1 - \varepsilon \frac{\lambda_2}{2}\right)(P_c - P) - \varepsilon \frac{\mu}{2}(P_c - P)^\beta \tag{2.102}$$

where $\varepsilon = 1$ for the liquid phase and $\varepsilon = -1$ for the vapour phase.

For a given temperature, the regression of the eight parameters P_c, γ, λ, ρ_c, λ_1, μ, x_c, λ_2 involved in these expressions (2.97 to 2.100) is possible from a set of coexistence points located below the critical point. In Figure 2.35, two examples are shown where this procedure is tested against an experimental data set. In order to reproduce the type of conditions prevailing in Monte Carlo simulations, the regression of the eight parameters has been applied to a subset of data corresponding to pressures significantly below the critical point. It can be seen that the critical density, critical pressure and critical composition are correctly extrapolated. Care must be taken to the selection of the data points in this procedure. Indeed, low pressure points must not be considered because the Taylor-like expansion of Eqs. (2.101) and (2.102) does not hold far away from the critical point.

C. *Vaporization Enthalpy*

The molar vaporization enthalpy, or heat of vaporization, is readily obtained as the difference of the molar enthalpies of the coexisting vapour and liquid phases in a Gibbs ensemble simulation. This leads to:

$$\Delta_{\text{vap}} h = N_a \left(\frac{\langle U^V \rangle}{\langle N^V \rangle} - \frac{\langle U^L \rangle}{\langle N^L \rangle} + P^V \left(\frac{\langle V^V \rangle}{\langle N^V \rangle} - \frac{\langle V^L \rangle}{\langle N^L \rangle} \right) \right) \tag{2.103}$$

Where U, N and V stand for potential energy, number of molecules and volumes, the superscript indicating the phase considered. This expression assumes that kinetic energy is the same in both phases, which is generally a good approximation because the kinetic energy is known to be equally distributed along all degrees of freedom.

A simplified expression can be used if the pressure is sufficiently low that the vapour behaves as a perfect gas:

$$\Delta_{\text{vap}} h = -N_a \frac{\langle U^L_{\text{ext}} \rangle}{\langle N^L \rangle} + RT \tag{2.104}$$

Where U_{ext} is the external potential energy. This simplified equation is convenient at temperatures lower than the boiling point, because the perfect gas law is then obeyed by the vapour (except some strongly associating compounds like acetic acid which form aggregates in the vapour phase at low pressure). However, it assumes that the internal energy of the molecules is the same in both phases. This may be untrue for long chain molecules (above 20 carbon atoms), because they may adopt a more condensed structure in the vapour phase than in the liquid.

D. Thermodynamic Derivative Properties

Thermodynamic derivative properties are second order derivatives of the thermodynamic potential, as opposed to variables like pressure, molar volume or enthalpy which are first order derivatives of the thermodynamic potential. The most common derivative properties are heat capacity, isothermal compressibility and isobaric thermal expansivity. They are useful to predict the behaviour of a system subject to a change in temperature, pressure or volume.

As indicated in Section 2.1.3, derivative properties can be determined in principle from the statistical fluctuations. As an example, it is straightforward to use Eq. (2.20) to determine the compressibility coefficient $\beta_T = -\frac{1}{\langle V \rangle}\left(\frac{\partial \langle V \rangle}{\partial P}\right)_T$ from Monte Carlo simulations.

On the opposite, heat capacity cannot be obtained so simply because its calculation involves the fluctuations of the *total* energy (Eq. 2.19), while Monte Carlo results provide only the *potential* energy. In a way similar to equations of state, Monte Carlo simulation allows only the determination of residual properties, i.e. the difference between the non-ideal and the ideal heat capacity. When a proper account of the various forms of energy is made in the partition function, the molar residual heat capacity c_P^{res} and the thermal expansivity α_P may be obtained from the following fluctuation formulae [Lagache *et al.*, 2001]:

$$\alpha_P = \frac{1}{\langle V \rangle}\left(\frac{\partial \langle V \rangle}{\partial T}\right)_P = \frac{1}{\langle V \rangle kT^2}\left(\langle V\hat{H}\rangle - \langle V \rangle\langle \hat{H}\rangle\right) \tag{2.105}$$

$$c_P^{\text{res}} = \frac{N_a}{kNT^2}\left(\langle U^{\text{ext}}\hat{H}\rangle - \langle U^{\text{ext}}\rangle\langle H\rangle\right) + \frac{N_a P}{kNT^2}\left(\langle V\hat{H}\rangle - \langle V\rangle\langle\hat{H}\rangle\right) - N_a k \tag{2.106}$$

where $\hat{H}$ is the configurationnal enthalpy (Eq. 2.111), U^{ext} the intermolecular potential energy, N_a the Avogadro number, N the number of molecules.

The total heat capacity at constant pressure is obtained by adding the ideal heat capacity $c_P^{\text{id}}(T)$ to the residual heat capacity. For a mixture, $c_P^{\text{id}}(T)$ can be easily calculated from the pure component ideal heat capacities $c_P^{\text{id}}(T) = \sum_i x_i c_{Pi}^{\text{id}}(T)$ where x_i is the molar fraction of each compound. The pure component ideal heat capacities $c_{Pi}^{\text{id}}(T)$ can be obtained from experimental correlations or from group contribution methods [Poling *et al.*, 2001; Benson, 1976].

The Joule-Thomson coefficient, i.e. the derivative of temperature versus pressure at constant enthalpy, is expressed by the following equation [Alberty and Silbey, 1997]:

$$\mu_{\text{JT}} = \left(\frac{\partial T}{\partial P}\right)_H = \frac{1}{c_P}\left[T\left(\frac{\partial v}{\partial T}\right)_P - v\right] = \frac{v}{c_P}\left[T\alpha_P - 1\right] \tag{2.107}$$

where $v = (N_a / N)\langle V \rangle$ is the molar volume. It therefore straightforward to obtain the Joule-Thomson coefficient once c_p and α_P are determined.

The molar heat capacity at constant volume, c_v, can be obtained from the Mayer equation:

$$c_P - c_V = \frac{\alpha_P vT}{\beta_T} \tag{2.108}$$

As a consequence of the above equations, a single Monte Carlo simulation in the isobaric-isothermal ensemble yields the various thermodynamic derivative properties of industrial interest that are under consideration in this book.

E. Excess Volumes and Excess Enthalpies

The usefulness of excess functions – either excess volume or excess enthalpy – is to quantify the deviation from ideal mixing in mixtures. They help in the quantitation of system behaviour resulting from changes in composition, e.g. the mixing of a fluid with a solvent [Vidal, 1997].

The excess volume V^E in a binary mixture of volume V is

$$V^E = V - x_1 V_1 - (1 - x_1) V_2 \tag{2.109}$$

where V_1 and V_2 are the volumes of the pure components with the same number of molecules as the mixture in the same P, T conditions, and x_1 is the mole fraction of component 1. This expression allows excess volume to be readily computed from Monte Carlo simulation results.

Similarly, the excess enthalpy in a binary system can be expressed as:

$$H^E = H - x_1 H_1 - (1 - x_1) H_2 \tag{2.110}$$

where $H = U_{\text{ext}} + U_{\text{int}} + K + PV$ is the enthalpy, U_{ext} is the intermolecular potential energy, U_{int} is the molar internal potential energy and K *is* the kinetic energy of the system, these properties being defined per mole of substance for either pure components or the mixture, at the same temperature and pressure.

The computation of excess enthalpy is delicate in principle, because the kinetic energy is not considered in Monte Carlo simulation, and because only a part of the internal potential energy, $\overline{U}_{\text{int}}$ is explicitly computed. For instance, the potential energy associated with bond stretching is ignored when bond lengths are kept constant. Monte Carlo results allow only to compute the configurational enthalpy:

$$\hat{H} = U_{\text{ext}} + \overline{U}_{\text{int}} + PV \tag{2.111}$$

The total enthalpy may be expressed as the sum of the molar configurational enthalpy $\hat{H}$ and some part of the ideal gas energy at the same temperature:

$$H = \hat{H} + E'^{\text{id}} \tag{2.112}$$

where $E'^{\text{id}} = E^{\text{id}} - \overline{U}_{\text{int}}$ the part of the ideal gas energy corresponding to the ignored potential energy and the kinetic energy. It can be assumed that in an ideal gas mixture, E'^{id} can be obtained from the mole average of the pure component contributions:

$$E'^{\text{id}} = x_1 E_1'^{\text{id}} + (1 - x_1) E_2'^{\text{id}} \tag{2.113}$$

The excess enthalpy is then:

$$H^E = \hat{H} - x_1\hat{H}_1 - (1-x_1)\hat{H}_2 \tag{2.114}$$

This expression allows to evaluate easily the excess enthalpy from three Monte Carlo simulations in the *NPT* ensemble. Its derivation does not assume that the average internal potential energy per molecule is the same in the pure components and in the mixture. This procedure is therefore adapted to large flexible molecules which may adopt different internal conformations according to their environment. As excess volume and excess enthalpy are extensive properties the corresponding molar properties are easily obtained through $v^E = N_a V^E / N$ and $h^E = N_a H^E / N$.

2.4.9 Computer Hardware and Software Considerations

A. Present Computer Capacity

Computer memory and speed has increased exponentially over the last thirty years, as summarized by the well-known Moore's law which states that speed increases by a factor of two every eighteen months at constant cost. Today's PC processors are as powerful as supercomputers were ten years ago. In consequence, computers are replaced frequently and any discussion of their use in simulation tends to become obsolete in a very short period of time. Nevertheless we will give some insights about present trends.

At first, supercomputers are no longer needed to address realistic systems. Although some of the results presented in this manual were generated on supercomputers, most of them could be performed on a modern PC workstation in one or a few days per simulation. Considering the moderate cost of workstations (less than 5 000 euros for an up-to-date biprocessor with a large screen), an operator may be given several processors to work with without material costs exceeding his or her salary. This means that several simulations can be carried out every day – largely meeting the engineer's capacity to prepare the runs, to analyze them and write reports!

B. Prospective Trends in Hardware

If we consider present trends in computer science, two factors are likely to increase computer capacity in the near future:

- the constantly increasing capacity of individual processors due to the diminishing size of individual elements on silicon chips,
- the increasing availability of reasonably priced parallel computers.

It is conjectured by major manufacturers that the frequency of individual processors will continue to rise for several years, so that a gain by a significant factor after 2005 is likely. They expect this improvement to slow down between 2010 and 2020 because the size of the elementary components on silicon chips cannot be reduced below 5 nanometers for fundamental reasons (with just a few atoms possible in this space!), compared with the size of approximately 100 nanometers today. Considering the major technological difficulties, it cannot be ruled out that the slow-down begins sooner than anticipated, i.e. before 2010.

As a result, the greatest increase in computer capacity will be provided either by a complete change of technology (although no reasonable alternative to silicon technology yet exists for large scale applications) or by using parallel processors (which is more likely). In large computing centres, it is common to see clusters of hundreds of processors, and clusters containing tens of thousands have been reported. The operational costs of such facilities are still high, but it is already possible to buy smaller integrated clusters at a reasonable price. It is likely that small parallel computers with 4, 8, or 16 processors will be increasingly affordable in the next few years. Altogether, an increase by a factor four in computer speed seems realistic in the next five years, without any drastic increase in cost.

C. Monte Carlo Software

The situation in Monte Carlo simulation is quite different from those of quantum mechanics or molecular dynamics for which free software packages (e.g. GAMESS or DL POLY) and commercial software (e.g. DMOL or GAUSSIAN) have been available for many years. Indeed, few multipurpose Monte Carlo packages are available, either from research groups or from scientific software companies. The main reason is probably that Monte Carlo moves are rather molecule- and ensemble-specific so researchers tend to write their own, custom-made tools. Commercial software is still quite limited in terms of intermolecular potentials, phase properties, multiphase equilibria, multicomponent systems, and account of polarisation, among others.

For the applications dealt with in this manual, we have mostly relied on the GIBBS software developed jointly by IFP, CNRS and University of Paris XI-Orsay started by Mackie *et al.* [1997]. This software includes several features, as shown in Table 2.5. Apart from the Monte Carlo code itself, it comprises preprocessors to compute energy grids for adsorption studies and post-processors to perform various types of result analysis (averages, thermodynamic derivatives, critical coordinates, etc.). It also features parallel tempering, which makes taking advantage of parallel computers possible. The architecture of the software is given in Figure 2.36.

Table 2.5 Features included in the Gibbs code.

	Features	**Specific aspects**
Statistical ensembles	*NPT*	May be used with parallel tempering
	NVT	May be used with parallel tempering
	Grand Canonical	Available with orthorhombic crystalline microporous adsorbents May be used with parallel tempering
	m-phase Gibbs ensemble	Available with imposed volume or imposed pressure
Monte Carlo moves	Internal	Translation, rigid body rotation, flip, CBMC regrowth, pivot, reptation,
	Transfers	With CBMC or reservoir bias
	Volume change	
	Insertions/deletions	With CBMC or reservoir bias

Table 2.5 Features included in the Gibbs code. *(continued)*

	Features	Specific aspects
Intermolecular energy	Dispersion - repulsion	Lennard-Jones 6-12 + combining rules of Lorentz-Berthelot, Kong or Waldmann-Hagler
	Electrostatic	Partial charges located on nuclei, or on AUA force centres, or on bonds, or at any location defined by neighbour atoms
Molecular structure	Fully rigid	No bending nor torsional energy
	Fully flexible	Linear, branched or cyclic structures with bending and torsion energy
	Partly rigid	Linear or branched structure with some bending angles or torsion angles fixed
Intermolecular potential	United Atoms	
	Anisotropic United Atoms	The AUA force centre may be either on the external bisector or on the bond with the next atom
	All Atoms	
Phase properties	Pressure	Virial equation
	Chemical potential	Widom test insertions with configurational bias or reservoir bias
	Energy, enthalpy	Separation of electrostatic from dispersion-repulsion contributions
	Thermodynamic derivative properties	Heat capacity, thermal expansion coefficient, compressibility coefficient, Joule-Thomson coefficient
Optimisation	Adjustable parameters $= \sigma, \varepsilon, \delta$	Derivatives versus parameters are obtained from statistical fluctuations

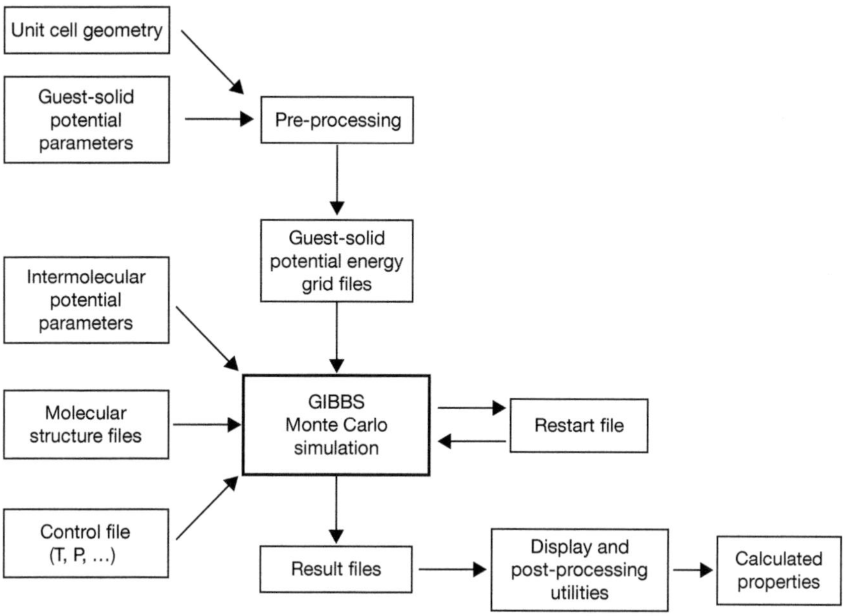

Figure 2.36 Organisation of the GIBBS Monte Carlo software.

3

Fluid Phase Equilibria and Fluid Properties

Over the generations, scientists have collected an enormous amount of data in the field of fluid phase equilibria so it might be considered that molecular simulation cannot provide major contributions for applications. Nevertheless, there are still gaps in our knowledge for various reasons, and well-tested simulations may be used to provide predictions here. For instance, the properties of many compounds of industrial interest have never been measured because they are too expensive, unavailable as pure chemicals or highly toxic. This is the case of hydrocarbons with more than 10 carbon atoms other than *n*-alkanes, and sulphur- and mercury-containing organic compounds which will be considered in this chapter. Also deficient is the coverage of phase equilibrium data for mixtures, because all possible combinations have not been investigated or because specific technical difficulties (e.g. high pressure and corrosion) may be encountered with some systems. The results that are described in this chapter aim at contributing to such topics.

In a first section, the capacity of molecular simulation to predict the equilibrium properties of pure hydrocarbons will be addressed. This will be done by developing an intermolecular potential that makes it possible to predict such properties for unknown compounds from their simple chemical structure (Section 3.1). The discussion will next turn to predicting derivative properties, such as heat capacity, isothermal compressibility or the Joule-Thomson coefficient (Section 3.2). Subsequently, consideration will be given to the modelling of equilibrium properties of organic polar compounds, such as sulphides, thiols, organo-mercury compounds, … which are important contaminants in natural gases and processed fuels (Section 3.3). The following section will be devoted to the prediction of the phase behaviour of mixtures, particularly those involving polar compounds which are difficult to predict with conventional models (Section 3.4). This section will be also the opportunity to illustrate the calculation of bubble points for multicomponent mixtures, and the solubility of high pressure gases in polyethylene. In the next section, the thermodynamic properties of natural gases will be discussed, with particular emphasis on the Joule Thomson coefficient which controls the thermal regime of high pressure gas production wells (Section 3.5). In the course of this discussion, a key point is how to represent the complex composition of a natural gas with a selected array of pure components, since pseudo-components cannot be readily used in molecular simulation in the same way as with equations of state. Finally, we will conclude this chapter by showing applications related to the development of a simplified process of acid gas removal from natural gas and further reinjection in deep reservoirs (Section 3.6). This will deal mainly with the phase equilibria of H_2S-rich systems involving methane, water and carbon dioxide although the volumetric and energetic properties of related fluids will be also considered.

3.1 PREDICTING THE PROPERTIES OF PURE HYDROCARBONS

3.1.1 General Strategy

The general aim in this section is the prediction of the thermodynamic properties of hydrocarbons that have not been investigated experimentally. With the exception of a few well-known families like the *n*-alkanes or *n*-alkylbenzenes, most hydrocarbons with more than 10 carbon atoms are unavailable in a pure form at reasonable prices. This explains why very few measurements have been made on branched alkanes, which exist in so many isomeric forms that the lack of commercially available products is easily understandable. In this carbon range, the same is true for hydrocarbons made of a polycyclic nucleus (either saturated or aromatic) substituted by alkyl groups. Molecular simulation is expected to provide better estimations than group contribution methods for hydrocarbons far above 10 carbon atoms. Among other features, it should be able to account for the exact molecular structure in a theoretical fashion, which is difficult to do with empirical group contribution methods. For instance, simulation should predict – at least qualitatively – the difference between alkane isomers differing just by the degree of branching, like 2-methylhexane and 2,4-dimethylpentane.

In order to achieve this, we need a transferable intermolecular potential that is sufficiently simple to be operated with large molecules. As indicated in Section 2.4.3, United Atoms models have been developed for this purpose. As the predictions obtained by this type of model have been abundantly covered in the literature, the primary goal in this section is not to provide a detailed review of United Atoms capabilities, but to review the features of an alternative model, the Anisotropic United Atoms potential. Nevertheless, the two methods will be compared for certain families, e.g. the olefins.

As mentioned in Sections 2.2 and 2.4, the Anisotropic United Atoms model appears as a good compromise because it is both a simplified method in which hydrogen atoms are not considered explicitly, and a realistic model in which the location of the force centre accounts for hydrogen atoms in some approximate way. In order to determine the potential parameters for each group, optimisation is based on a few reference compounds, as explained in Section 2.4.4. The purpose of this section is not to discuss the details of the optimisation procedure for all the groups (the alkanes, olefins and aromatics) but rather to present an overview. The optimisation procedure for the *n*-alkanes and olefins is described with more details in the corresponding sections.

The general flowchart of Figure 3.1 shows on which reference components optimisation of the AUA intermolecular potential has been based for the various groups (the resulting parameters are indicated in Appendix 1). There are three initial independent optimisation processes:

- optimisation of the CH_2 and CH_3 group parameters of linear alkanes, which is based on ethane, *n*-pentane and *n*-dodecane [Ungerer *et al.*, 2000],

- optimisation of the saturated cyclic CH_2 group parameters, which has been based on cyclohexane and cyclopentane [Bourasseau *et al.*, 2002a],
- optimisation of the aromatic CH group parameters, which has been based on benzene [Contreras-Camacho *et al.*, 2004a];

Optimisation of the other groups is not independent, since it is based on the use of previously determined group parameters:
- optimisation of the CH group in branched alkanes makes use of the *n*-alkane-based CH_3 group parameters to describe the methyl groups of the reference molecule, i.e. isobutane [Bourasseau *et al.*, 2002 b],
- optimisation of the $=CH_2$, $=CH-$ and $=C<$ groups in olefins makes use of the *n*-alkane-based CH_3 and CH_2 group parameters to describe the saturated part of the reference olefins [Bourasseau *et al.*, 2003],
- optimisation of the C group in substituted aromatics makes use of the *n*-alkane-based CH_3 and the benzene-based aromatic CH group parameters to describe the reference xylene molecule [Contreras-Camacho *et al.*, 2004b].

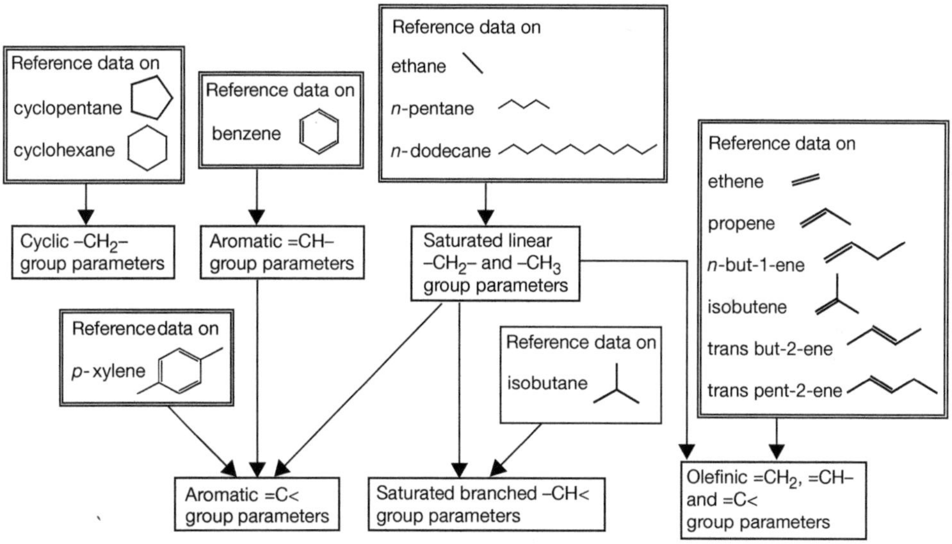

Figure 3.1 General scheme of the optimization of the Anisotropic United Atoms (AUA) intermolecular potential parameters for hydrocarbons showing the reference molecules used for the various groups.

It is thus clear that the AUA intermolecular potential has been developed in the same way as group contribution methods. Implicitly, it has been assumed that the properties assigned to a given group are independent on its environment or on the location of the group in the molecule. This is a very bold hypothesis! Indeed, the resulting potential comprises only nine different groups for all hydrocarbons, involving 25 non-zero parameters for Lennard-Jones interactions.

Before comparing experimental measurements with predicted properties, it is interesting to ask whether optimised potential parameters follow regular trends. As illustrated by Table 3.1, the optimised position of the AUA force centre is always located close to the geometric centre of the carbon and the related hydrogens, which is its expected location. In Figure 3.2, the Lennard Jones parameters σ and ε are displayed for the various groups. It may be seen that there is a regular evolution of σ and ε with decreasing number of hydrogen atoms in the group within a given series. For instance, there is a regularly decreasing trend of σ in the sequence CH_4 (3.73 Å), CH_3 (3.607 Å), CH_2 (3.461 Å) and CH (3.362 Å). The AUA potential of alkanes therefore makes it possible to correct a significant inconsistency of the classical United Atoms potentials, in which the Lennard Jones diameter is larger for CH_2 and CH_3 groups (3.75 to 3.95 Å, see Table 2.3) than for methane.

Table 3.1 Comparison of the positions of AUA force centers and geometrical centers of the atoms of the group. It has been assumed that the C–H bond distance was 1.14 Å, sp^3 bond angles 109° and sp^2 bond angles 120°.

Group	Optimized δ_{AUA} (Å)	Carbon-geometric center distance (Å)	Difference (Å)
Saturated CH_3	0.216	0.285	0.07
Saturated CH_2	0.384	0.44	0.06
Saturated CH	0.646	0.57	0.09
Cyclic $–CH_2–$	0.336	0.44	0.1
Olefinic =CH	0.295	0.38	0.09
Olefinic =CH–	0.414	0.57	0.16
Aromatic CH	0.407	0.57	0.16

Similarly, the Lennard-Jones diameters of the π-bonded groups indicated in Figure 3.2 form a consistent sequence, since the olefinic σ_{CH2} appears larger than σ_{CH} and than σ_C. The optimised energetic parameters are regularly spaced, and the AUA force centres are close to the geometrical centre of the optimised group. Moreover, the comparison between parameters of π-bonded groups and parameters of saturated groups shows that the entire set of AUA parameters is internally consistent. In contrast to these results, the United Atoms parameters as optimised by various authors [Spyriouni *et al.*, 1999; Wick *et al.*, 2000; Nath *et al.*, 2001] do not form logical sequences, e.g. olefinic CH_2 groups appear smaller than CH groups in these models.

These observations tend to indicate that the AUA potential parameters have a sound physical basis. This feature is expected to provide a good extrapolation capability to molecules other than those considered in optimisation procedure, as will be investigated in the following sections.

3.1.2 Predicting the Properties of Linear Alkanes

This section will be devoted to linear alkanes containing two carbon atoms or more. The case of methane is specific, because it is supercritical in the context of the applications considered in this manual. Therefore the simulation of methane will be treated in Sections 3.2 and 3.3.

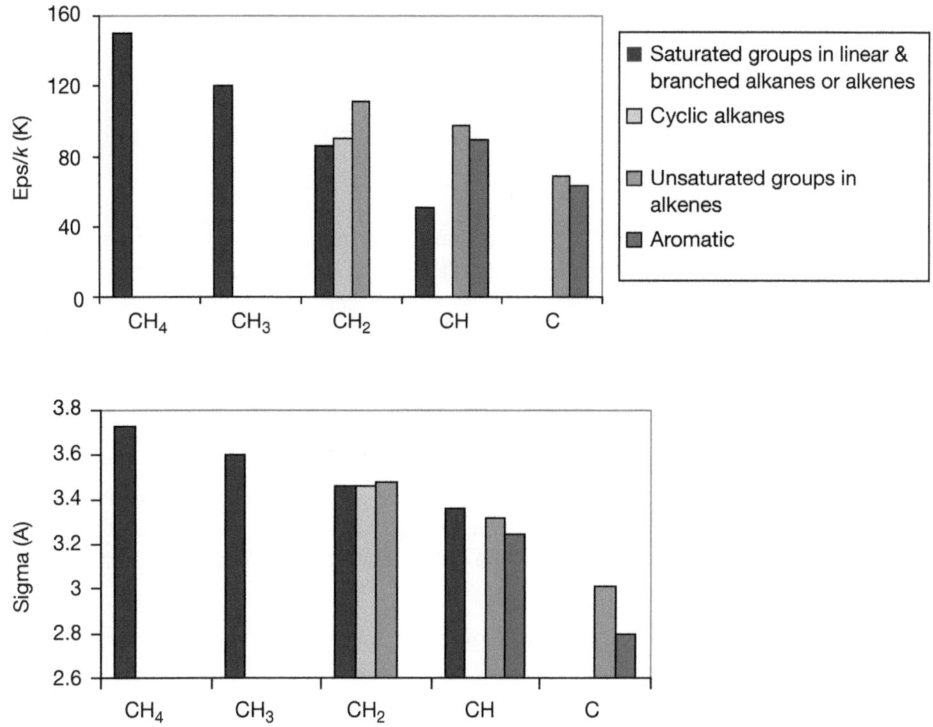

Figure 3.2 Lennard-Jones diameters σ (bottom) and energetic parameters ε/*k* (top) of the various groups considered in the AUA intermolecular potential of hydrocarbons.

A. Optimisation of Potential Parameters

The reference compounds selected for optimisation of the CH_2 and CH_3 groups of linear alkanes are ethane, *n*-pentane and *n*-dodecane [Ungerer *et al.*, 2000]. Thus they range from an exclusively CH_3-based compound, ethane, to a dominantly CH_2-based compound, *n*-dodecane. For each reference compound, two reference temperatures were used: a temperature intermediate between the normal boiling point and the critical temperature which was simulated using the Gibbs ensemble technique, and a temperature intermediate between the freezing and boiling temperatures, which was simulated in the *NPT* ensemble.

The higher temperature makes it possible to carry out an accurate simulation in the Gibbs ensemble (it is neither subject to near-critical density fluctuations nor to low acceptance of transfer moves). Three independent results derived from this simulation are introduced in the definition of the error criterion (Eq. 2.81) for comparison with experimental data: saturated vapour pressure (Eq. 2.14), vaporization enthalpy (Eqs. 2.103 or 2.104) and saturated liquid density.

At the lower temperature, only two independent results may be used in the error criterion: the liquid density and the vaporization enthalpy. The advantage of complementing the Gibbs ensemble simulation by this low temperature *NPT* simulation is that the potential is then likely to provide prediction in the largest possible temperature range.

Altogether, there are consequently 5 independent experimental data per reference component, i.e. 15 data in the case of *n*-alkanes. The number of reference data is therefore largely superior to the number of optimised parameters for the CH_2 and CH_3 groups, which is 6 (one set of ε, σ and δ per group).

A delicate point in optimisation is the selection of the starting point for ε, σ and δ. In the case of *n*-alkanes, a good starting point was provided by the AUA potential of Toxvaerd [Toxvaerd, 1997]. For this point, the partial derivatives of every reference result versus every parameter must be determined, as discussed in Section 2.4.4. The statistical uncertainties s_i appearing in the error criterion (2.81) are estimated from the block average method (see Section 2.4.7). Then, the optimum is estimated by solving the system (2.85). As can be seen in Table 3.2, a substantial improvement is found from this single optimisation step. The average dimensionless error is reduced from $F = 4.2$ with Toxvaerd's parameters to $F = 0.53$ with the optimised parameters. Having the dimensionless error lower than 1 means that on average, the 15 reference data are matched with an accuracy slightly better than the estimated statistical uncertainty. In such cases it is unlikely that the optimisation can be significantly improved.

Table 3.2 Summarized table of the optimization of the AUA potential with all reference data points. AUA3 stands for the potential published by Toxvaerd [1997] and AUA4 stands for the optimized AUA parameters.

Reference compound	T (K)	Simulation conditions	Type of data	Experimental data	Estimated statistical uncertainty	AUA3	AUA4
Ethane	240	GEMC	P_{sat}(kPa)	966	100	968	945
			ΔH_{vap}(kJ/mol)	12.0	0.5	12.31	11.97
			ρ_l(kg/m^3)	472	10	513	471
	140	*NPT*, $P = 0$	ΔH_{vap} (kJ/mol)	16.36	0.5	15.97	15.91
			ρ_l(kg/m^3)	598	5	652	600
nC_5	385.82	GEMC	P_{sat}(kPa)	775	100	866	765
			ΔH_{vap}(kJ/mol)	20.25	0.5	19.14	20.36
			ρ_l(kg/m^3)	521	10	512	521
	232.3	*NPT*, $P = 0$	ΔH_{vap}(kJ/mol)	29.6	0.5	28.4	29.55
			ρ_l(kg/m^3)	683	5	697	689
nC_{12}	550	GEMC	P_{sat}(kPa)	355	100	600	350
			ΔH_{vap}(kJ/mol)	37.45	1	30.73	37.57
			ρ_l(kg/m^3)	530	10	486	531
	305.57	*NPT*, $P = 0$	ΔH_{vap}(kJ/mol)	60.15	1	55.10	59.37
			ρ_l(kg/m^3)	740	5	723	746

B. Evaluation of the Potential

In Figure 3.3, the saturated vapour pressures as determined using the optimised AUA parameters are compared with reference data on C_2 to C_{30} *n*-alkanes. For this purpose, saturation pressures above 1 bar have been computed with the Gibbs ensemble technique, while lower pressures have been computed by thermodynamic integration, as explained in Section 2.3.8. Although statistical uncertainties tend to cause larger discrepancies for the heavier *n*-alkanes, it is clear from this graph that Monte Carlo simulation is able to provide high quality predictions for much heavier alkanes than those used as reference, and for a very large range of temperatures as well.

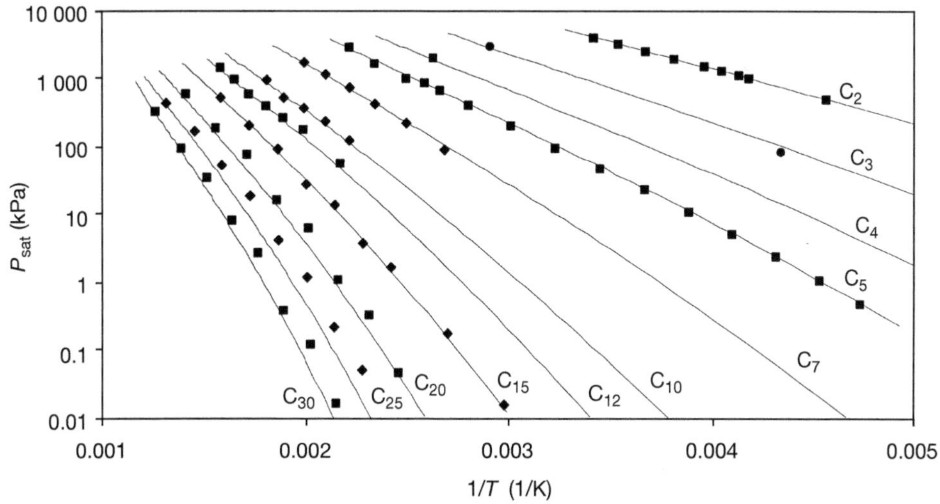

Figure 3.3 Saturated vapour pressure of *n*-alkanes obtained from simulations using the AUA potential (symbols) compared with experimental correlations from the DIPPR databank (continuous lines).

The liquid densities predicted for *n*-alkanes are shown in Figure 3.4. Here again, a satisfactory prediction is found. It may be noted however that some systematic differences are observed for heavy hydrocarbons in this family. This might be due to the difficulty of exploring the whole configuration space within a reasonable computer time, since these simulations were achieved without using the reptation moves presented in Section 2.3.2. The vaporization enthalpies are compared with experiments on Figure 3.5 and show similar agreement with experimental data.

At first glance, it may seem surprising that saturation pressure predictions extrapolate well to very low pressures, considering that there is just one saturation pressure at high temperature per compound among the reference points. This can be explained by the fact that in the Clapeyron equation (Eq. 2.72), the derivative of the saturation pressure versus $1/T$ is mainly controlled by the vaporization enthalpy. As this property is well predicted (Fig. 3.5), the slope of the Clapeyron plot is well reproduced, explaining why the whole trend of saturation pres-

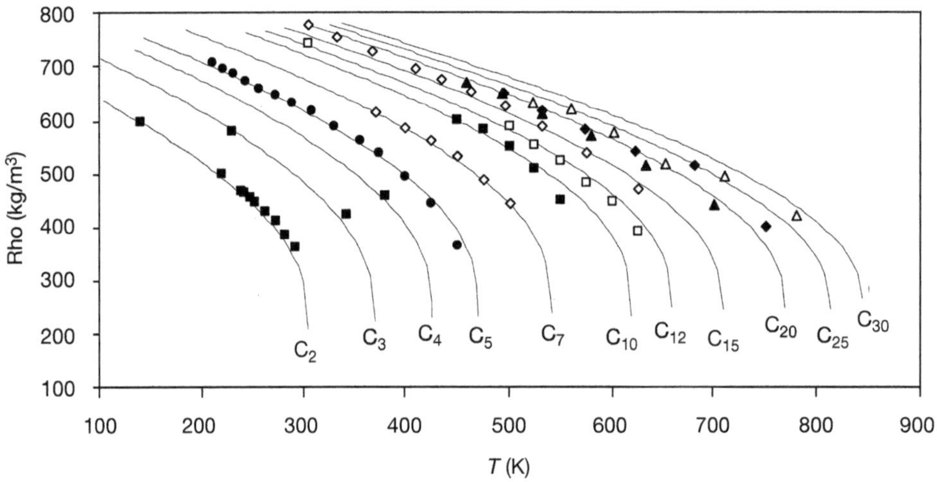

Figure 3.4 Saturated liquid density of *n*-alkanes obtained from simulations using the AUA potential (symbols) compared with experimental correlations from the DIPPR databank (continuous lines).

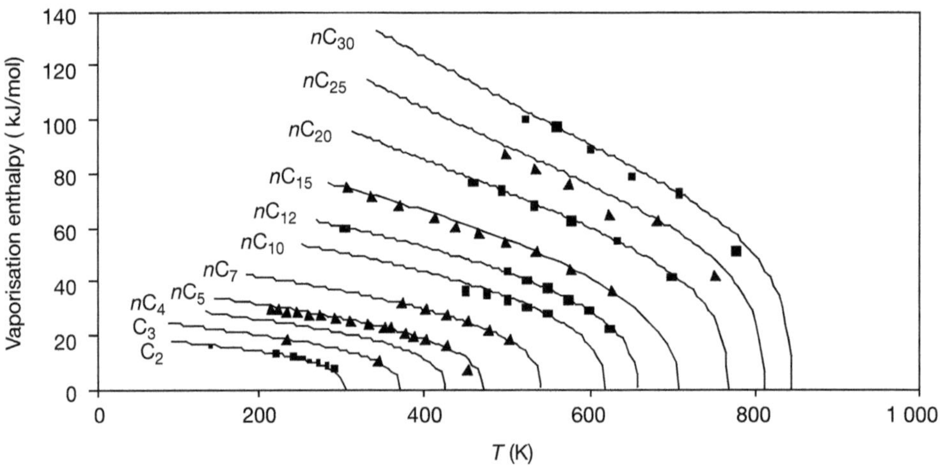

Figure 3.5 Vaporization enthalpy of *n*-alkanes obtained from simulations using the AUA potential (symbols) compared with experimental correlations from the DIPPR databank (continuous lines).

sures is correctly predicted. This justifies the introduction of vaporization enthalpy among the reference data for the optimisation of the intermolecular potential.

Critical points may be also determined from simulation results, using the procedure discussed in Section 2.4.8. The critical temperatures and critical densities obtained by this method for *n*-alkanes are given in Table 3.3. It shows that the deviations from critical densi-

Table 3.3 Comparison of critical parameters and boiling temperatures determined by simulation with the AUA potential and experimental measurements. Values in parentheses are likely results of extrapolations rather than true measurements.

Compound	Critical temperature (K)		Critical density (kg/m^3)		Normal Boiling temperature (K)	
	Exp.	Simulation	Exp.	Simulation	Exp.	Simulation
Ethane*	305.4	311	202	210	184	nd
n-pentane*	469.7	469	237	218	307.7	307
n-heptane	540.3	547	232	231	372	376
n-decane	617.7	616	236	225	447	445
n-dodecane*	658.2	652	239	217	489	481
n-pentadecane	(708)	nd	(250)	nd	543.1	541
n-eicosane	(768)	nd	(251)	nd	617	609
Isobutane*	407.8	nd	229	nd	261.4	268
2-methylhexane	530.4	537	227	238	362	362
2,4-dimethylpentane	519.8	529	240	247	354	357
Cyclopentane*	511.8	507	271	275	322.4	311
Cyclohexane*	553.6	559	273	271	353.9	352
Cyclooctane	647.2	647	283	303	424.3	nd
Ethylene*	282.3	280	214	212	169.2	166
Propene*	365	373	223	226	225.1	224
1-butene*	419.5	420	233	236	267.1	262
1-pentene	464.7	462	234	232	303.3	298
1-hexene	504	499	237	241	336.6	328
1-octene	567.1	566	241	240	394.4	388
Cis-2-butene	435.6	448	240	238	276.8	279
Trans-2-butene*	428.6	439	236	234	274	265
Isobutene*	417.9	426	234	232	266.2	260
Butadiene	425	422	245	237	268.6	264
Trans-2-pentene*	475	480	234	243	309.5	316
Benzene*	562.2	558	302	302	353.2	350
Toluene (methylbenzene)	591.8	580	292	307	383.8	376
p-xylene* (1,4 dimethylbenzene)	616.2	598	280	303	411.5	403
o-xylene (1,2 dimethylbenzene)	630.3	610	289.1	311	417.2	401
1,3,5 trimethylbenzene	637.4	622	280.2	285	437.8	415
n-propylbenzene	638.2	627	273	281	432.4	421
Naphthalene	(748)	716	(315)	323	487.8	491

* reference compounds.

ties are generally below 10 K, the average being about 6 K. This is typical of the uncertainty that can be expected from Monte Carlo simulation as described here. Among the heavier alkanes (C_{12} and above), the experimental critical parameters are generally obtained by extrapolation of the vapour pressure curves because the pyrolysis kinetics are too fast for accurate direct measurements. Nevertheless, there is a good agreement between the critical coordinates obtained by simulation and by this extrapolation.

In addition to the quantitative representation of macroscopic properties, molecular simulation is good at showing the structure of the alkane chains at the molecular level. As illustrated by the configuration of liquid *n*-pentadecane at 306 K (Fig. 3.6) the alkane molecules are far from being linear and have very tortuous shapes. Nevertheless, their conformation is by no means random. Firstly, the bending angles do not vary by more than +/– 20° either side of the average, as shown in the distribution in Figure 3.7. Secondly, the torsion angles are not random – otherwise we would not see several molecules displaying regular zig-zag segments over 10 atoms or so, as in Figure 3.6. Indeed, these zig-zag segments correspond to 'trans' configurations, i.e. portions of the molecule where successive torsion angles are close to the minimum of the torsion potential (Fig. 3.8). It has been found that this tendency is particularly strong at lower temperatures [Bourasseau *et al.*, 2002b]. This shows that the *n*-alkanes in the liquid phase at low temperature display some common features with the solid phase in which all torsion angles are in the 'trans' configuration.

It is interesting to compare the performance of the AUA potential with those of other intermolecular potentials from the literature using the classical United Atoms concept (see

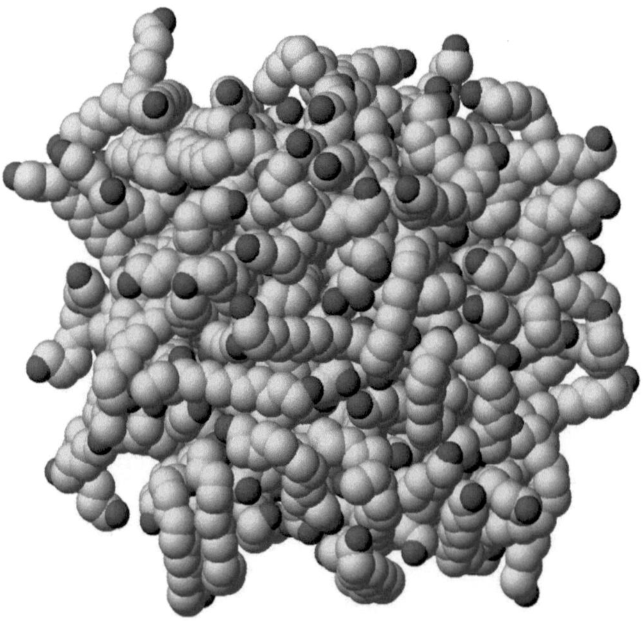

Figure 3.6 Example of configuration of a *n*-pentadecane liquid phase at 305.85 K at saturation pressure.

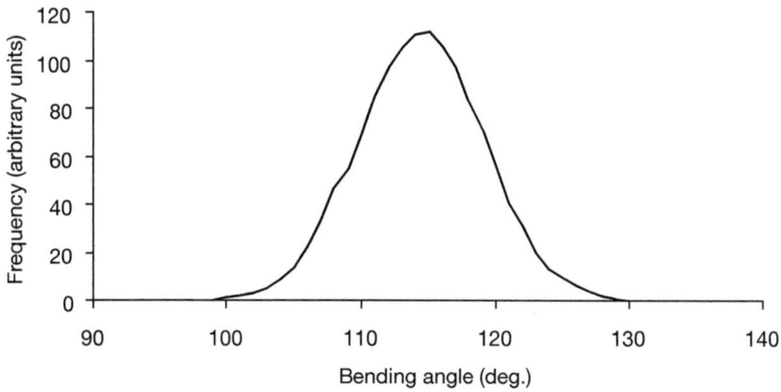

Figure 3.7 Example of the distribution of bending angles CH_2–CH_2–CH_2 in *n*-dodecane at 423 K, showing variations of approximately 15° from the equilibrium value of 114°.

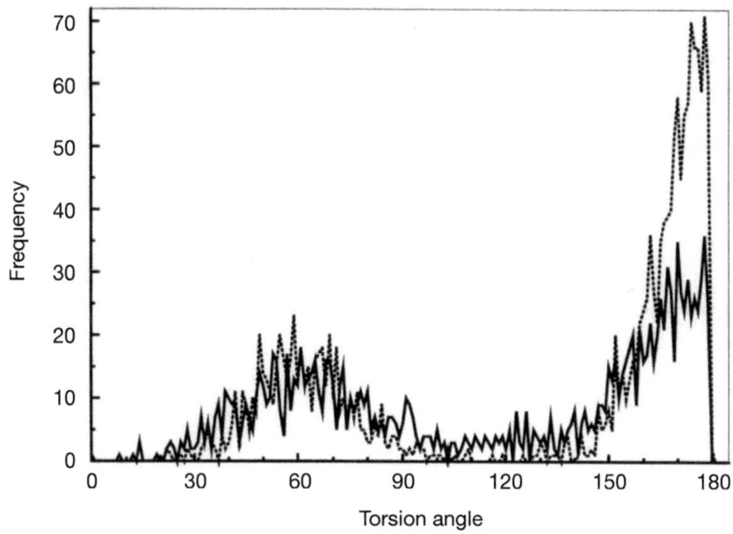

Figure 3.8 Distribution of the dihedral angle φ in *n*-eicosane at 350 K (dotted line) and at 700 K (full line), illustrating the preference for the trans conformation (φ = 180°) against the *cis* conformations (φ = 60°), reprinted with permission from Bourasseau *et al.* [2002], © Taylor & Francis Ltd (http://www.tandf.co.uk/journals).

Section 2.2.6). Compared with the United Atoms potentials TraPPE [Martin and Siepmann, 1998] and NERD [Nath *et al.*, 1998], it seems that the optimised AUA potential described here provides a similar account of density and a better account of vaporization enthalpy [Ungerer *et al.*, 2000]. As a consequence, accurate vapour pressure predictions are obtained with the AUA model at low temperatures, below the normal boiling point.

More advanced models like the exp-6 potential [Errington and Panagiotopoulos, 1999] or the hydrogen-explicit model [Chen and Siepmann, 1999] are probably more accurate than the AUA model for *n*-alkanes but they have not been applied to a sufficiently large variety of chemical families to provide a general alternative to UA and AUA potentials.

3.1.3 Branched Alkanes

As mentioned in Section 3.1.1, branched alkanes are modelled by using the same CH_2 and CH_3 groups as with *n*-alkanes, the CH group being optimised on the basis on the properties of isobutane. As this reference basis is small, it is important to evaluate the potential on other molecules. Although the experimental data are far less complete on branched alkanes than on *n*-alkanes, there is a sufficient amount of data on branched alkanes to allow such tests [Bourasseau *et al.*, 2002b and 2004]. We show here evaluations based on two branched alkanes with 7 carbon atoms (2-methylhexane and 2,4-dimethylpentane) and one branched alkane with 19 carbon atoms (pristane or 2,6,10,14-tetramethylpentadecane). Together with these evaluations, we show the predictions of two branched alkanes with 15 carbon atoms for which no data is available, 2-methyltetradecane and 2,6-dimethyltridecane.

As shown in Figure 3.9, the saturation pressures of the various branched alkanes are well predicted. It is particularly satisfactory to see that the small difference between the vapour pressure curves of 2-methylhexane and 2,4-dimethylhexane is well predicted, because these hydrocarbons have the same chemical formula and differ only in terms of the degree of their branching. More precisely, Monte Carlo simulation is able to predict the slight decrease in critical temperature with increasing branching on the basis of the sole chemical structure. The predicted vapour pressure pattern of pristane is also consistent with experimental measurements in the temperature range for which they are available. This validation of a heavy branched alkane makes us confident that the predicted vapour pressures of 2-methyltetrade-

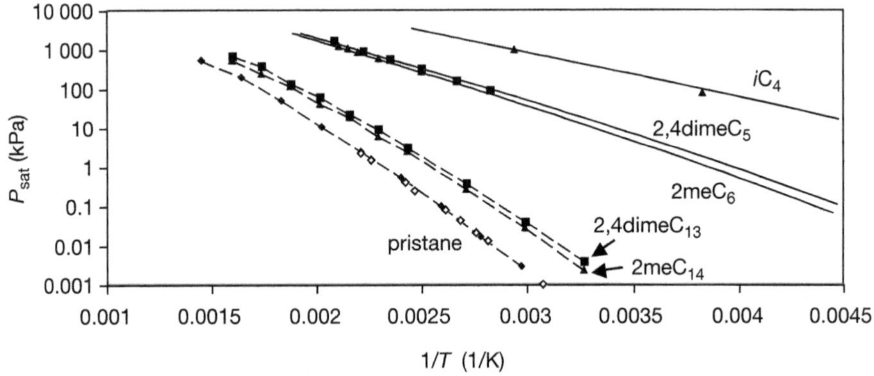

Figure 3.9 Saturated vapour pressure of branched alkanes obtained from simulations using the AUA potential (symbols and dotted lines) compared with experimental correlations from the DIPPR data bank (continuous lines). Pristane is 2,6,10,14 tetramethylpentadecane.

cane and 2,4-dimethyltridecane are accurate, although no experimental measurements are available for these compounds. The variation of vapour pressure with the number of branched methyl groups is qualitatively similar to the C_7 branched alkanes. Comparison of these two isomers shows that combined use of the Gibbs ensemble technique and thermodynamic integration is sufficiently accurate to predict consistently the behaviour of slightly different hydrocarbons at temperatures much lower than the normal boiling point, where experimental measurements are difficult because of very low vapour pressure.

In Figure 3.10, liquid density predictions are shown for the same compounds. A good agreement is observed between experimentally-based correlations and simulation results for the reference molecule (isobutane) as well as for the C_7 branched alkanes. Although statistical uncertainties are of similar amplitude as the differences between isomers, it seems that the slightly higher density of 2-methylhexane with respect to 2,4-dimethylhexane is qualitatively reproduced. In the case of pristane, the predicted trend is consistent with the only experimental data available [Fermeglia and Torriano, 1999].

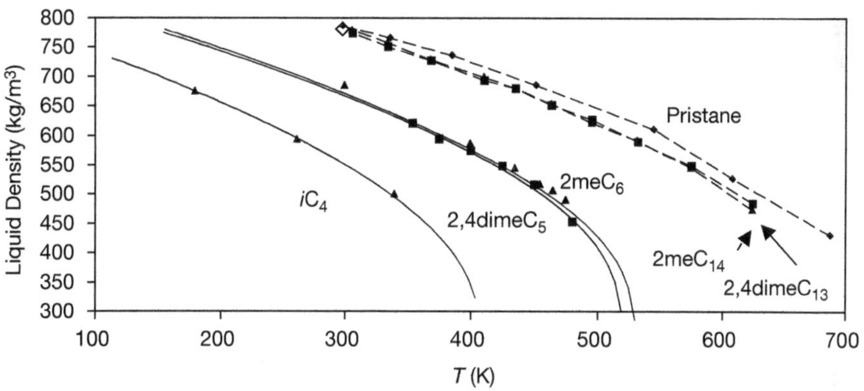

Figure 3.10 Saturated liquid density of branched alkanes obtained from simulations using the AUA potential (symbols and dotted lines) compared with experimental correlations from the DIPPR data bank (continuous lines). The diamond indicates the liquid density of pristane (2,6,10,14 tetramethylpentadecane) measured by Fermeglia and Torriano [1999].

Finally, the vaporization enthalpy of the investigated branched alkanes is also close to the truth (Fig. 3.11). For this property, larger differences are found between isomers than for vapour pressure or liquid density.

As a whole, all three properties are consistently predicted for much heavier branched alkanes than the reference molecules, a felicitous observation. The accuracy of molecular simulation *vis-à-vis* branching effects in alkanes makes it a much more useful predictive tool for this family than any known group contribution method.

The comparison with the TraPPE United Atoms models of branched alkanes [Martin and Siepmann, 1999] seems to indicate that the AUA model has an equivalent prediction capacity. For instance, the TraPPE model is also capable of predicting reasonably the properties of

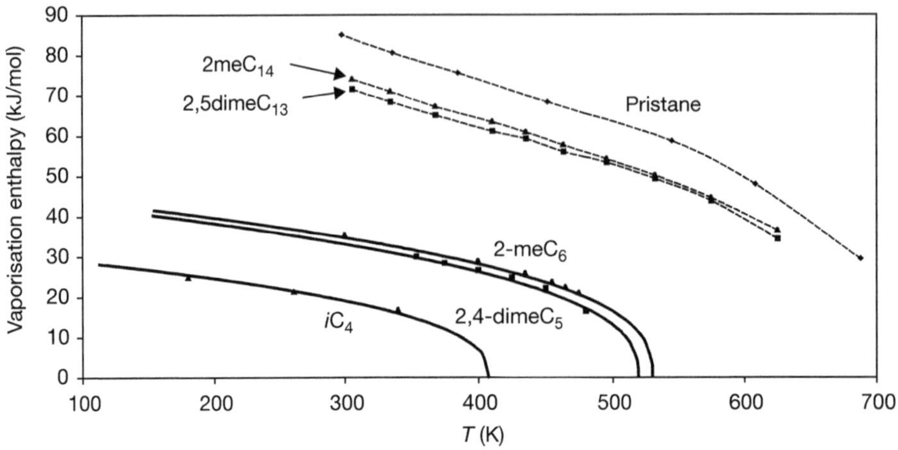

Figure 3.11 Molar Vaporisation enthalpy of branched alkanes obtained from simulations using the AUA potential (symbols and dotted lines) compared with experimental correlations from the DIPPR data bank (continuous lines). Pristane is 2,6,10,14 tetramethylpentadecane.

squalane, a C_{30} branched alkane [Zhuravlev and Siepmann, 1997; Zhuravlev *et al.*, 2002]. The NERD potential is shown to provide very good predictions on light branched alkanes [Nath and de Pablo, 2000] but it is based on more parameters, as the CH_3 group is not the same for all positions in this model.

3.1.4 Cyclic Alkanes

The intermolecular potential for the cyclic CH_2 group has been optimised for cyclopentane and cyclohexane, because five-membered and six-membered rings are by far the most common saturated cyclic structure in crude oils and natural gas components. Nevertheless, we will check here that the model is still applicable to larger rings by investigating cyclooctane as well.

These molecules are flexible, as illustrated by the various different conformations of the cyclohexane molecule (Fig. 3.12). This flexibility is modelled using the same framework as linear or branched alkanes, by introducing bending and torsion energy (see Section 2.2.5). The C–C bond lengths are taken as identical to those in a *n*-alkane, in addition to bending and torsional parameters (see Appendix 1). The predominance of the "chair" conformation over the other conformations ("boat" or "skewed") is well reproduced by simulation [Bourasseau *et al.*, 2002a].

As mentioned in Section 2.3.6, the main difference in simulating phase equilibrium of a cyclic molecule is that a specific Monte Carlo movement, the reservoir bias transfer, is used to respect the constraint of closing the ring. Apart from this specific move, the Gibbs ensemble technique and the thermodynamic integration method have been applied in the same way as for linear and branched alkanes.

The comparison of saturated vapour pressures (Fig. 3.13) reveals that the AUA potential predicts not only the behaviour of cyclopentane and cyclohexane (which is expected as they

Figure 3.12 Typical conformation of cyclohexane molecules (450 K, 0.8 MPa).

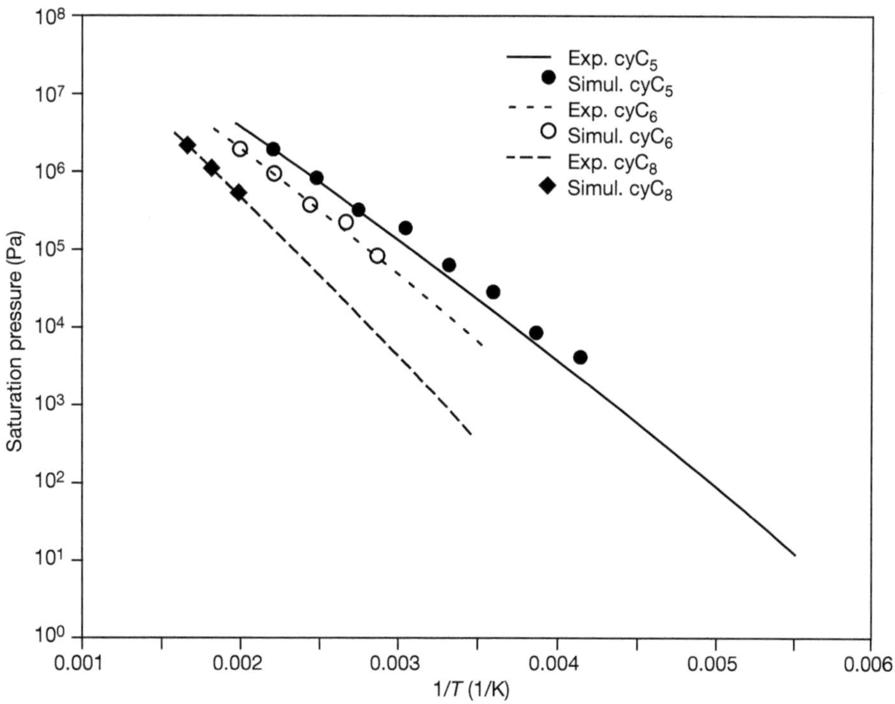

Figure 3.13 Saturated vapour pressure of cyclopentane, cyclohexane and cyclooctane obtained from simulations using the AUA potential (symbols) compared with experimental correlations from the DIPPR data bank (lines). Reprinted with permission from "Prediction of Equilibrium Properties of Cyclic Alkanes by Monte Carlo Simulation". J. Phys. Chem. B 106: 5483-5491. © 2002, American Chemical Society.

are the reference molecules) but also the cyclooctane pattern. Similarly, liquid density and vaporization enthalpy (Figs. 3.14 and 3.15) are also well reproduced for the reference molecules and for cyclooctane, although with a significant deviation for cyclooctane density.

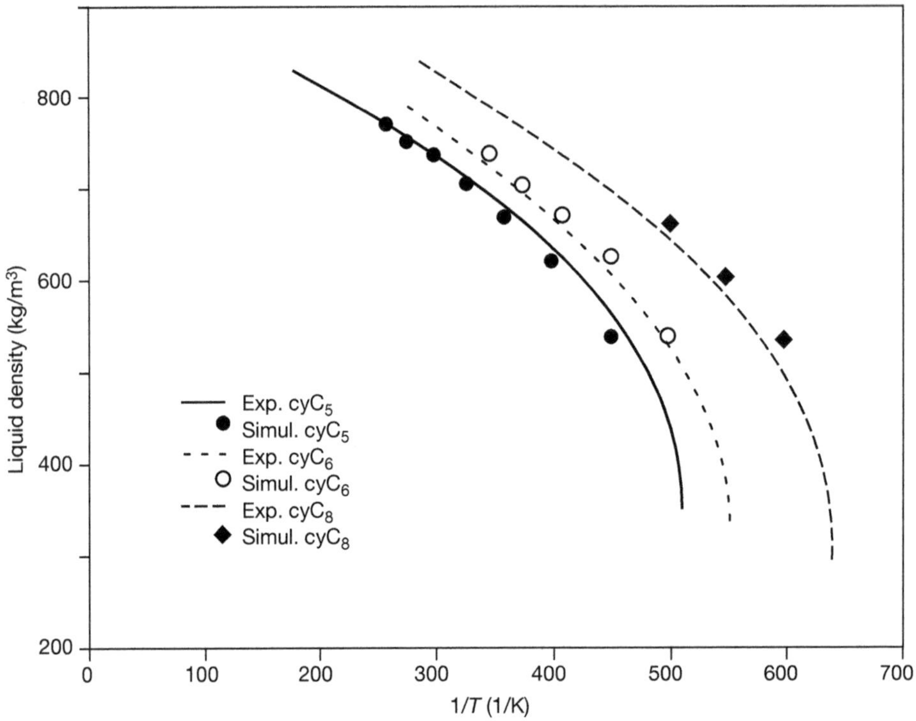

Figure 3.14 Saturated liquid density of cyclopentane, cyclohexane and cyclooctane obtained from simulations using the AUA potential (symbols) compared with experimental correlations from the DIPPR data bank (lines). Reprinted with permission from "Prediction of Equilibrium Properties of Cyclic Alkanes by Monte Carlo Simulation". J. Phys. Chem. B 106: 5483-5491. © 2002, American Chemical Society.

It is not surprising that a single set of parameters does not represent the behaviour of all the cyclic alkanes investigated. Indeed, it is known that five-membered and eight-membered rings are subject to greater internal tension than the six-membered ring of cylcohexane. Also, it can be seen that, in the limit case of an infinitely large ring, the CH_2 parameters for linear alkanes should be closer to the truth than those determined for small rings. Nevertheless, the predictions obtained for the cycloalkanes in question are adequate because the cyclic CH_2 parameters are close to those for linear alkanes (see Table 2.3). Compared with previous intermolecular potentials of cyclic alkanes [Errington and Panagiotopoulos, 1999; Neubauer *et al.*, 1999] the proposed potential offers the advantage of a larger scope because it extends to five-and eight-membered rings and takes a larger temperature range into account.

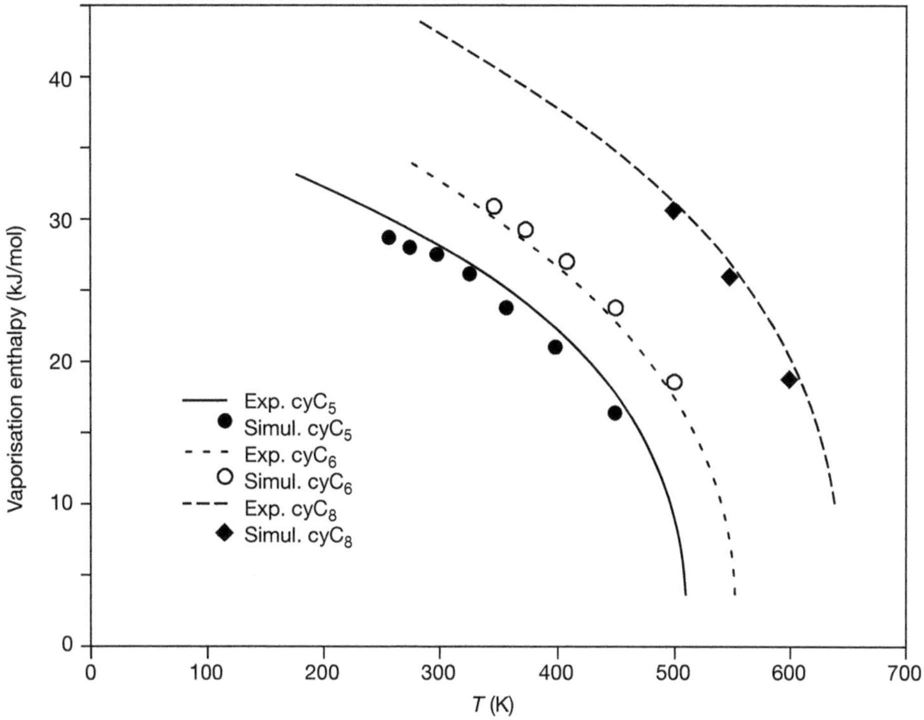

Figure 3.15 Molar vaporization enthalpy of cyclopentane, cyclohexane and cyclooctane obtained from simulations using the AUA potential (dots) compared with experimental correlations from the DIPPR data bank (lines). Reprinted with permission from "Prediction of Equilibrium Properties of Cyclic Alkanes by Monte Carlo Simulation". J. Phys. Chem. B 106: 5483-5491. © 2002, American Chemical Society.

3.1.5 Olefins

We detail below (Subsection A) the optimisation procedure followed by Bourasseau *et al.* [2003] to generate intermolecular potential parameters for olefins, since the procedure illustrates certain difficult situations in which unrealistic or inaccurate results may be obtained when attention is not specifically devoted to the problem. Evaluation of the intermolecular potential is discussed in Subsection B.

Geometry and torsional flexibility require specific models for olefins. As a result of the sp^2 hybrid orbital of the two carbons participating to the double bond, the other bonded atoms belong to a common plane in the most stable conformation. The C=C bond length is smaller (1.33 Å) than the C–C bond (1.535 Å on average), and the equilibrium bond angles C=C–H or C=C–C are close from 120°, with small variations. For the sake of simplicity, the torsional flexibility around a double bond can be neglected, i.e. the torsion angle Φ is fixed to either 0° (as in trans-2-butene) or to 180° (as in *cis*-2-butene). The torsion potential for the C=C–C–C sequence may be taken from the work of Jorgensen and Madura [1984]. For the C=C–C=C sequence which appears for instance in butadiene, the torsion potential parameters have been fitted with Eq. (2.47) to the energies determined from quantum chemistry calculations [Szalay *et al.*, 1989; Kofranek *et al.*, 1992]. The corresponding parameters are indicated in Appendix 1.

A. Optimisation of the Potential

In a first optimisation stage, three low molecular weight alpha-olefins have been used as reference molecules: ethylene, propene, and isobutene. The CH_2 π-bonded group appears four times, CH and C π-bonded groups only once. In the same way as *n*-alkanes, five reference data are considered for each molecule to build the maximum likelihood criterion $F(y_1)$ of Eq. (2.81): one saturation pressure, two liquid densities and two vaporization enthalpies. The initial parameters have been selected by analogy with the previously optimised groups for linear and branched alkanes. The outcome of this first optimisation is satisfactory, i.e. the criterion is significantly lower than 1 and the optimum is stable with respect to sensitivity tests. With the resulting set of optimised parameters, the properties of the three reference molecules used in the optimisation process are well represented but the predictions are not completely successful for other olefins, particularly beta-olefins (but-2-ene, pent-2-ene), or butadiene. The explanation of this disappointing behaviour is that only alpha-olefins have been included as reference in this first stage. The CH_2 and CH π-bonded group parameters were thus optimised to describe molecules containing exclusively CH_2=CH–, or CH_2=CH_2 sequences. These sequences do not appear in beta-olefins, where the –CH=CH– sequence prevails. It is therefore necessary to enlarge the experimental database to include other types of olefins in a second optimisation stage.

The six reference molecules in the second stage are indicated in Table 3.4. We can note that CH_2 and CH π-bonded groups are included in equivalent proportions, and appear on different types of sequences. A branched olefin – isobutene – is included to incorporate the π-bonded C group in the reference database. As there are five reference data points per molecule, the error criterion is composed of 30 reference data points and it depends on 8 parameters. For this second optimisation stage, the initial parameters are those obtained from the first stage.

Table 3.5 shows the successive parameter sets in both optimisation stages.

Table 3.4 Reference molecules used in the second optimization trial. For each molecule two temperatures have been included in the reference database.

		π-bonded CH_2	**π-bonded CH**	**π-bonded C**	**Low T**	**High T**
Ethylene	CH_2=CH_2	2	–	–	150	225
Propene	CH_2=CH–CH_3	1	1	–	200	290
Isobutene	CH_2=C–$(CH_3)_2$	1	–	1	260	330
1-butene	CH_2=CH–CH_2–CH_3	1	1	–	250	350
Trans-but-2-ene	CH_3–CH=CH–CH_3	–	2	–	302.1	380
Trans-pent-2-ene	CH_3–CH=CH–CH_2–CH_3	–	2	–	310	430

Table 3.5 Lennard-Jones parameters and carbon to force center distances of AUA π-bonded groups in olefins during the two successive stages of the optimization of the AUA potential for olefins. Note that due to the increased number of reference components from 3 (first stage) to 8 (second stage), the error criterion *F* is not defined in the same way in the two stages.

	π-bonded CH_2			**π-bonded CH**			**π-bonded C**		***F***
	ε (K)	**σ (Å)**	**δ (Å)**	**ε (K)**	**σ (Å)**	**δ (Å)**	**ε (K)**	**σ (Å)**	
Initialization	96.3	3.56	0.3	81.7	3.3	0.4	52.8	3.16	6.98 ↓
First stage	111.0	3.48	0.295	97.5	3.32	0.414	68.4	3.02	0.53
Second stage	111.1	3.48	0.295	90.6	3.32	0.414	61.9	3.02	2.05 ↓ 1.33

In this second trial, the use of the fluctuation method based on Eq. (2.86) makes it possible to evaluate the partial derivatives of the error criterion with respect to every potential parameter. This requires 12 simulations, while 58 simulations would be required with the finite difference method to evaluate the same derivatives. Although the fluctuation method requires longer runs for satisfactory convergence, it provides an overall speed-up by a factor three approximately.

When all the 8 parameters are optimised simultaneously, it is found that some of the optimised parameters y_1 have unrealistic values, e.g. the optimised value of ε(C) appears excessively low (3 K), indicating that the optimisation problem is ill defined. As discussed in Section 2.4.4, two ways are possible in this type of case to define the optimisation problem: reducing the number of parameters being optimised simultaneously, or increasing the number of experimental data points under consideration. The second way is unlikely to succeed because significant effort has already been put in to extend the reference database from three to six components. In contrast, the first option appears to be appropriate since the initial set of parameters is close to the desired optimum. This was implemented by optimising only three parameters over the eight parameters taken from the first optimisation trial: ε(CH_2), ε(CH) and ε(C). Table 3.5 shows the resulting parameters, which represent the last stage of the optimisation procedure for the three groups.

Before evaluating these parameters by a comprehensive evaluation of a variety of different olefins, it is worth performing a series of sensitivity tests as discussed in Section 2.4.4. The results lead to the conclusion that a small variation of experimental data – within estimated experimental uncertainties – produces a variation of no more than 5% in any of the optimised parameters. Similarly, variations of simulation results change the optimised parameters by 2-3%, and a variation of about 10% in partial derivatives causes less than 1% variation in all optimised parameters. In each case, the minimised error criterion is not significantly changed during these sensitivity tests. It is thus likely that the set of optimised parameters issued from the second optimisation stage is reasonably stable and meaningful. As mentioned above, these parameters have also a good physical sense, especially when compared to classical United Atom potentials.

B. Predicting the Equilibrium Properties of Olefins

Figures 3.16 to 3.18 show respectively vapour pressures, coexistence densities and vaporization enthalpies for the alpha-olefins with between two and eight carbon atoms used as reference molecules (ethylene, propene, 1-butene) and other olefins (1-pentene, 1-hexene and 1-octene). On average, the optimised AUA potential predicts vapour pressures as or more accurately than the United Atoms potentials (shown for comparison). The same applies to liquid densities (Fig. 3.17). Comparison with United Atoms potentials is not possible for vaporization enthalpies because their authors have not determined this property in their work. Nevertheless, the AUA potential accounts very well for vaporization enthalpies, including the three heavier alpha-olefins that were not considered in optimisation process.

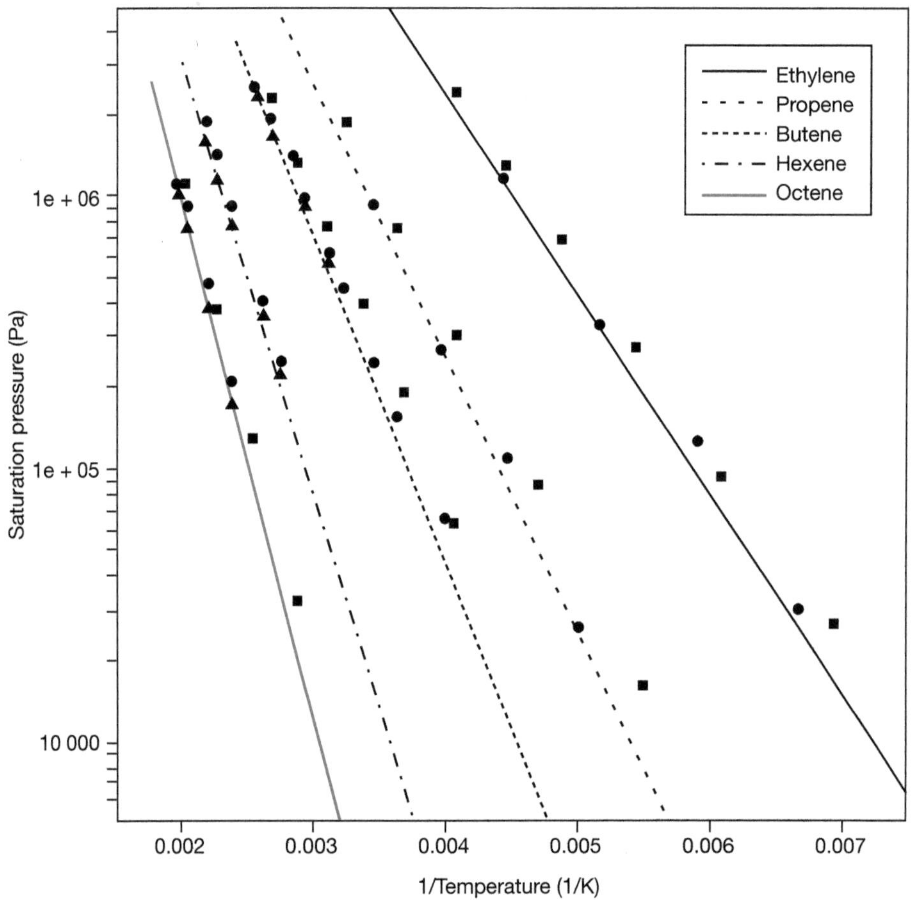

Figure 3.16 Saturated vapour pressure of alpha-olefins from ethylene to 1-octene after Bourasseau *et al.* [2003]. Lines are correlations from the Dortmund Data Bank and symbols are simulation results with three intermolecular potentials: AUA (dots), Spyriouni *et al.* [1999] (triangles), Wick *et al.* [2000] (squares).

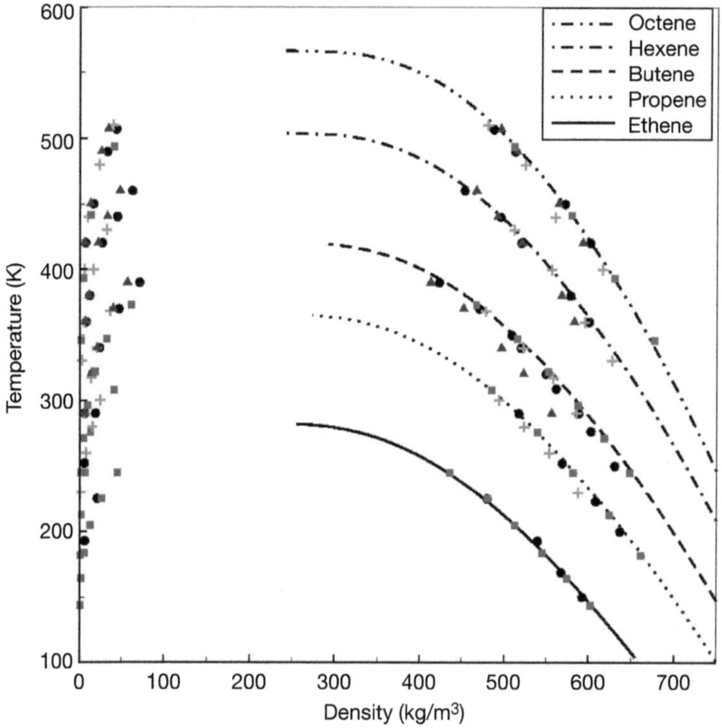

Figure 3.17 Saturated liquid and vapour density of alpha olefins after Bourasseau *et al.* [2003]. Lines are correlations from the Dortmund Data Bank and symbols are simulation results with three intermolecular potentials: AUA (dots), Spyriouni *et al.* [1999] (triangles), Wick *et al.* [2000] (squares), Nath *et al.* [2001] (crosses).

Figures 3.19 to 3.21 show simulation results for olefins with four or five carbon atoms (isobutene, butadiene, *cis*-2-butene, *trans*-2-butene, 1-pentene, *trans*-2-pentene). These graphs are aimed at investigating the capability of the AUA potential to represent the differences in properties between isomers differing by the structure (branched versus linear, *cis* versus *trans* conformers). Equilibrium properties of these molecules are close and the separation between some of them is a challenging test of our potential. For instance, the gap between the two extreme critical temperatures (1-butene and *trans*-2-pentene) is only 55 K (Table 3.3).

Changes in properties with temperature are reasonably reproduced. In Figure 3.20, the sequence 1-butene, isobutene, butadiene, *trans*-but-2-ene and *cis*-but-2-ene is observed when liquid density increases at a fixed temperature. Moreover, the evolution of saturated liquid densities when temperature increases is consistent for each isomer. Similar observations can be made from the saturation pressures of Figure 3.19 or from the vaporization enthalpies of Figure 3.21. As three of these molecules were not taken as reference in optimisation, these predictions may be considered very satisfactory. However, the AUA potential shows some

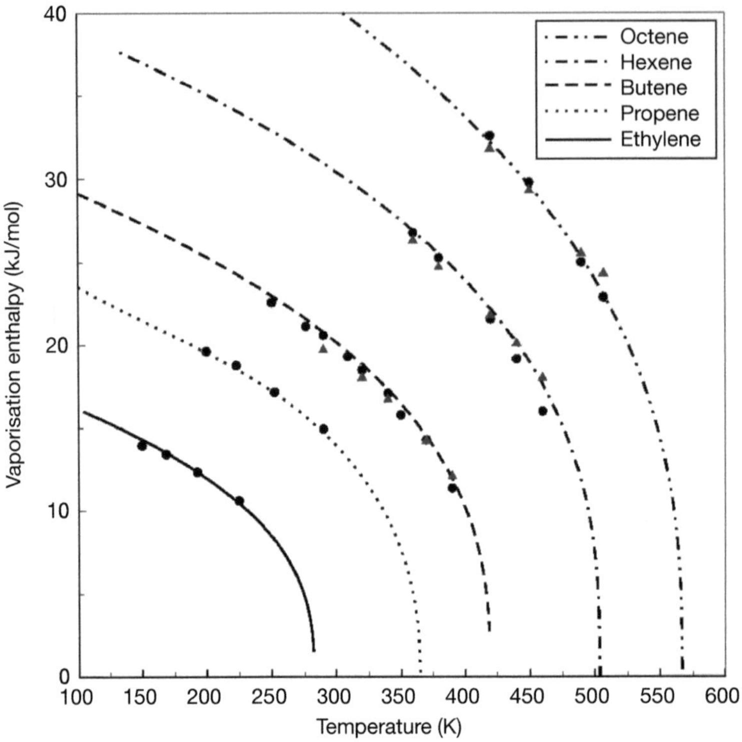

Figure 3.18 Molar vaporization enthalpy of alpha olefins after Bourasseau *et al.* [2003]. Lines are correlations from the Dortmund Data Bank and symbols are simulatyion results with three intermolecular potentials: AUA (dots), Spyriouni *et al.* [1999] (triangles).

intrinsic limitations in separating the properties of some close isomers. This is particularly the case of *trans*-2-butene and *cis*-2-butene, whose vaporization enthalpies are predicted similar while experimental data indicate significant differences between these isomers (Fig. 3.21). As many assumptions have been made in deriving the potential (negligible electrostatics, fixed C–C=C angles, planar geometry of the C–C=C–C sequence, ...) it is not surprising that the potential is subject to such limitations. Nevertheless, it may be concluded that the relative average errors for the olefins shown in Figures 3.16 to 3.21 are 1.5% on liquid densities, 2% on vaporization enthalpies and 10% on saturation pressures.

It is therefore not surprising that the AUA potential makes it possible also an accurate calculation of critical temperatures, critical densities and normal boiling point temperatures for the olefins considered (Table 3.3). Critical temperatures are estimated with a relative error of about 2%, and critical densities with a relative error of about 3%. Compared with United Atoms potentials of olefins [Spyriouni *et al.*, 1999; Wick *et al.*, 2000; Nath *et al.*, 2001], the AUA model presents an accuracy at least equivalent on alpha-olefins and a wider range of application, as Spyriouni's and Nath's model are restricted to alpha-olefins and Wick's work does not apply to diolefins.

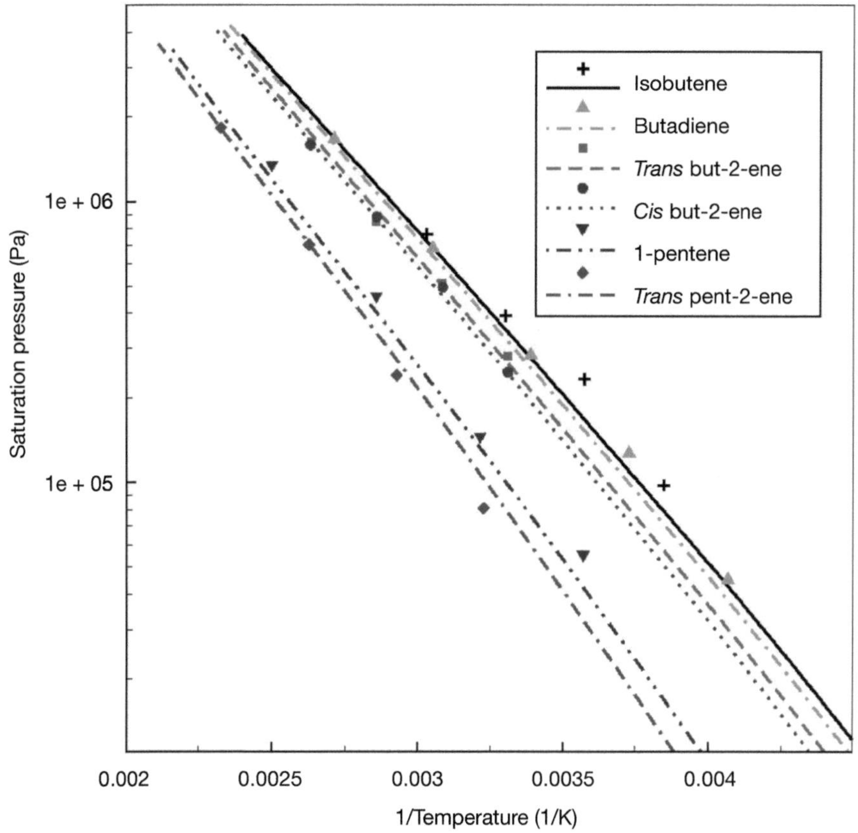

Figure 3.19 Saturated vapour pressure of C_4 and C_5 olefins obtained from simulations using the AUA potential (symbols) compared with experimental correlations from the DIPPR data bank (lines).

3.1.6 Aromatics

As indicated on Figure 3.1, the parameters of the CH aromatic group have been optimised on the basis of benzene, while the substituted aromatic C group has been optimised with reference to *p*-xylene [Contreras-Camacho *et al.*, 2004b] or to naphthalene [Ahunbay *et al.*, 2005]. The location and diameter of the CH group are very close to a previous AUA benzene model [Friedrich and Lustig, 2002] in which the force centre is located 0.29 Å away from the carbon.

In Figures 3.22 to 3.24, the properties of six aromatics from 6 to 10 carbon atoms are shown, including the two reference molecules and four new aromatic molecules. It order to compute the properties of new alkylbenzenes (like toluene or propylbenzene for instance), the sp^3-bonded CH_3 and CH_2 group parameters have been taken identical to alkanes, in the same way as it has been done when simulating *p*-xylene. Beyond the good agreement between simulation results and experimental data, it is particularly instructive to observe that molecular

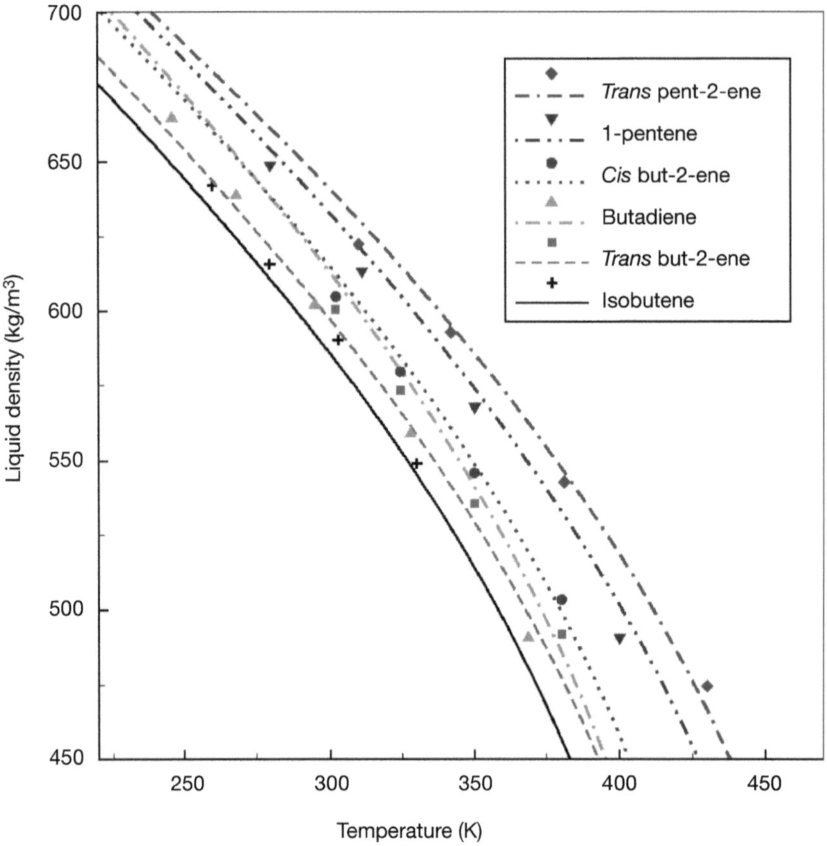

Figure 3.20 Saturated liquid density of C_4 and C_5 olefins obtained from simulations using the AUA potential (symbols) compared with experimental correlations from the DIPPR data bank (lines).

simulation is able to capture the major part of the interplay between aromatic and saturated groups. For instance, the addition of an alkyl chain in propylbenzene tends to decrease liquid density compared with benzene, while the addition of a second aromatic ring (as is the case with naphthalene) results in a strong increase in liquid density (Fig. 3.22). This is the consequence of the lower carbon-carbon bond distance in aromatics (1.40 Å) compared to saturated chains (1.535 Å), among other causes. Nevertheless, the vapour pressure is predicted to increase continuously with the length of the alkyl chain which is substituted to the aromatic ring (Fig. 3.23). Similarly, the vaporization enthalpy (Fig. 3.24) is well depicted on average, although its reveals that the AUA potential underestimates the critical temperature of methyl-substituted monoaromatics.

The very good behaviour of the AUA potential on benzene may be attributed to its good physical sense, as revealed by the reasonable account of the pairwise radial distribution function (Fig. 3.25). This distribution curve is known from experimental diffusion measurements

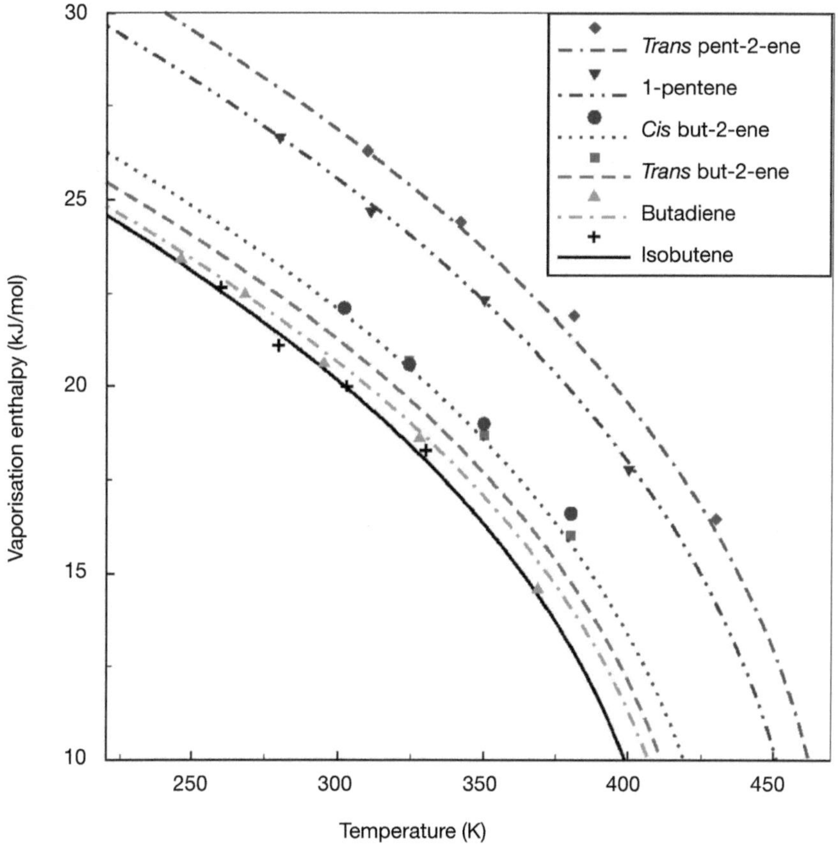

Figure 3.21 Molar vaporization enthalpy of C_4 and C_5 olefins obtained from simulations using the AUA potential (symbols) compared with experimental correlations from the DIPPR data bank (lines).

[Narten, 1977]. The first shoulder corresponds mainly to stacked molecules, a configuration for which the C–C distance is approximately 4 Å on average. It may be seen that the AUA potential predicts the fundamental liquid structure with an accuracy equivalent to the All Atoms potential of Jorgensen [Jorgensen and Severance, 1990]. This is probably why the AUA potential manages to depict rather well the properties of aromatics in spite of its intrinsic simplicity.

However, significant deviations with thermodynamic properties are found for some aromatic molecules. This is particularly the case of the aromatics bearing several methyl groups, like *p*-xylene and trimethylbenzene. It is possible that the AUA potential is somewhat too simple to capture all the structural differences between these molecules. Indeed, it neglects the possible influence of electrostatic forces, while it is known that the aromatic ring is negatively charged in its central region and positively charged at its periphery. This results in dipole moments for asymmetric molecules (toluene) and higher-order moments for symmetric molecules (benzene, *p*-xylene), whose influence is neglected in this potential. A special mention

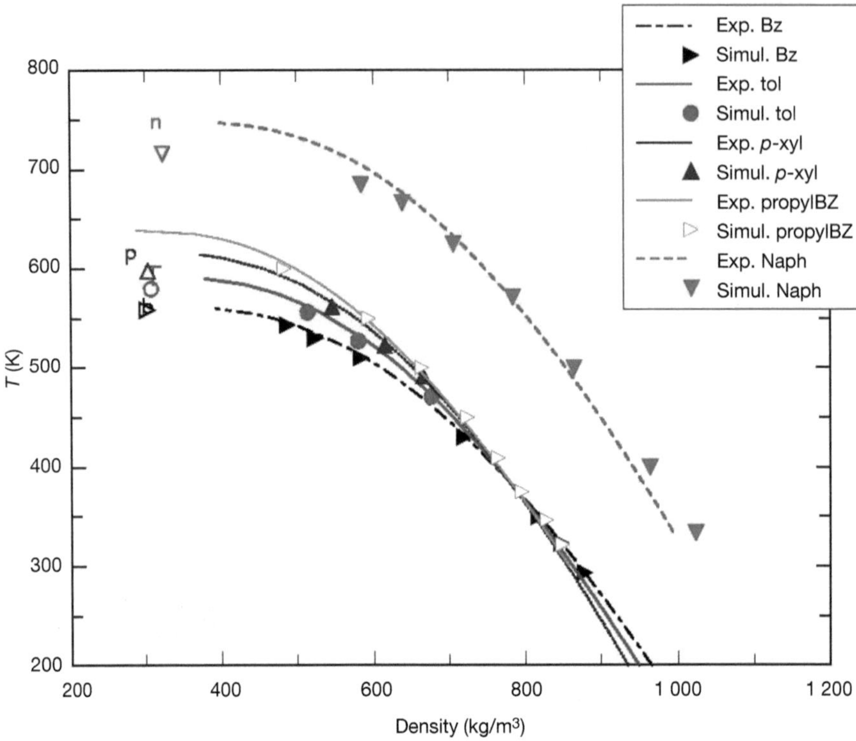

Figure 3.22 Liquid density of aromatic hydrocarbons obtained from simulations using the AUA potential of Contreras-Camacho *et al.* [2004a and 2004b] (symbols) compared with experimental correlations from the DIPPR data bank (lines).

must be made about naphthalene however. It is indeed doubtful that the properties of this molecule can be reliably measured in the critical region, because it undergoes rapid pyrolysis above 700 K as most hydrocarbons. Therefore, the critical temperature is obtained by extrapolation, and the comparison between the measured and predicted critical coordinates of naphthalene is not a meaningful test of the potential. On the opposite, the correspondence between the measured and predicted boiling temperatures (Table 3.3) is meaningful, indicating that the properties of bicyclic aromatic compounds other than naphthalene are likely to be well predicted.

Compared with other intermolecular potentials for aromatics in terms of thermodynamic property prediction, the AUA model discussed in this section appears slightly superior to the TraPPE United Atoms model [Wick *et al.*, 2000]. For instance, it provides a better account of the vaporization enthalpies of benzene [Contreras-Camacho *et al.*, 2004a]. It is also clearly superior to the All Atoms model of Jorgensen [Jorgensen and Severance, 1990] for temperatures higher than ambient. The AUA model is as accurate as Friedrich's AUA model of benzene [Friedrich and Lustig, 2002]. The only more accurate model for benzene is an exp-6 model [Errington and Panagiotopoulos, 1999]. However Errington's and Friedrich's models have not yet been extended to other aromatic molecules than benzene.

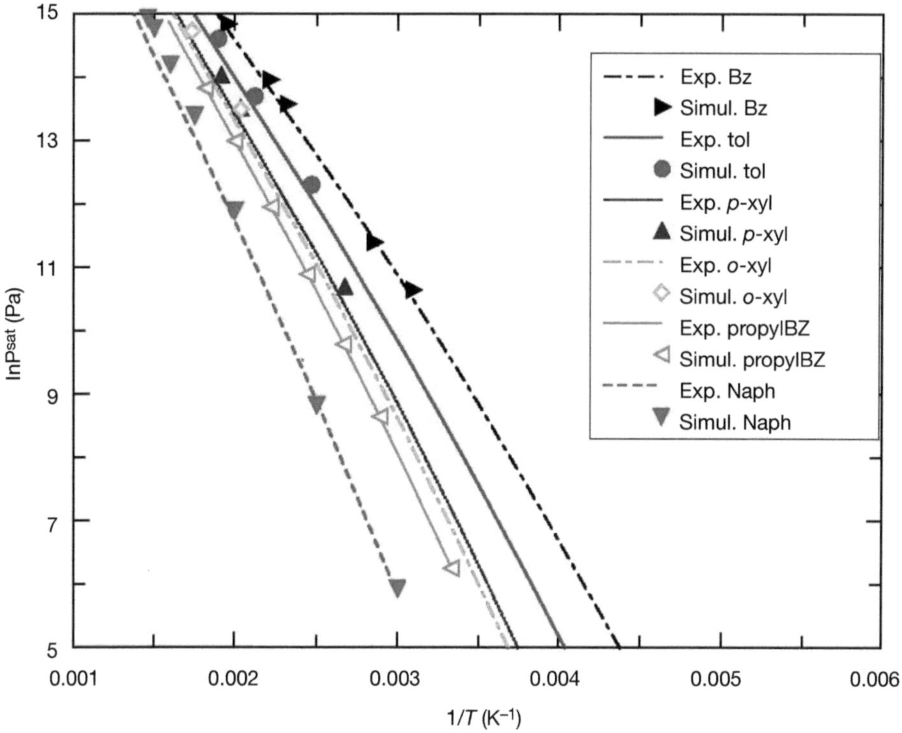

Figure 3.23 Saturated vapour pressure of aromatic hydrocarbons obtained from simulations using the AUA potential of Contreras-Camacho *et al.* [2004a and 2004b] (symbols) compared with experimental correlations from the DIPPR data bank (lines).

3.1.7 Perspectives

A challenging aspect of AUA potentials is that there are still few well-tested computer programs in which they are implemented, compared with UA potentials. Also, the configurational bias algorithms are somewhat less efficient with AUA than with UA potentials, as it was discussed in Section 2.3.6. Nevertheless, the results described in Sections 3.2 to 3.6 show that current algorithms are efficient enough to simulate a large set of molecular structures with the AUA potential efficiently.

The main advantage of the proposed AUA intermolecular potential over classical UA potentials is that it combines an ability to make accurate predictions with the simplicity of United Atoms methods. Also, the AUA has been tested over a wider range of properties, since the UA potentials from the literature have not been compared systematically with measured vaporization enthalpies and do not show a comparable accuracy *vis-à-vis* this property. As a result, the AUA potential can be safely extrapolated to temperatures much lower than the normal boiling point. Here again, UA potentials have been less tested in these conditions, because

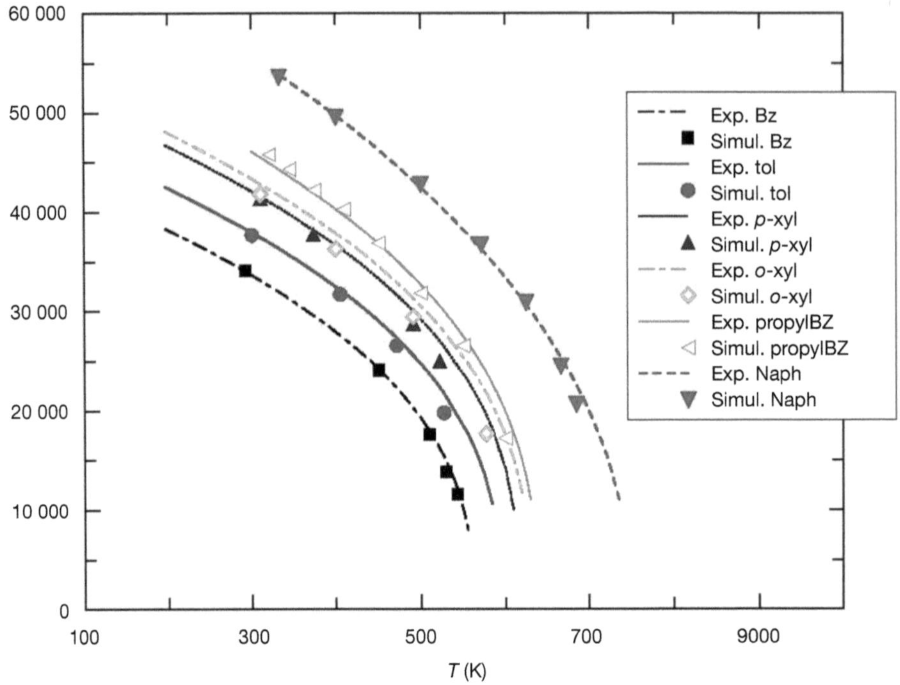

Figure 3.24 Molar vaporization enthalpy of aromatic hydrocarbons obtained from simulations using the AUA potential of Contreras-Camacho *et al.* [2004a and 2004b] (symbols) compared with experimental correlations from the DIPPR data bank (lines).

they have generally focused comparison to temperatures higher than the normal boiling point for which phase equilibrium properties are easier to evaluate by molecular simulation.

Another advantage of the AUA potentials is the low number of independent parameters that have been calibrated. Altogether, 25 optimised parameters are involved for all groups, and all the hydrocarbon properties predicted in the previous sections are based on the simple input of molecular structure together with these 25 parameters. Compared to a UA potential like NERD, the AUA potential requires a lower number of effective parameters to address branched alkanes.

Recent work, not covered in this book, has convincingly shown that the AUA potential could be used to address many types of hydrocarbons with successful results. Ahunbay *et al.* [2004] have investigated alkyl-substituted cycloalkanes, polycyclic alkanes like decalin, and naphthenoaromatics like tetralin. The parameters of aromatic carbon rings carrying some substituent have been extrapolated from naphthalene by Ahunbay *et al.* [2005]. This is a minor change however, because this carbon does not interact as much as the CH aromatic group, and because these parameters have not changed much from the values proposed by Contreras *et al.* [2004b]. Logically, these parameters appear to provide a better account of the properties of polyaromatics like phenanthrene and anthracene [Ahunbay *et al.*, 2005]; interestingly,

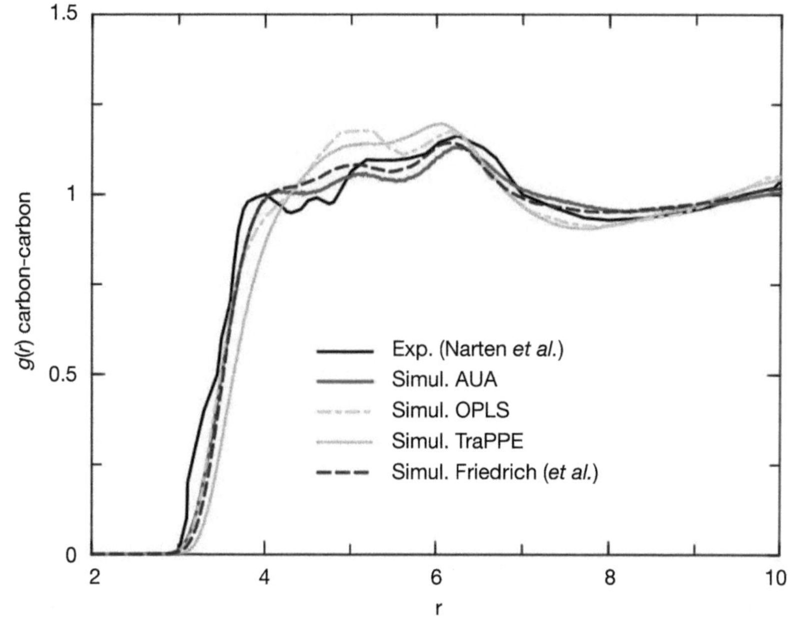

Figure 3.25 Carbon-carbon intermolecular pair distribution function of liquid benzene at 298 K computed with the AUA potential (AUA-AROM), and the intermolecular potentials of Jorgensen *et al.* (OPLS-AA), Wick *et al.* (TraPPE), Friedrich and Lustig (AUA-FL), compared with experimental data. Reprinted with permission from [Contreras *et al.*, 2004a], © 2004 American Chemical Society.

they improve the accuracy of predictions for the alkylbenzenes as well. The related parameters are given in Appendix 1, where the AUA parameters for all groups are reviewed.

With the exception of this small reparametrization, it can be stated that the AUA potential discussed in this section is validated and can be used without additional testing for predicting the properties of all hydrocarbons. Thus simulation can be used as a tool to generate an appropriate database which will serve to assign thermodynamic properties to the heavy hydrocarbons identified by gas chromatography and mass spectrometry. This is no longer a dream and it is already being carried out for C_{10}-C_{20} hydrocarbons. Further application to heavier hydrocarbons will become possible when more powerful computers are available.

3.2 THERMODYNAMIC DERIVATIVE PROPERTIES OF LIGHT HYDROCARBONS

In Section 2.4.8, thermodynamic derivative properties were defined as second order derivatives of the thermodynamic potential. These comprise, among others, isothermal compressibility, isobaric thermal expansivity and heat capacity. These properties can be evaluated only from the statistical fluctuations, i.e. from the amplitude of the variations of parameters such as volume or energy around the average, or from their correlation functions (Sections 2.1.3 and 2.4.8). By comparison, the properties discussed in Section 3.1 (density, vapour pressure, enthalpy) are directly linked to first order derivatives of the thermodynamic potential, which can be obtained from simple thermodynamic averages.

Thermodynamic derivative properties are interesting for industrial applications in many respects. Isothermal compressibility defines how fluid density will change with pressure, an important property when it comes to evaluating how oil density changes in a reservoir when the pressure is lowered during production. Thermal expansivity which characterises the density response to a change in temperature is useful in evaluating liquid density in various processes. Heat capacity is an important property too, as it quantifies the amount of heat that must be provided to a fluid to raise its temperature and thus controls the energy budget of many operations.

Another property that we will consider in this section is the Joule-Thomson coefficient, which characterises the change in temperature of a fluid when it is submitted to a change in pressure at constant enthalpy. This is typically the case of a flowing fluid that experiences hydraulic resistance (such a valve, a porous medium or a narrow tube) without exchanging heat with its surroundings. This property is derived from the heat capacity and the thermal expansivity (Eq. 2.107), so that it is natural that it be discussed along with derivative properties. As it controls the sign and amplitude of temperature variations in various phases of industrial processes, the Joule-Thomson coefficient is of great importance in the oil and gas industry.

A fundamental reason to investigate thermodynamic derivative properties by molecular simulation is the difficulty of classical equations of state to predict these properties, e.g. as is the case for heat capacity of pure hydrocarbons [Barreau *et al.*, 1993]. In contrast, molecular simulation appears now as a feasible and reliable route to evaluate the conditions of Joule-Thomson inversion in CO_2 and other gases [Colina and Müller, 1997; Chacin *et al.*, 1999; Escobedo and Chen, 2001; Colina *et al.*, 2002]. In Section 3.2.1 we will discuss how thermodynamic derivative properties can be predicted quantitatively for light hydrocarbon gases from methane to butane, as investigated by Lagache *et al.* [2001]. A particular emphasis will be put on the near-critical behaviour of thermodynamic derivative properties in Section 3.2.2. Molecular simulation is not able to simulate the critical point itself but it has been shown to predict well non-classical behaviour in the near-critical region [Wilding, 1995]. No analytical equation of state can do this, neither cubic nor more complicated versions [Levelt-Sengers, 1970]. It is therefore interesting to test the ability of simulation to predict the behaviour of thermodynamic derivative properties in the near-critical region, as they systematically diverge at the critical point.

3.2.1 Predictions at High Pressure

This section is devoted to predicting the derivative properties of methane, ethane and *n*-butane at high pressure, on the basis of the work of Lagache *et al.* [2001].

If a quantitative prediction of thermodynamic derivative properties is expected for light hydrocarbons, a preliminary requisite is an accurate intermolecular potential. As seen in Section 3.1, the AUA potential provides an accurate representation of volumetric properties as well as equilibrium pressure and vaporisation enthalpy for the light alkanes from ethane on (that this section will focus on). As this intermolecular potential involves a moderate number of centres of force, it is ideal for the calculation of derivative properties, which require longer simulations for convergence than other properties.

In the case of methane, Lennard-Jones potential parameters have been determined by Möller *et al.* [1992] on the basis of vapour-liquid equilibrium data. This potential provides a very satisfactory prediction of the volumetric behaviour at high pressure, as will be seen in Section 3.5, and it will be adopted therefore in the calculation of the derivative properties considered in this section.

A compilation of the thermodynamic properties is available for *n*-butane [Sychev *et al.*, 1995], providing a convenient data set for the comparison of predicted and experimental values for various derivative properties.

The first derivative property that will be considered here is the thermal expansivity. As shown in Figure 3.26, a quantitative prediction of this property is achieved for *n*-butane at 380 K, either in the vapour or in the liquid phase. Below the saturation pressure (1.7 MPa at this temperature) butane is in the vapour state and its thermal expansivity increases with increasing pressure. For very low pressures, the thermal expansion would be expected to be $1/T = 1/380 = 0.00263\ K^{-1}$, following on from the perfect gas law which must be obeyed at low pressure. It may be seen from Figure 3.26 that the experimental and simulation trends are in agreement with this low pressure limit. It is worth noticing that significant departures from perfect gas behaviour are observed in the vapour, e.g. thermal expansivity almost doubles between 0 and 1.5 MPa due to increasing intermolecular forces. Above the saturation pressure, thermal expansivity relates to the liquid phase. Although the error bars are significantly greater in this region, simulation results clearly indicate that thermal expansivity is dropping, in agreement with the experimental data.

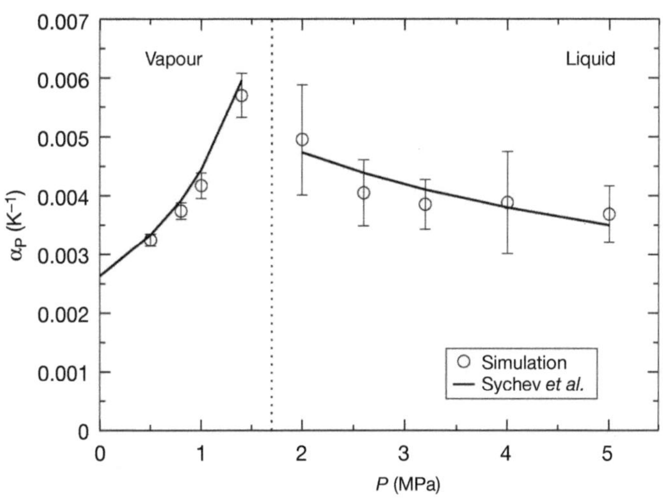

Figure 3.26 Thermal expansivity of *n*-butane at 380 K after Lagache *et al.* [2001], reproduced by permission of the PCCP Owner Societies. The continuous lines are the correlations from Sychev [1995], open dots are simulation results.

The second derivative property considered here is isothermal compressibility, which is given for n-butane in the same conditions as thermal expansivity (Fig. 3.27). In the vapour state, isothermal compressibility is undefined (ie. infinite) for $P = 0$ and decreases continuously with increasing pressure. A strong discontinuity is found at the saturation pressure since isothermal compressibility in the liquid phase is decreased by two orders of magnitude compared with the coexisting vapour state. Here again, significant error bars are found – particularly in the liquid state – but the Monte Carlo results are in good agreement with the experimental evidence.

Heat capacity provides the third example of derivative property, again illustrated by the behaviour of n-butane at 380 K (Fig. 3.28). More precisely, the variable plotted in Figure 3.28 is the *residual* heat capacity, i.e. the difference between total heat capacity and ideal heat capacity (see Section 2.4.8). As the ideal heat capacity is the total heat capacity of a gas in the limit of very low pressure, the residual heat capacity is zero for $P = 0$. It increases with increasing pressure as long as pressure is lower than the saturation pressure, and shows a slightly decreasing trend for higher pressures in the liquid state. At the temperature of 380 K, the ideal heat capacity of n-butane is 120.3 $J \cdot mol \cdot K^{-1}$ [Sychev *et al.*, 1995], and the residual heat capacity lies between 0 and 80 $J \cdot mol \cdot K^{-1}$ depending on pressure. Thus the total heat capacity ranges between 120 and 200 $J \cdot mol \cdot K^{-1}$ in the pressure range considered, the residual heat capacity representing up to 40% of the total heat capacity. The significant contribution of the residual heat capacity at high pressure reflects the important role of intermolecular interactions in the determination of total heat capacity.

Correct thermal expansivity, isothermal compressibility and residual heat capacity predictions for n-butane at high pressure may be interpreted as the result of the accurate modelling of molecular interactions in the vapour as well as in the liquid state, thanks to the AUA intermolecular potential discussed in Section 3.1. Now other light hydrocarbons will be considered to extend the scope of this comparison.

In the case of methane, comparisons with experimental data will be performed at two temperatures, 294.2 K and 377.5 K, on the basis of the compilations of IUPAC [Wagner and Reuck, 1996]. In these conditions, no vapour-liquid transition is observed because methane is supercritical. As a result, the trend of residual heat capacity with pressure does not exhibit any discontinuity (Fig. 3.29). At the critical temperature (191 K), the residual heat capacity of methane would exhibit a sharp peak flanking the divergence at the critical pressure (4.5 MPa). At 294.2 K, a well-defined peak is observed around 20 MPa, but no divergence is found. At 377.5 K the maximum of residual heat capacity is still noticeable around 40 MPa but far less developed. This drift and attenuation of the heat capacity maximum is predicted by Monte Carlo simulation (within the error bars representing the statistical uncertainties), indicating that the predicted residual heat capacity is reliable for supercritical methane. In the temperature range under consideration, the ideal heat capacity of methane lies between 36 and 39 $J \cdot mol \cdot K^{-1}$ [Wagner and de Reuck, 1996]. Thus, the residual heat capacity is a minor yet significant part of total heat capacity (0 to 35%), as was the case with n-butane. This confirms the value of Monte Carlo simulation when it comes to generating reliable estimates of heat capacity, although the ideal heat capacity must be provided by experimental data or by correlations since it cannot be accessed by simulation.

Predicting the Joule Thomson coefficient of methane at high pressure is also a challenging way to test Monte Carlo simulation, as its calculation through Eq. (2.107) involves molar vol-

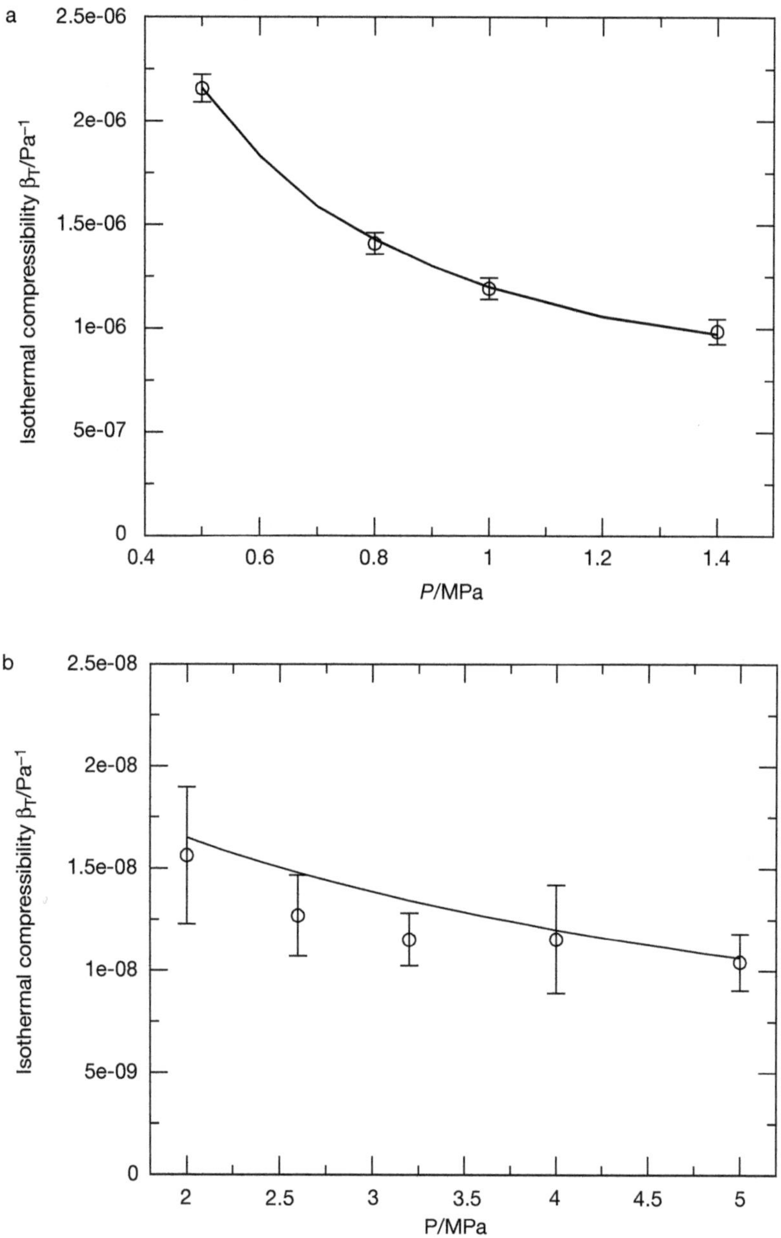

Figure 3.27 Isothermal compressibility of *n*-butane at 380 K after Lagache *et al.* [2001], reproduced by permission of the PCCP Owner Societies (a) vapour phase (b) liquid phase. The continuous lines are the correlations from Sychev [1995].

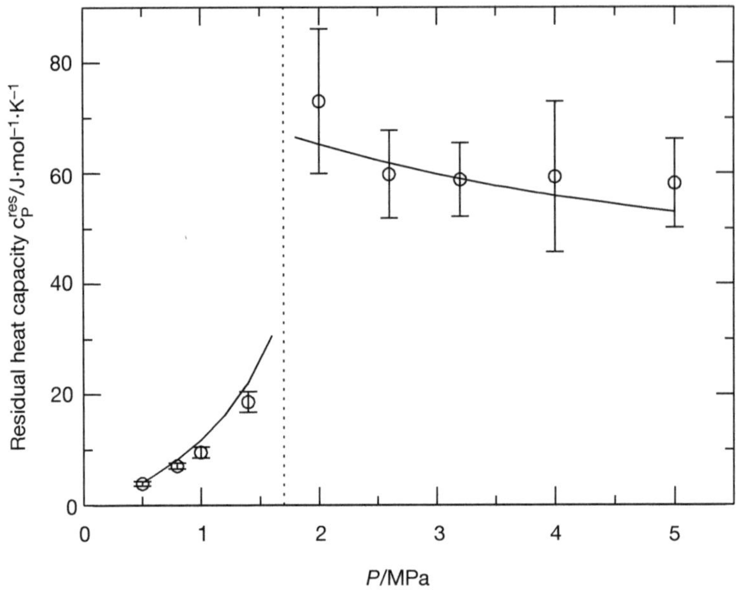

Figure 3.28 Residual heat capacity of *n*-butane at 380 K after Lagache *et al.* [2001], reproduced by permission of the PCCP Owner Societies. The continuous lines are the correlations from Sychev [1995], open symbols are simulation results.

ume, heat capacity and thermal expansivity. As illustrated in Figure 3.30, the Joule-Thomson coefficient tends toward a non-zero value (approximately 2.5 $K \cdot MPa^{-1}$) at very low pressure. Obtaining a non-zero value in the low pressure limit may seem surprising because methane should be behaving like a perfect gas with a thermal expansivity of 1/T. According to Eq. (2.107) the Joule-Thomson coefficient should be zero, a conclusion that is obviously contradicted by the experimental data shown in Figure 3.30. In fact, it may be shown that the convergence toward a non-zero value at P = 0 can be explained by the influence of second order terms on thermal expansivity. It is therefore highly satisfactory that the simulation results are consistent with the limiting value of 2.5 $K \cdot MPa^{-1}$ at P = 0, although statistical uncertainties increase significantly when low pressures are addressed. The increase in statistical uncertainty at low pressure, which is specific to the prediction of the Joule-Thomson coefficient, may be understood when looking at the terms appearing in Eq. (2.107). Indeed, the Joule Thomson coefficient is the product of two terms, one of them (v/c_p) tending to infinity, while the other $[T\alpha_p - 1]$ approaches zero in the limit of low pressures. A small relative uncertainty on the thermal expansivity α_p causes then a large uncertainty on the Joule-Thomson coefficient because it is multiplied by the very large term (v/c_p). At higher pressures, this is no more the case and the statistical uncertainties are moderate. This allows a good determination of the inversion pressure of the Joule-Thomson effect, i.e. the pressure at which heating is observed instead of cooling in the Joule-Thomson expansion. As seen in Figure 3.30, the predicted inversion pressure (approximately 52 MPa) is in very good agreement with experimental observations.

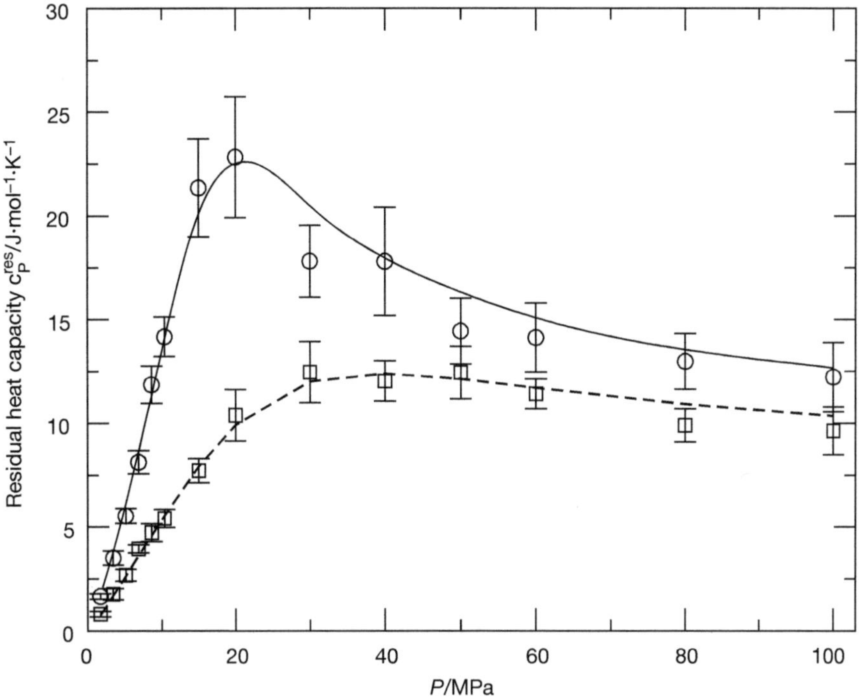

Figure 3.29 Residual heat capacity of methane computed by simulation at 294.2 K (dots) and 377.5 K (squares) after Lagache *et al.* [2001], reproduced by permission of the PCCP Owner Societies. The lines indicate the correlations of Wagner and Reuck [1996].

If heavier hydrocarbons are considered, the presence of the liquid-vapour phase transition causes an important discontinuity of the Joule-Thomson coefficient, as illustrated by the case of *n*-butane (Fig. 3.31). In most cases, the Joule-Thomson coefficient for pure liquids is known to be negative, so that the case of *n*-butane at 380 K shown in Figure 3.31 is rather unusual as the liquid phase existing above 1.7 MPa exhibits a positive Joule-Thomson coefficient. Nevertheless, the behaviour of the Joule-Thomson coefficient is quantitatively predicted by simulation. If higher pressures were investigated, the Joule-Thomson would be negative as it is observed for methane.

Finally, this investigation of light hydrocarbons will address a mixture of methane and ethane. In a general way, the simulation of mixtures is delicate because the sampling of all possible system conformations is always more difficult for a mixture than for a pure compound. In this respect, the prediction of the Joule-Thomson coefficient of the methane-ethane mixture at high pressure shows that it is possible to keep the uncertainty to a level that is comparable to that of the pure components involved (Fig. 3.32), with reasonable agreement with experimental measurements. In this comparison, the methane mole fraction was 75%, i.e. within the usual range of methane concentrations observed in natural gases (65 to 95%).

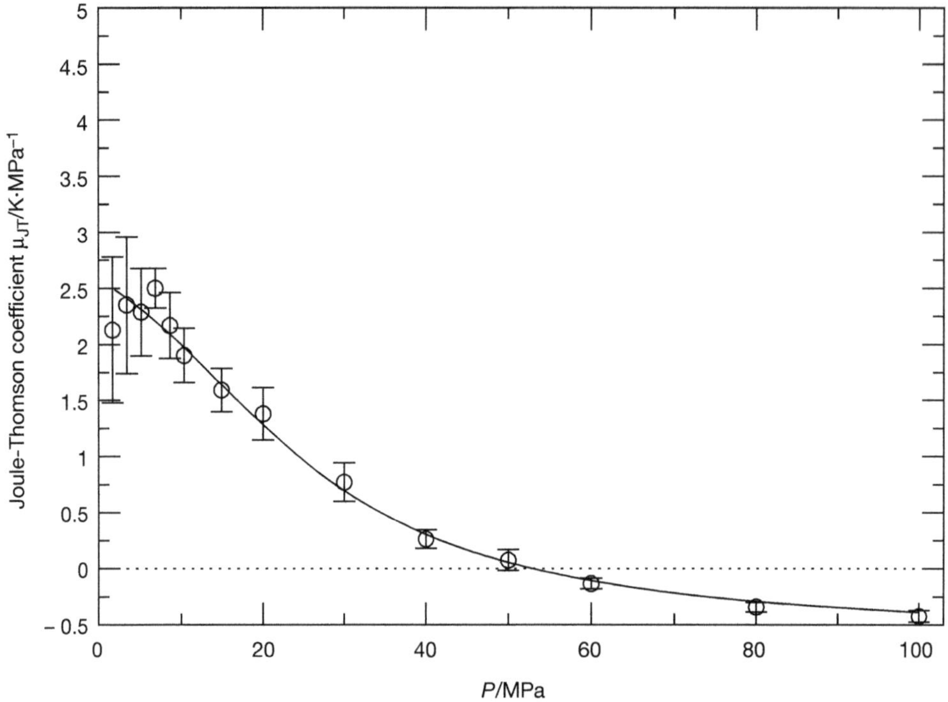

Figure 3.30 Joule-Thomson coefficient of methane obtained by simulation (dots) at 377.5 K after Lagache *et al.* [2001], reproduced by permission of the PCCP Owner Societies. The lines indicate the correlations of Wagner and de Reuck [1996].

However, it is still a very simplified model compared to the composition of real gases. In Section 3.5, we will see that it is possible to determine the Joule-Thomson coefficient of natural gases containing a variety of hydrocarbons other than methane.

As a concluding remark about the evaluation of derivative properties, it can be said that the general agreement between predicted values and experimental data (or quasi-experimental correlations) is excellent. This quantitative agreement is firstly evidence of the practical efficacy of the proposed fluctuation method when it comes to evaluating derivative properties. Secondly, it provides striking confirmation of the intermolecular potential of Möller *et al.* [1993] for methane and of the AUA potential of alkanes. The performance of intermolecular potentials is especially impressive when it is considered that they are not based on any derivative property, but only on standard properties as discussed in Section 3.1. It is also striking that butane properties are well represented, while no data about *n*-butane was considered in developing the AUA potential. However, it may be argued that the error bars are often large, and the proposed method might be considered insufficiently accurate for practical use in the industry. This would be too pessimistic, because computer capacity has increased since these investigations [Lagache *et al.* 2001] were performed on computers that were four times

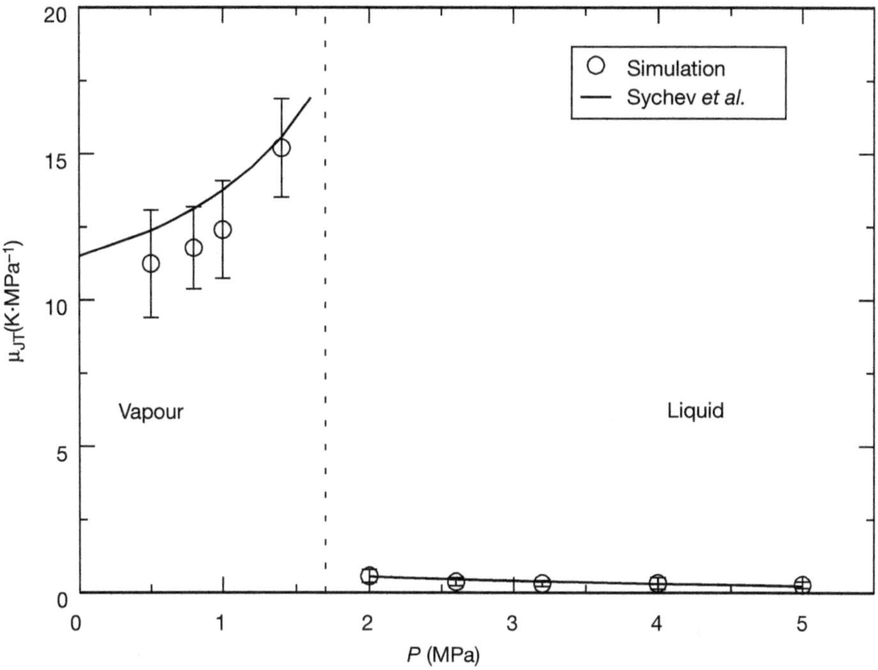

Figure 3.31 Joule-Thomson coefficient of *n*-butane at 380 K from simulations (dots) and from the correlation of Sychev (lines), after Lagache *et al.* [2001], reproduced by permission of the PCCP Owner Societies. The dashed vertical line separates the vapour region (left) and the liquid region (right).

less rapid than today's workstations. Simulations four times as extensive could be run in the same time with only half the statistical uncertainty. With further increases in speed and with parallel computers, there is no doubt that it will be possible to obtain smooth trends for derivative properties in the near future in the same way as density or saturation pressures are obtained today.

3.2.2 Prediction of Derivative Properties in Near-critical Conditions

This investigation is aimed at investigating the divergence of derivative properties in the near-critical region, using ethane and methane as examples.

Some derivative properties are said to exhibit strong divergence (or anomalies) at the critical point, i.e. they tend to infinite values at the critical point in classical theories as well as in non-classical theories involving scaling laws like Eq. (2.95) [Levelt-Sengers *et al.*, 1983]. Isothermal compressibility, thermal expansivity and heat capacity at constant pressure are

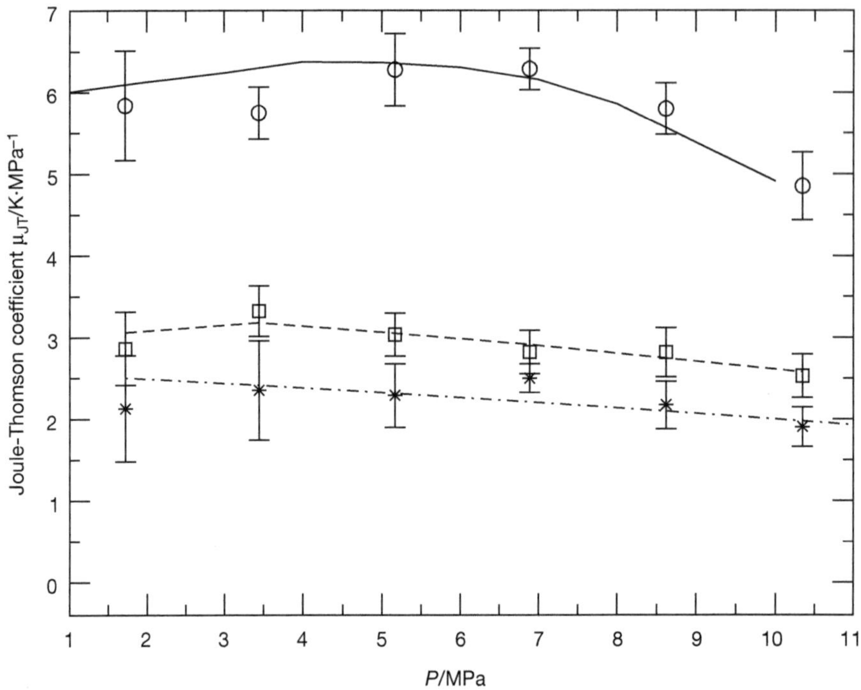

Figure 3.32 Joule-Thomson coefficient in the methane-ethane system at 377.5 K, after Lagache *et al.* [2001], reproduced by permission of the PCCP Owner Societies: pure methane (stars), 25% ethane (squares), pure ethane (dots). The lines are the recommended values from Landolt and Bornstein [1961].

examples of such strongly diverging properties at the critical point. A first objective of this section is to show that such anomalies can be represented, at least qualitatively, although specific difficulties are encountered as the critical point is approached.

According to classical analytical models, other derivative properties do not diverge at the critical point but, in practice, they are actually observed to diverge weakly, i.e. a very sharp maximum is found in the vicinity of the critical point. This is particularly true of heat capacity at constant volume. Such weak divergences can only be described with non-classical theories. Another aim in this section is to examine if the behaviour of a weakly divergent property can be reproduced by molecular simulation.

Before investigating derivative properties in the near-critical region, it is worth plotting ethane's liquid-vapour coexistence diagram as generated by Gibbs ensemble simulations for 1 000 molecules (Fig. 3.33), using the AUA potential discussed in Section 3.1. The highest temperature simulated is 293 K, because density fluctuations cause excessive fluctuations above this temperature. In order to simulate coexistence in the Gibbs ensemble at higher temperatures, the simulated system should comprise a larger number of molecules. The critical temperature obtained by fitting Eqs. (2.95) and (2.96) on simulated points yields a critical temperature of 311 K, overestimating by approximately 5 K the experimental critical temperature. In the following development, the derivative properties will be computed along three isotherms corresponding to reduced temperatures $T/T_c = 1$, $T/T_c = 1.058$ and $T/T_c = 1.14$, where $T_c = 305.4$ K is the actual critical temperature.

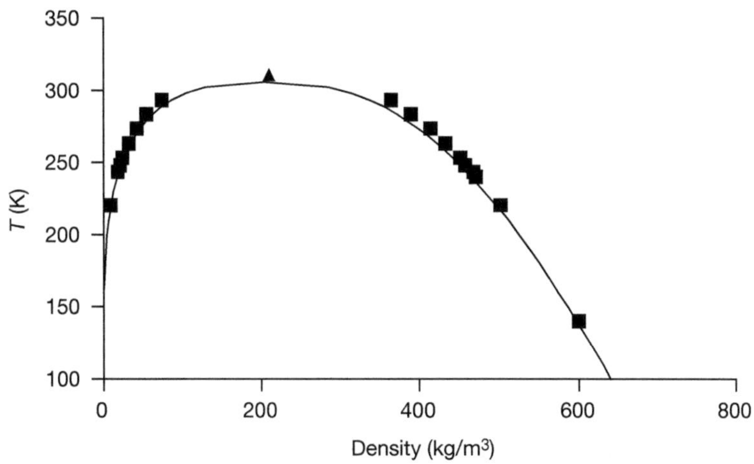

Figure 3.33 Coexistence curve of ethane obtained by simulation (dots) and from the quasi-experimental equation of state of Friend *et al.* [1991].

The thermal expansivity coefficient shows the typical behaviour of a strongly diverging derivative property (Fig. 3.34). The experimental data related to the first isotherms ($T/T_c = 1$) exhibits a true divergence characterized by a narrow peak centrered on the critical pressure. The position of the peak shifts to larger pressures when temperature is increased, and the peak height is diminished. Qualitatively, this behaviour is well reproduced by simulation results obtained by the fluctuation method in the *NPT* ensemble. However, large statistical uncertainties are found for those points close to the divergence. The amplitude of the peak is not well reproduced for $T/T_c = 1$: this may be explained by the difference between the critical temperature of the model and the actual critical temperature. In fact, the simulation performed at $T/T_c = 1$ is subcritical, and it is not surprising that the anomaly is less than observed. On the opposite, the condition $T/T_c = 1.058$ (i.e. $T = 323$ K) is somewhat closer from the critical temperature of the model (311 K) than from the experimental critical temperature, so its not surprising that it overestimates the anomaly of the thermal expansivity.

The heat capacity at constant pressure (C_p) of ethane is shown for the same isotherms (Fig. 3.35), for which quantitative comparison is made with the experimental data of Bier *et al.* [1976]. Here again, the strong divergence is qualitatively predicted by simulation. As found with thermal expansivity, the anomaly of C_p is somewhat underpredicted for $T/T_c = 1$ and overpredicted for $T/T_c = 1.058$, the explanation being the overestimation of the critical temperature by the AUA potential used in the study.

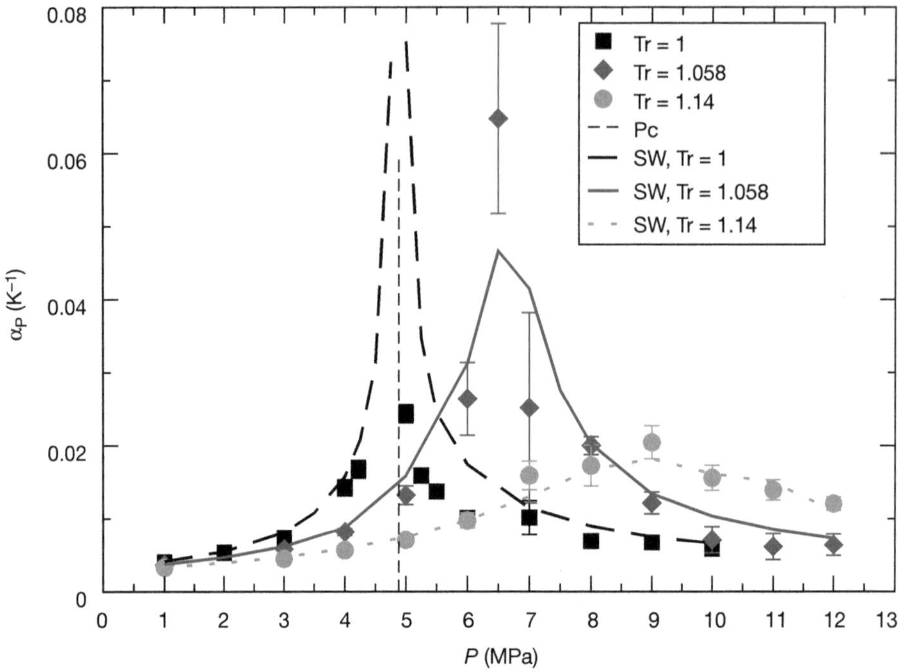

Figure 3.34 Thermal expansivity of ethane in the near-critical region from simulation at three reduced temperatures T/T_c, compared with the quasi-experimental equation of state of Friend *et al.* [1991].

When considering C_V, the heat capacity at constant volume (Fig. 3.36), it can be seen that the weak divergence is also well reproduced by simulation for ethane along these three isotherms. Although there are no direct measurements to compare with, it is clear that C_V displays qualitatively the same behaviour as the thermal expansion coefficient and C_p. Another illustration that simulation predicts correctly the anomaly of C_V is provided by the example of methane (Fig. 3.37) at the temperature of 210 K, i.e. $T/T_c = 1.1$. Here, C_V is plotted versus density, and the maximum of C_V at 150 kg/m^3 is predicted in reasonable agreement with experimental correlations.

From these comparisons, it may be concluded that anomalies can be investigated by molecular simulation in a semi-quantitative way, including the case of a weakly diverging property like C_V. In this study, the use of a simulation box containing 1 000 ethane molecules was chosen to be able to come close enough to the critical point. Although larger than the system routinely used for liquid-vapour equilibria, this is still moderate in size and there is little doubt that it will become possible to investigate near-critical anomalies in more complex systems in the near future. Nevertheless, it is worth to stress the importance of having an accurate intermolecular potential for such predictions. In the case of ethane, discrepancies between simulation and experiment are indeed likely to be caused by the 5 K deviation of the AUA potential on the critical temperature rather than by statistical uncertainties. This statement is not specific to derivative properties, but it is particularly sensitive in this case because such striking divergences are only observed within a few degrees of the critical point.

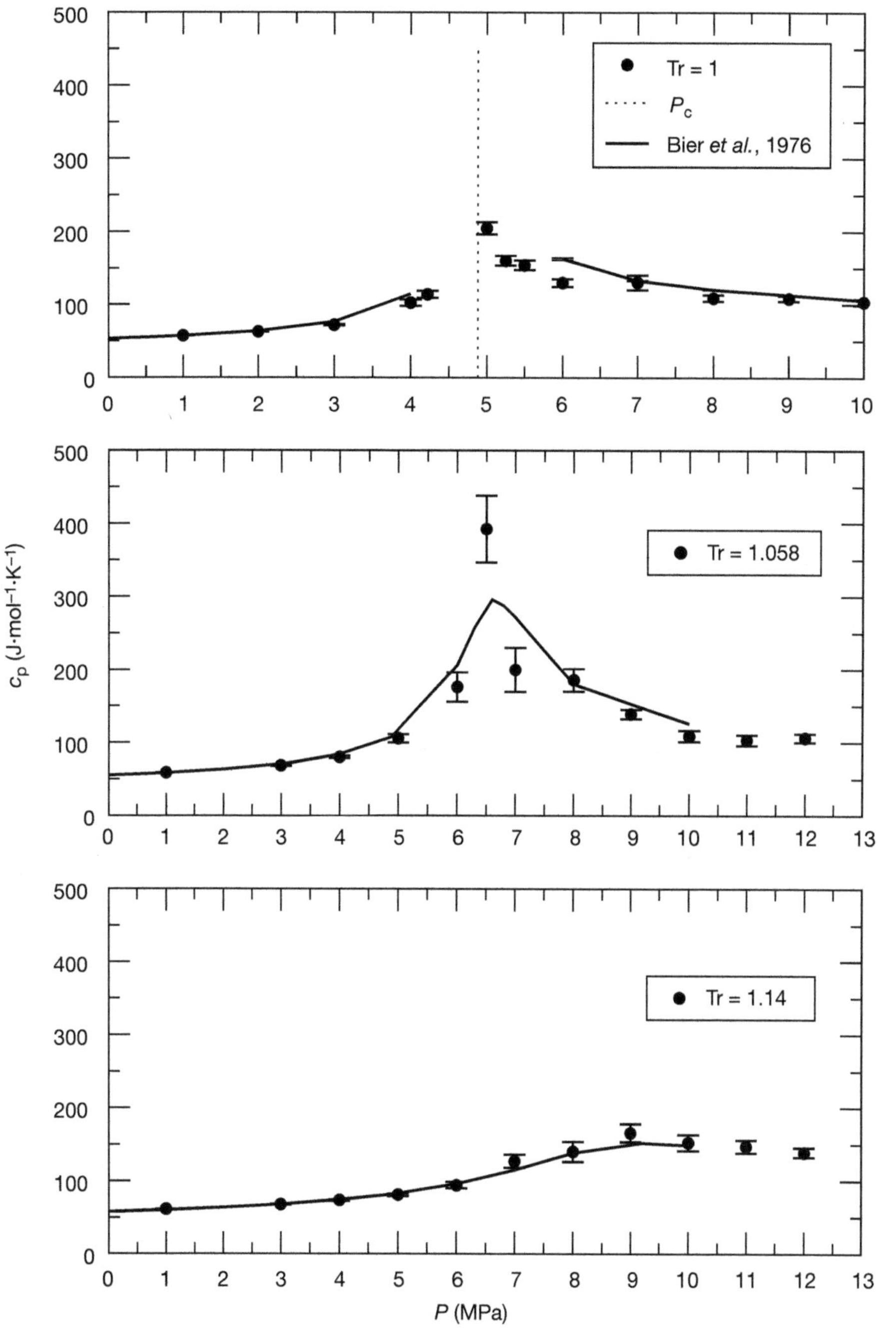

Figure 3.35 Molar heat capacity at constant pressure of ethane in the near-critical region from simulation at three reduced temperatures T/T_c, compared with the quasi-experimental equation of state of Friend *et al.* [1991].

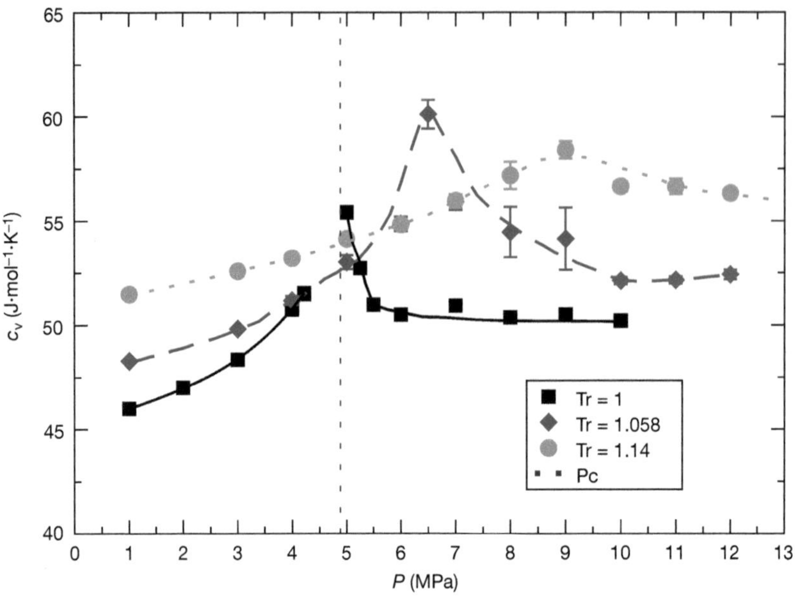

Figure 3.36 Molar heat capacity at constant volume of ethane in the near-critical region from simulation at three reduced temperatures T/T_c, compared with the quasi-experimental equation of state of Friend *et al.* [1991].

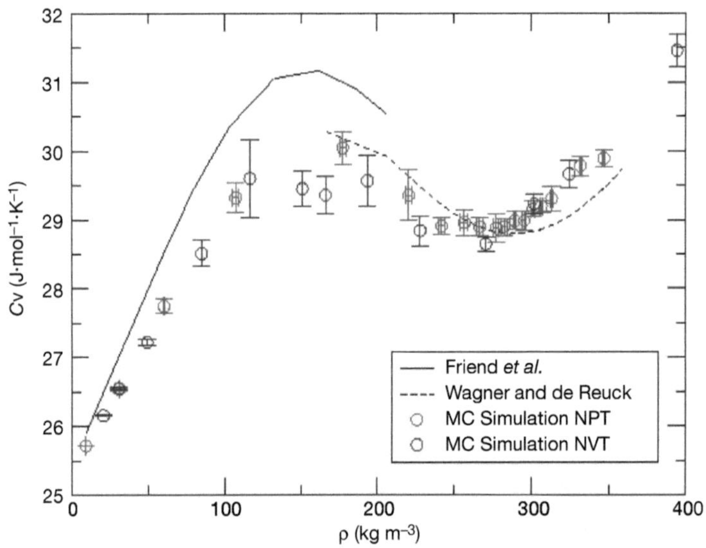

Figure 3.37 Molar heat capacity at constant volume of methane versus density computed from *NPT* and *NVT* simulations at a reduced temperature of 1.1 (i.e. 210 K). The continous line is the quasi-experimental equation of Friend *et al.* [1991] and the dashed lines are recommended values from Wagner and de Reuck [1996].

3.3 PROPERTIES OF POLAR ORGANIC COMPOUNDS

In many instances, the oil and gas industry needs to handle polar organic compounds, i.e. compounds in which polar groups such as –SH, =O or –OH are fixed onto a saturated or aromatic skeleton. Sulphur-containing compounds are often present in significant amounts in natural gases and crude oils and they have to be removed for several reasons: they smell noxious, they are poisonous and their combustion releases sulphur oxides which are converted into sulphuric acid in the atmosphere of large cities. They comprise mainly sulphides (R–S–R'), thiols (R–SH), and thiophene derivatives (i.e. a five-membered aromatic ring containing four carbon atoms and one sulphur atom). Oxygenated compounds are also important as additives in gas production (e.g. methanol is often used to prevent hydrate formation in pipelines) and to fuels (etheroxides are systematically used to raise octane number). Organo-mercuric compounds sometimes accompany mercury in natural gases as trace compounds, and their fate in gas processing must be known because they are corrosive and extremely toxic.

Our understanding of thermodynamic behaviour of polar organic compounds in industrial processes is often limited by a lack of experimental data, e.g. many sulphur-containing compounds have been investigated as pure compounds but there is little thermodynamic data about mixtures of sulphur compounds with hydrocarbons or common solvents. Also, there is little information about the adsorption of these compounds in zeolites, even though adsorption-based processes might provide a good way of removing these contaminants. For all these reasons, there is a need for reliable ways to simulate the thermodynamic behaviour of such compounds.

These molecules are sufficiently polar that their simulation must take electrostatic energy into account, as discussed in Section 2.2. This is done by introducing electrostatic point charges which are defined on the basis of ab initio calculations, following the general scheme shown in Figure 3.38. Several force fields have been developed with this approach for polar organic compounds [Sun, 1998; Jorgensen 1986a and 1986b]. However, these intermolecular potentials have been designed to match liquid properties in ambient conditions rather than vapour pressures or critical properties. It is therefore suitable to develop special potentials to provide accurate predictions of thermodynamic properties. The purpose of this section is to illustrate this approach in the context of a few examples: organic sulphides and thiols, organo-mercuric compounds, ketones and aldehydes and alcohols. It is however still far from complete. Many high quality contributions have been made to the simulation of vapour-liquid equilibria based on etheroxides [Lisal *et al.*, 1999] and alkanols [Chen *et al.*, 2001], as well as other polar organic compounds; the reader is referred to the extensive related literature for information about other systems than those discussed below.

3.3.1 Organic Sulphides and Thiols

The following section is a mainly an account of the work of Delhommelle *et al.* [2000b] on the development of an original intermolecular potential for organic sulphides (R'–S–R"), disulphides (R'–S–S–R") and thiols (R–SH) where R, R' or R" are alkyl groups. Its objective is to device a general AUA potential that would be consistent with the alkane potential used in Section 3.1. to describe the liquid-vapour coexistence properties of these compounds.

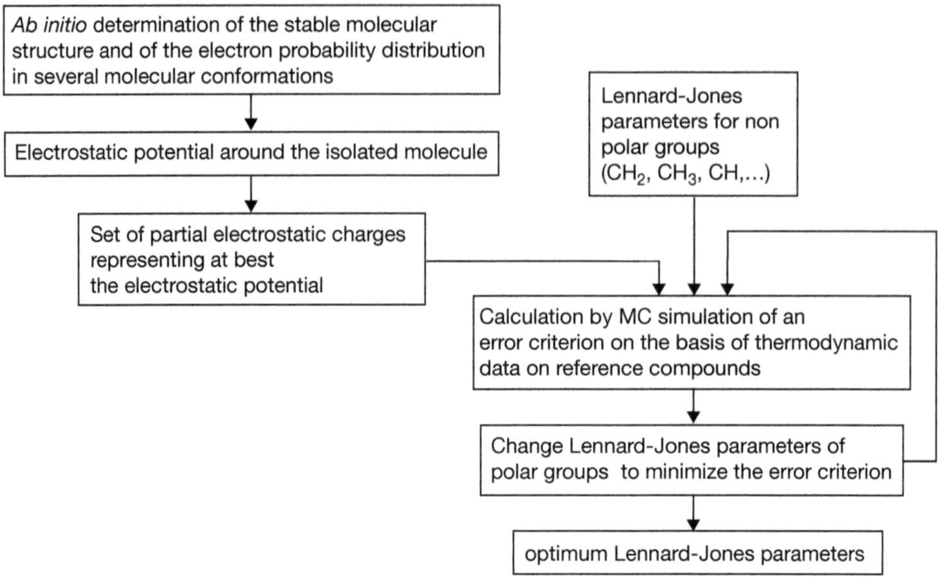

Figure 3.38 General procedure followed for the sequential determination of the electrostatic partial charges and the Lennard-Jones parameters for polar organic molecules.

In a first step, a set of partial electrostatic charges is determined to account for the electrostatic field around the isolated molecule, based on ab initio calculations. In a second step, the Lennard-Jones parameters of the sulphur-bearing group are fitted onto selected reference compounds, the influence of the alkyl chain(s) being computed with the same parameters for CH_2 and CH_3 as for the *n*-alkanes (Fig. 3.38).

As discussed in Section 2.4.4, the determination of electrostatic charges for these compounds raises a specific difficulty with the bending and torsional flexibility of the alkyl chains meaning that multiple conformations may be possible. If the usual procedure is followed, a separate set of charges is obtained for each different conformation. In Delhommelle's work, the application of a recent method [Lévy and Enescu, 1998] presents the advantage that the atomic charges depend little on the conformation, as illustrated by Figure 3.39 in the case of dimethyl disulphide. As can be seen on the figure, the normal least-squares fit yields charges that depend strongly on the dihedral angle which characterises torsion in this molecule. On the opposite, the eigenvalue method of Lévy *et al.* yields charges that are less dependant on the conformation. The charges determined for selected sulphides and thiols are indicated in Table 3.6. As a general feature, the hydrogens bear small positive charges between 0.15 and 0.25, carbon atoms bear negative charges (– 0.7 to – 0.4) as well as sulphur (0 to – 0.2). The distribution of these charges causes a dipole moment in those molecules which are asymmetric, in qualitative agreement with observed dipole moments. For instance, the simulated dipole moment of methanethiol is 1.92 D, while experimental measurements indicate a value of 1.52 D. It seems that dipole moments based on ab initio calculations tend to be overestimated.

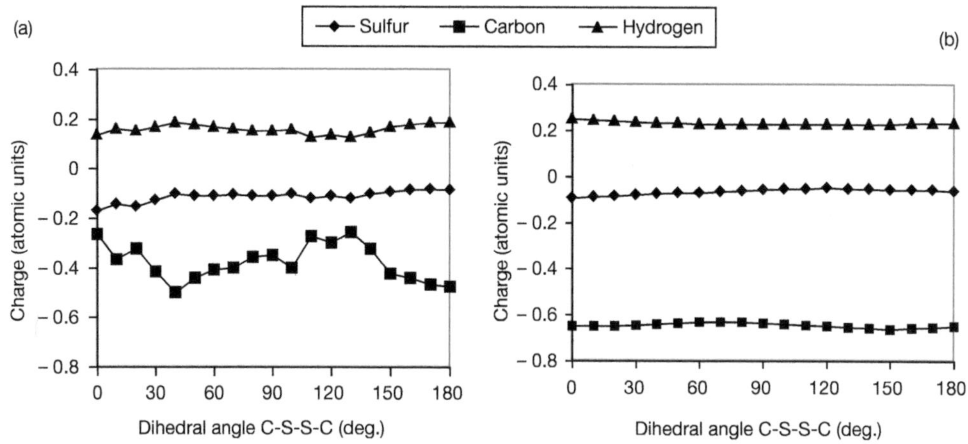

Figure 3.39 Dependency of the electrostatic charges of dimethyldisulfide as a function of the dihedral angle, after Delhommelle *et al.* [2000b], reprinted with permission, © 2000 American Chemical Society. (a) charges obtained by normal least-squares fit. (b) charges obtained by eigenvalue analysis.

Table 3.6 Partial charges of sulfides and thiols determined by eigenvalue analysis on the basis of ab initio calculations [Delhommelle *et al.*, 2000b].

Compound	S	H(SH)	C(CH_3)	C(CH_2)	H(CH_3)	H(CH_2)
CH_3–S–CH_3	– 0.054	–	– 0.673	–	0.234	–
CH_3–CH_2–S–CH_3	– 0.043		– 0.587	– 0.454	0.202	0.231
CH_3–CH_2–S–CH_2–CH_3	– 0.010		– 0.500	– 0.496	0.180	0.231
CH_3–S–S–CH_3	– 0.090		– 0.654		0.211	
CH_3–SH	– 0.281	0.195	– 0.454		0.242	
CH_3–CH_2–SH	– 0.208	0.150	– 0.495	– 0.477	0.189	0.231
CH_3–CH_2–CH_2–SH	– 0.267	0.162	– 0.491	– 0.417	0.165	0.233

Once the charges have been determined, the Lennard-Jones parameters of the sulphur-bearing group can be determined in the same way as non-polar hydrocarbons by the method exposed in Section 2.4.4. In the case of organic sulphides, the reference molecule is the dimethylsulphide CH_3–S–CH_3, the thermodynamic properties of which at 375 K and 298 K define the error criterion. As the sulphur group consists of a unique atom, the Lennard-Jones centre is assumed to be the sulphur nucleus, so that only two parameters are optimised, i.e. $\sigma = 3.60$ Å and $\varepsilon/k = 160$ K. This set of parameters is found to represent correctly the coexistence curve of dimethylsulphide (Fig. 3.40a). In the case of thiols, the reference molecule is ethanethiol (C_2H_5–SH), and the SH group is represented as an anisotropic united atom, with a force centre located on the axis between the sulphur and the hydrogen atoms. The optimised parameters of the SH group are found to be $\sigma = 3.6547$ Å, $\varepsilon/k = 220.62$ K and $\delta = 0.0588$ Å. This set of values gives a better representation of the coexistence curve of ethanethiol than the OPLS potential [Jorgensen, 1986] as shown in Figure 3.40b. When compared with the H_2S potential of Kristof and Liszi discussed in Section 3.3 ($\sigma = 3.73$ Å, $\varepsilon/k = 250$ K), it may be observed that both σ and ε increase in the sequence S, SH and H_2S. It appears thus that the addition of hydrogen atoms to the sulphur increases the Lennard-Jones diameter and energy, in the same logical way as it was observed for the various CH_x groups in Section 3.1. The logical position of the AUA force centre of the SH group, located close to the sulphur nucleus, is an additional confirmation that the optimised parameters have a good physical sense.

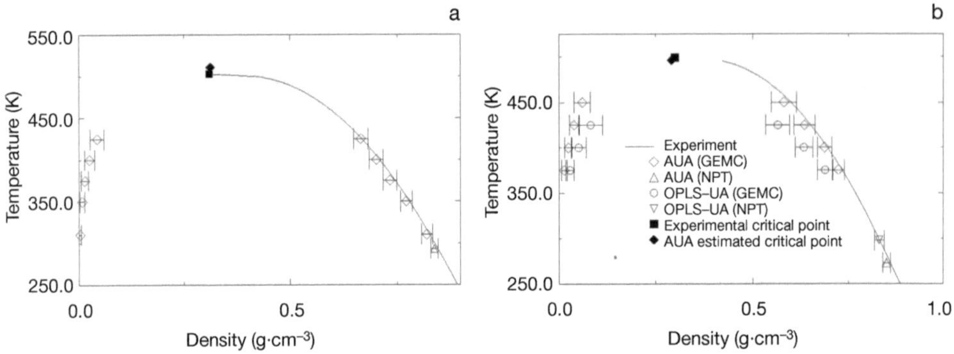

Figure 3.40 Liquid-vapour coexistence curves of dimethylsulphide (a) and ethanethiol (b), after Delhommelle *et al.* [2000b], reprinted with permission, © 2000 American Chemical Society.

Application of these potential parameters to other sulphides than dimethylsulphide is illustrated by Figure 3.41, which shows that the vapour pressures of diethylsufide and ethylmethylsulphide are also predicted correctly. It has been shown that liquid densities and vaporization enthalpies are also correctly represented for these molecules [Delhommelle *et al.*, 2000b]. Similarly, the potential has been applied to methanethiol, indicating again a good agreement with experimental liquid densities (Fig. 3.42). These results indicate that the influence of the alkyl chain is well represented by the alkane-based potential. However, the saturation pressure of methanethiol appears to be significantly underestimated.

In order to reduce the number of electrostatic charges (which is rather high in Delhommelle's model), an attempt has been made to use only three point charges, located on the sulphur, the neighbouring carbon and the hydrogen of the SH group. This has not only the advantage of reducing computer time, but also makes the model more consistent with the AUA alkane model in which there is no electrostatic charge on the alkyl chain. With a negative charge of -0.296 on the sulphur atom, and positive charges of $+0.112$ on the hydrogen and $+0.184$ on the carbon, the potential provides a better account of the dipole moment of methanethiol, ethanethiol and propanethiol. The Lennard-Jones parameters of the SH group have been readjusted accordingly on the equilibrium properties of ethanethiol. The resulting new parameter set ($\sigma = 3.6046$ Å, $\varepsilon = 258.27$ K, $\delta = 0.5601$ Å) is somewhat different from the previous one, but not inconsistent. It can be checked that the properties of ethanethiol and propanethiol are described in the same way as with the detailed electrostatic model, while the

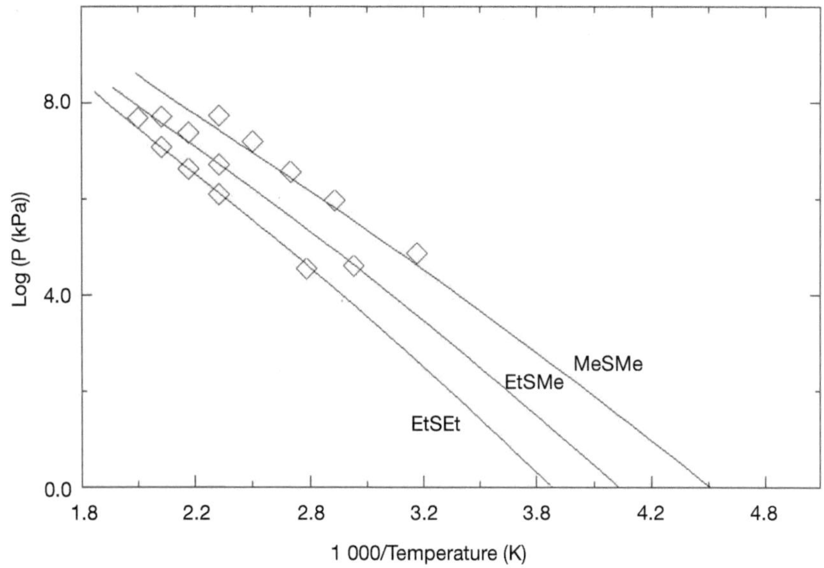

Figure 3.41 Saturated vapour pressure of organic sulphides, after Delhommelle *et al.* [2000b], reprinted with permission, © 2000 American Chemical Society.

EtSEt: diethylsulphide; EtSMe: ethylmethylsulphide; MeSMe: dimethylsulphide

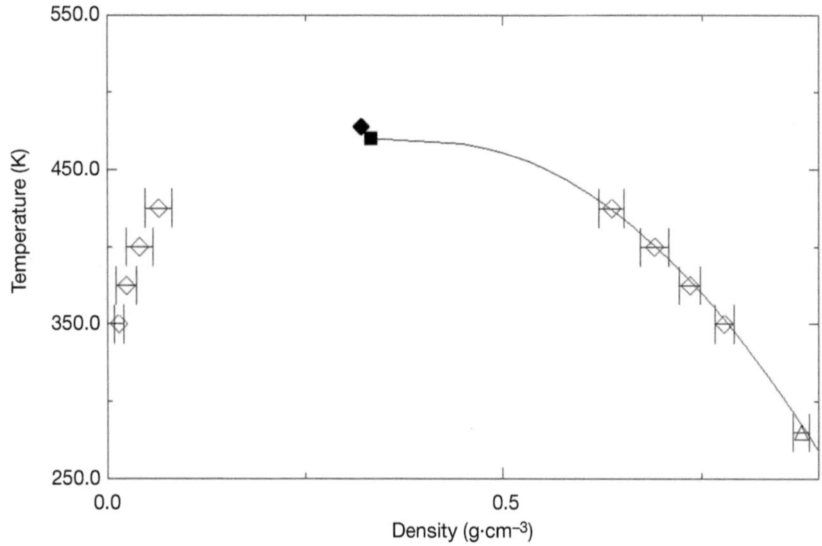

Figure 3.42 Liquid-vapour coexistence curve of methanethiol, after Delhommelle *et al.* [2000b], reprinted with permission, © 2000 American Chemical Society.

properties of methanethiol are slightly less satisfactory (Table 3.7). This example is interesting because it shows that the distribution of electrostatic charges in polar organic molecules may be sometimes simplified without adversely affecting property prediction.

Table 3.7 Properties of ethanethiol in the conditions selected as reference for the optimization of the Lennard-Jones parameters with atomic charges, i.e. nine electrostatic charges [Delhommelle *et al.*, 2000b] and with a simplified set of three charges on the sulfur atom and its two immediate neighbours.

Temperature	Property	Simulation Model 1 atomic charges	Simulation Model 2 3 charges	Experiment (DIPPR)
400 K	P_{sat} (kPa)	1 211	1 190	1 131
	$\Delta_{vap}h$ (kJ/mol)	21.35	20.97	20.95
	ρ_{liq} (kg/m^3)	687.0	688.8	693.5
276.16 K	$\Delta_{vap}h$ (kJ/mol)	27.3	28.11	28.4
	ρ_{liq} (kg/m^3)	849.0	850.5	857.2
	Dipole Moment (D)	1.86	1.69	1.58 [a]

[a] Mc Cleallan [1963].

This opens the way to future studies to investigate the behaviour of sulphides and thiols in liquid-vapour phase equilibria or in adsorption-based separation processes. Applying these potentials to adsorption will be discussed in Chapter 4.5.

3.3.2 Organo-Mercuric Compounds

Organo-mercuric compounds are of the form R–Hg–R where the R radical may be a methyl, ethyl or benzyl group if only the stable members of this family are considered. Here we will focus on dimethylmercury and diethylmercury compounds. Because of the extreme toxicity of these compounds, very few experimental measurements have been made and molecular simulation is a way of profiting from existing measurements in order to predict phase equilibria [Lagache *et al.*, 2004b]. Vapour pressures are documented between 260 and 360 K for dimethylmercury [Long and Cattanach, 1961; Thomson and Linnett, 1936], but only two measurements are available for diethylmercury (made by Thomson and Linnett). Densities are documented only from the compilation of Pascal [1957]. Similarly, the only information on the structure of these compounds is that their structure is linear with an estimated average C–Hg bond length of 2.094 Å [Rao *et al.*, 1960].

A first step in the simulation of these compounds is to check their equilibrium structure and to determine bending and torsional potentials to describe their conformations. This is done by applying a quantum chemistry method, using specific features to represent a very heavy atom like mercury. This study confirms that the C–Hg–C structure is linear. The stable conformations of both molecules are shown on Figure 3.43. The C–Hg distance on the basis of ab initio calculations on dimethylmercury (2.086 Å) turns out to be very close to Rao's experimental measurement. In diethylmercury, the C–Hg distance is not very different (2.101 Å) and the average C–C–Hg angle is 113.4°, i.e. similar to the average C–C–C angle in the *n*-alkanes (114°).

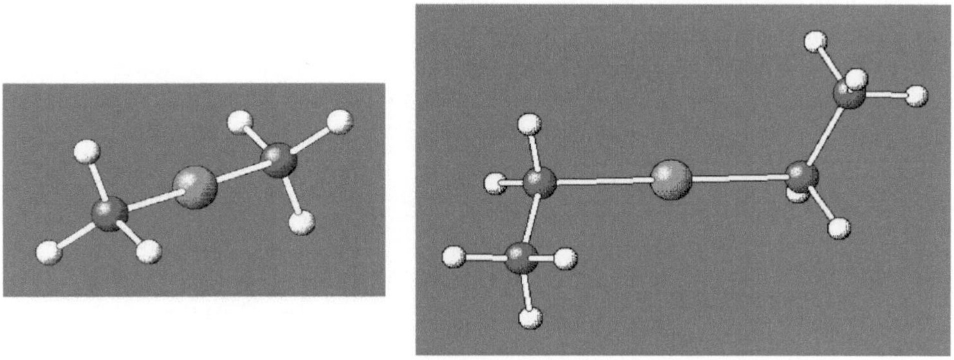

Figure 3.43 Structure of dimethylmercury and diethylmercury.

In order to determine the potential energy associated with bending deformations, ab initio calculations can be performed at various C–Hg–C bending angles, as shown on Figure 3.44 for dimethylmercury. On the basis of these results, the parameters of the bending potential can be determined unambiguously (see Appendix 1).

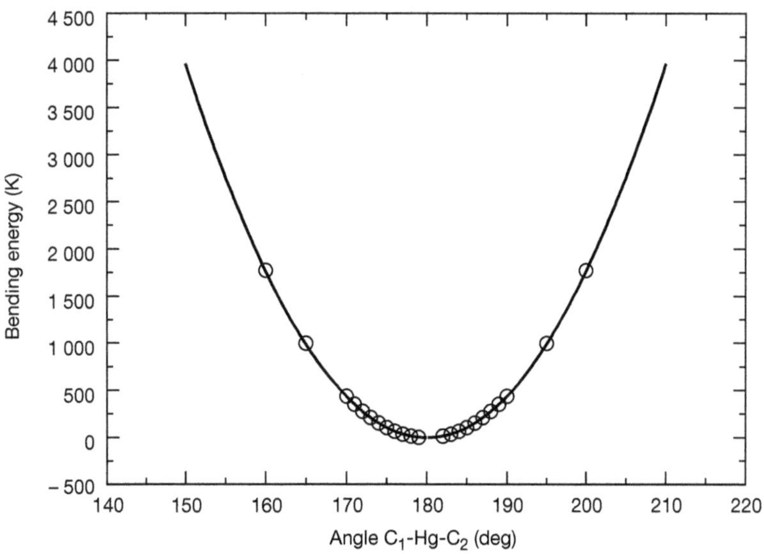

Figure 3.44 Bending potential energy C–Hg–C in organic sulfides, as determined from ab initio simulations (dots) and correlated by the standard bending potential function, after Lagache *et al.* [2004b], reprinted with permission, © 2004 American Chemical Society.

Dimethylmercury (which contains only three centres of force) has no torsion potential but diethylmercury does. However, the dihedral angle defined by four successive atoms C–C–Hg–C is undetermined in the stable conformation, because three of the atoms are aligned. Therefore it is decided to introduce the *supertorsion angle*, which is defined as the dihedral angle formed by the four carbon atoms. Ab initio results indicate that the potential energy of diethylmercury varies by less than 200 K when the supertorsion angle is varied progressively from 0 to 180° (Fig. 3.45). As this energy barrier is significantly lower than ordinary temperatures, it is not surprising that the supertorsion is mostly free, i.e. that no particular supertorsion angle is privileged when Monte Carlo simulation is performed.

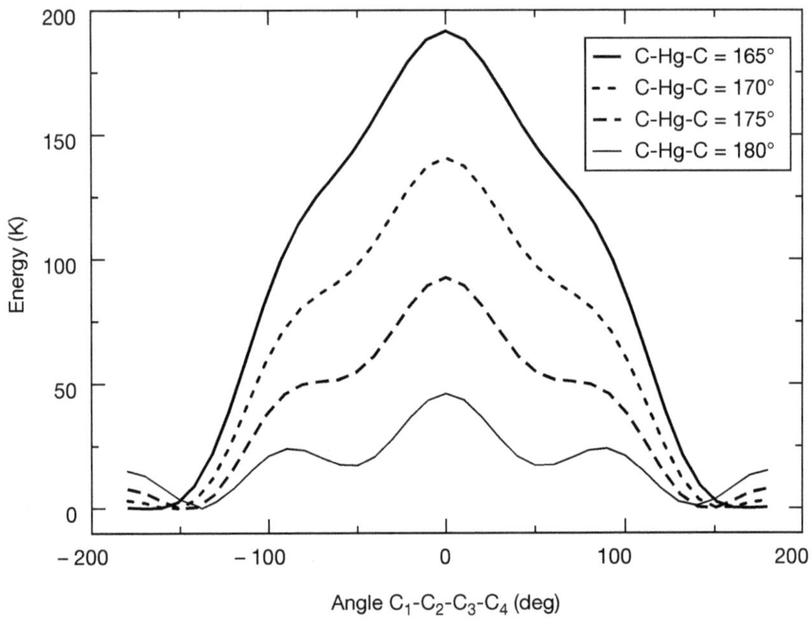

Figure 3.45 Potential energy associated with supertorsion (C–C–C–C dihedral angle) in the diethylmercury. Ab initio results (dots) are correlated by the same type of expression as torsion potential, after Lagache *et al.* [2004b].

The electrostatic partial charges on dimethylmercury and diethylmercury (Table 3.8) are determined with the method described above. In the example of dimethylmercury, a good representation is obtained when atomic charges are considered (rrms deviation lower than 10%), while the deviation exceeds 20% when 3 or 5 charges per molecule are considered. However, it turns out that electrostatic energy is only a minor part (less than 0.4%) of the whole interaction energy in the liquid phase, for both pure dimethylmercury and pure diethylmercury. Why is this so? The first reason is the absence of a significant dipole moment, so that diethylmercury and dimethylmercury display only quadrupole moments due to the symmetry of the molecule. Another reason is the magnitude of the Lennard-Jones potential of the mercury atom, which makes the electrostatic interactions relatively small in comparison with other molecules displaying significant quadrupole moments, such as carbon dioxide.

Table 3.8 Electrostatic charges of the dimethylmercury and diethylmercury determined by Lagache [2004b], following the eigenvalue analysis method of Lévy and Enescu [1998]. The charges are given in atomic units, i.e. as a fraction of the unit proton charge.

	Hg	C	H	CH_2 (AUA)	CH_3 (AUA)	q_{add} [a]
CH_3–Hg–CH_2 (9 charges)	0.542	– 0.769	0.166	–	–	–
CH_3–Hg–CH_3 (3 charges)	0.206	–	–	–	– 0.103	–
CH_3–CH_2–Hg–CH_2–CH_3 (5 charges)	0.184	–	–	– 0.071	– 0.02	–
CH_3–CH_2–Hg–CH_2–CH_3 (7 charges)	0.176	–	–	– 0.091	0.194	– 0.191

[a] Charges located on the axis of the C_1-C_2 bond, 0.446 Å from the external carbon C_1 toward the outside of the molecules.

In consequence, in the course of simulations of these molecules as pure compounds or mixed with non-polar fluids, electrostatic interactions can be ignored without incurring a significant error. Taking into account these charges is expected to be necessary only in systems where organic mercury compounds are interacting with highly polar compounds, such as alcohols or water.

The liquid density and saturation pressures of dimethylmercury and diethylmercury are indicated in Figures 3.46 and 3.47. The liquid density at ambient temperature is reproduced with a deviation of 1% for dimethylmercury and 3% with diethylmercury. This result may be considered satisfactory, because the absence of multiple experimental measurements prevents us from knowing the uncertainty. The saturation pressure of dimethylmercury is represented with a good accuracy, but it is overestimated by 30 to 50% in the case of diethylmercury. This shows the extrapolation capability of the intermolecular potential is only qualitative for organic mercuric compounds. If predictions are to be made for diethylmercury, it is recommended that potential parameters are optimised to match its saturation pressures.

3.3.3 Ketones and Aldehydes

Ketones and aldehydes are solvents which owe their polar character to the carbonyl C=O group, in which the oxygen is negatively polarised (Fig. 3.48). In this paragraph, we give a brief account of the work achieved by Kranias *et al.* [2003] who have extended the Anisotropic United Atoms potential to model the phase equilibrium behaviour of this important family of compounds.

In a first step, atomic electrostatic charges are determined from ab initio simulations of the isolated molecule, following the Lévy and Enescu method outlined in Section 2.4.4. For each

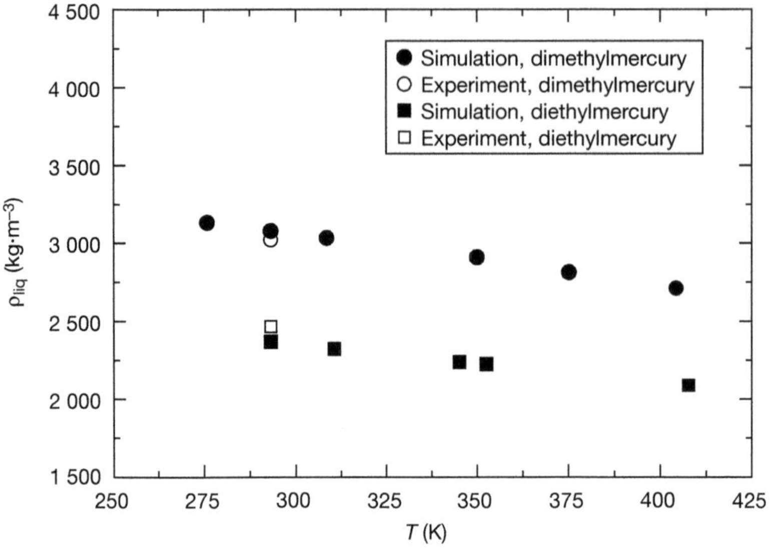

Figure 3.46 Density of dimethylmercury and diethylmercury [Lagache *et al.*, 2004b], reprinted with permission, © 2004 American Chemical Society.

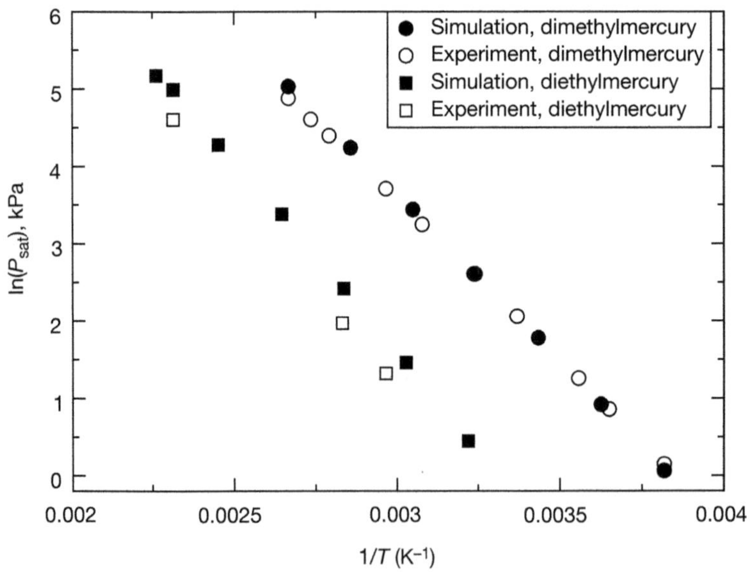

Figure 3.47 Saturation pressure of dimethylmercury and diethylmercury [Lagache *et al.*, 2004b], reprinted with permission, © 2004 American Chemical Society.

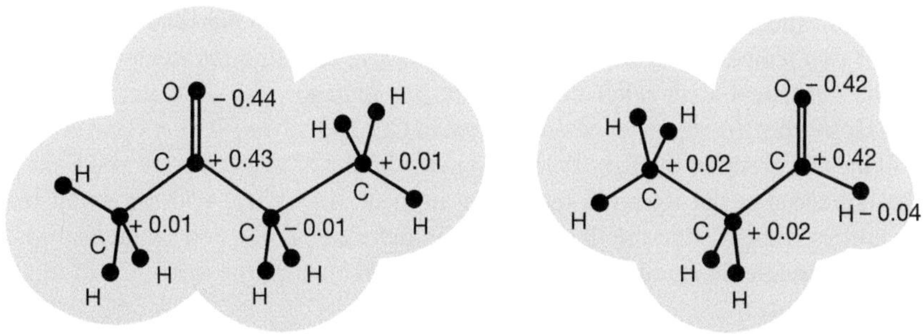

Figure 3.48 Schematic structure of ketones and aldehydes showing the location of partial electrostatic charges.

ketone or aldehyde, this is done for several molecular conformations. These simulations yield a C=O distance of approximately 1.21 Å, in agreement with experimental measurements [Vollhardt and Schore, 1999]. As illustrated by Table 3.9, the partial charge located on the oxygen is varying little, from – 0.45 to – 0.4 elementary electronic charges, for the three molecules considered. The charge located on the neighbouring carbon is more or less opposite. The non-polar part of the molecule, i.e. the CH_2 and CH_3 groups, bear only very small charges compared to the atoms involved in the carbonyl group. Although this electrostatic model is quite simple, it is sufficient to reproduce the electrostatic field around the molecule with a rrms deviation lower than 10%. The dipole moments of these molecules, which are in the 2.5-2.75 D range irrespective of the hydrocarbon skeleton, are also reproduced with a good accuracy by these sets of atomic charges. Compared with the case of organo-mercuric compounds discussed in the previous section, the case of ketones and aldehydes appears simpler because the strongly polarised C=O bond is relatively insensitive to the nature of the non-polar hydrocarbon structure to which it is attached.

Table 3.9 Electrostatic charges of acetone, 2-butanone and propaldehyde determined by Kranias *et al.* [2003], following the eigenvalue analysis method of Lévy and Enescu [1998]. The charges are given in atomic units, i.e. as a fraction of the unit proton charge.

Location of charge	Acetone	2-Butanone	Propaldehyde
$C(CH_3)$	0.013	0.011	0.015
$C(CH_2)$	–	– 0.010	0.018
C	0.422	0.435	0.424
O	– 0.448	– 0.443	– 0.418
H	–	–	– 0.039
$C(CH_3)$	0.013	0.007	–

The Lennard-Jones parameters of the oxygen and carbon of the C=O group have been fitted on the equilibrium properties of acetone and 2-butanone, while the properties of the terminal hydrogen next to the C=O group in aldehydes has been fitted on propaldehyde. As in previous optimizations, this fit considers vapour pressures, liquid densities and vaporization enthalpies at two temperatures for each compound. The resulting parameters (Appendix 1) appear to be consistent with other compounds. For instance, the diameter of the oxygen (3.04 Å) is close from the diameter of the oxygen in CO_2, which is 3.033 Å [Harris and Yung, 1995] or of the OH group of methanol (3.03 Å) in the model of van Leeuwen and Smit [1995]. It may be noticed that the terminal hydrogen atom of aldehydes, with its small diameter (1.7 Å) is almost embedded in the field of the neighbouring carbon and oxygen atoms. Also, the energetic parameter is small ($\varepsilon/k = 35$ K). As a consequence, the influence of this hydrogen is essentially local.

The comparison of experimental and predicted vapour pressures for acetone, butanone and pentanone indicates that simulation predicts nicely the influence of the length of the alkyl chain of the molecule (Fig. 3.49). Similarly, the vaporization enthalpies are well represented (Fig. 3.50). In ketones, the critical temperature is increased by approximately 100 K, compared with the *n*-alkanes having the same number of carbon atoms. Why does the model of Kranias *et al.* account for this effect? The answer is that the cohesive energy of the liquid is significantly greater than in the *n*-alkanes, due to the additional Lennard-Jones site on the oxygen and to significant electrostatic interactions. Compared with sulphides and thiols, the electrostatic interactions play a larger role in ketones as a result of their larger dipole moments.

In the case of the aldehydes, the prediction of thermodynamic properties is shown in Figure 3.51. Although the butanal and heptanal molecules shown in this graph are not part of

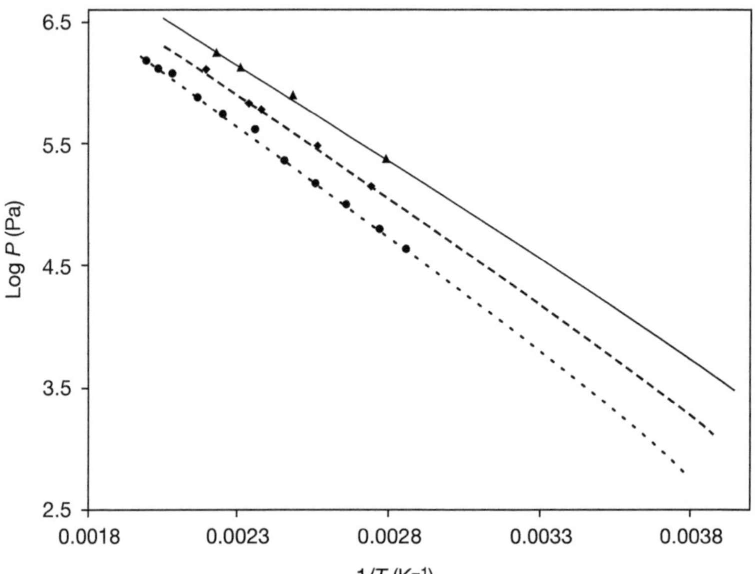

Figure 3.49 Saturation pressure obtained by simulation for acetone (trinagles), 2-butanone (diamonds) and 2-pentanone (dots), after Kranias *et al.* [2003], reproduced by permission of the PCCP Owner Societies. The lines correspond to the correlations taken from the Dortmund Data Bank.

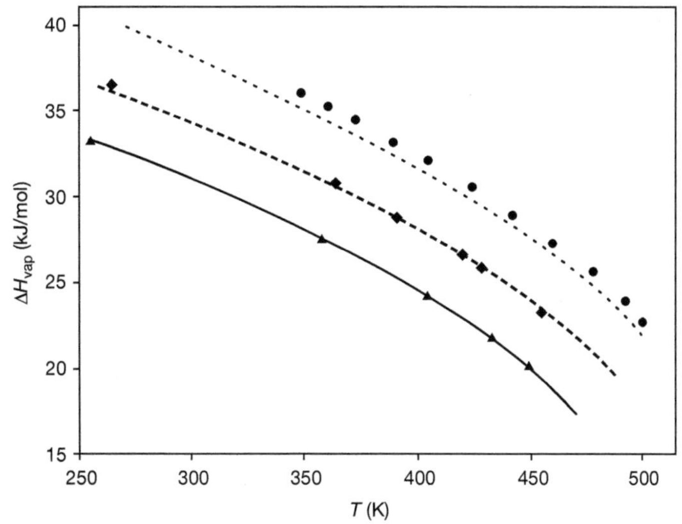

Figure 3.50 Molar vaporization enthalpy obtained by simulation for acetone (triangles), 2-butanone (diamonds) and 2-pentanone (dots), after Kranias *et al.* [2003], reproduced by permission of the PCCP Owner Societies. The lines correspond to the correlations taken from the Dortmund Data Bank.

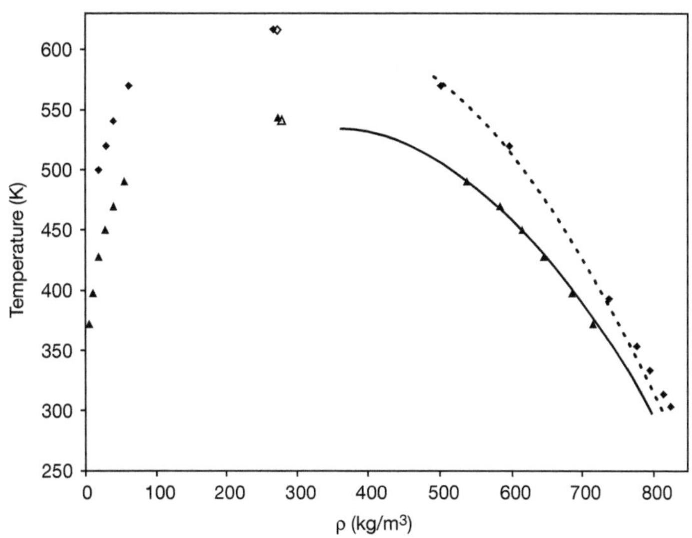

Figure 3.51 Liquid-vapour coexistence curve obtained by simulation for butanal (triangles) and heptanal (diamonds), after Kranias *et al.* [2003], reproduced by permission of the PCCP Owner Societies. The lines correspond to the saturated liquid density correlations taken frm the Dortmund Data Bank.

the reference molecules on which the Lennard-Jones parameters are fitted, their coexistence curves are correctly predicted. The critical temperatures determined from simulation results agree with experimental measurements within less than 6 K. Kranias and coworkers have also shown that the vapour pressures and vaporization enthalpies are well predicted for these compounds.

Whether ketones or aldehydes are considered, it is satisfactory to observe that the influence of chain length is reproduced with the same Lennard-Jones parameters as in the n-alkanes for the CH_2 and CH_3 groups. This confirms the good transferability of the AUA intermolecular potential that was already noticed in Section 3.3.1 about sulphides and thiols.

3.3.4 Alcohols

Alcohols are specific because they form hydrogen bonds with other alcohol molecules or with the molecules of other hydrogen-bonding fluids like water. This strong attractive interaction explains why alcohols display much greater critical temperatures and boiling temperatures than alkanes of similar molecular weight. For instance, ethane and methanol have similar molecular weights but their critical temperatures differ by approximately 200 K.

Although the existing intermolecular potentials do not include a detailed account of the hydrogen bonding interactions, they describe its effects by the same kind of electrostatic point charges reasonably well, as seen for ketones, sulphides, etc. However these charges are greater. In the case of the potentials of van Leeuwen and Smit [1995] and of Chen *et al.* [2001], the oxygen bears a strong negative charge (– 0.7 e) while the neighbouring carbon and hydrogen bear positive charges (+ 0.265 e and + 0.435 e respectively), the whole molecule being neutral. As a result, the hydrogen is strongly attracted by the oxygen of other alcohol molecules but it is repelled by their hydrogen and CH_3 groups. Thus there is a limited angle in which a strong negative electrostatic energy can develop (Fig. 3.52). In this way, the electrostatic model accounts approximately for the preferred orientation of hydrogen bonds in the bisector plane of the C–O–H angle of the first alcohol molecule, opposite to its H and C atoms.

Depending on the intermolecular potential, the C–O–H angle is considered rigid (van Leeuwen's model) or flexible (Chen's model), with an adapted intermolecular potential. The Lennard-Jones parameters of the OH group have been fitted on the liquid-vapour coexistence properties of either methanol alone [van Leeuwen and Smit, 1995] or methanol and ethanol [Chen *et al.*, 2001]. These parameters are given in Table 3.10. As illustrated by Figure 3.53, the coexistence curve of methanol is very well reproduced by the United Atom models investigated. It has been shown by Chen *et al.* [2001] that vapour pressures are correctly reproduced for methanol and ethanol as well with his model.

Recently, a new AUA intermolecular potential has been proposed for alcohols, involving four electrostatic charges per OH group [Bourasseau, 2003]. This model seems to provide equivalent results as Chen's potential on methanol and ethanol, but it has not yet been tested thoroughly. Its parameters are given in Appendix 1.

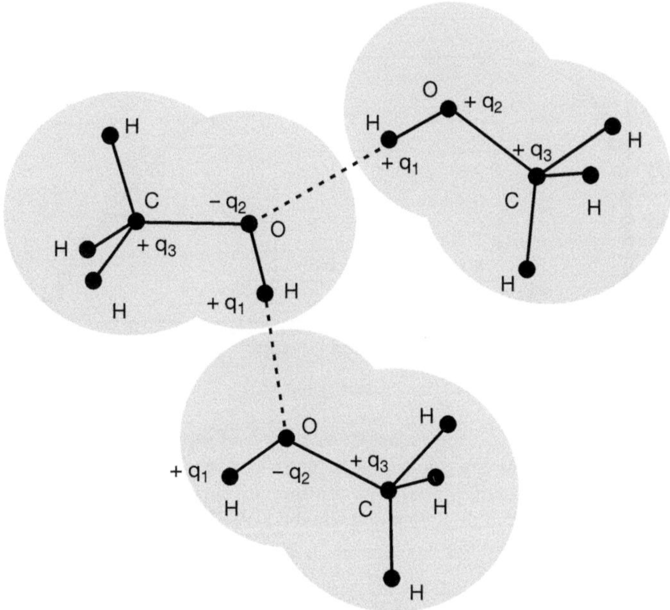

Figure 3.52 Schematic structure of methanol, based on the potential models of van Leeuwen and Smit [1995] and Chen *et al.* [2001], showing the location of electrostatic charges and the hydrogen bonding (dotted lines).

Table 3.10 Lennard-Jones parameters and electrostatic charges of the rigid methanol model of van Leeuwen and Smit [1995] and of the flexible United Atoms model of Chen *et al.* [2001].

		Bond length (r) bond angles (α) bond constant (k_θ/k)	**Lennard-Jones parameters**			**Electrostatic charge q/e**
			Site	**ϵ/k (K)**	**σ (Å)**	
Van Leeuwen and Smit [1995]	r_{C-O}	1.4246 Å	CH_3	105.2	3.74	0.265
	r_{O-H}	0.9451 Å	O	86.5	3.03	− 0.7
	α_{C-O-H}	108.53 Å	H	0	0	0.435
Chen *et al.* [2001]	r_{C-O}	1.43 Å	CH_3	98	3.75	0.265
	r_{O-H}	0.945 Å	O	93	3.02	− 0.7
	α_{C-O-H}	108.5 Å	H	0	0	0.435
	$(k_\theta/k)_{C-O-H}$	55 400 K				

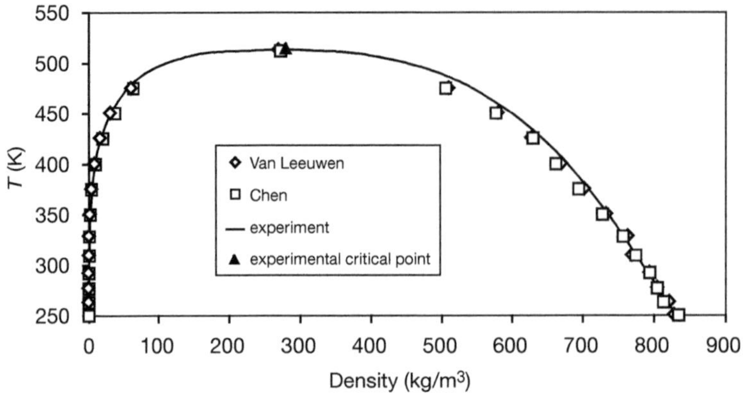

Figure 3.53 Liquid-vapour coexistence curve of methanol obtained by simulation with the intermolecular potentials of van Leeuwen and Smit [1995] and Chen *et al.* [2001]. The continuous line indicates the recommended values from de Reuck and Craven [1993].

3.4 PHASE BEHAVIOUR OF MIXTURES

When considering the phase equilibria of mixtures, the main industrial reason for using simulation is to avoid fitting specific interaction parameters when a new binary system is encountered. This section will mainly deal with comparing predictions with experimental data for various binary systems of interest to the oil and gas industry, in order to illustrate the predictive capacity of today's molecular simulation techniques. These comparisons will focus on systems which need specific calibration of interaction parameters when classical thermodynamic models are used and which therefore represent rather severe tests of the predictive capacity of intermolecular potentials.

However, we must be aware that the prediction of phase behaviour may be sensitive to the combining rule used to compute unlike interactions, especially when there are significant size differences between the various groups [Potoff *et al.*, 1999]. These authors have shown that better predictions of fluid phase equilibria are achieved with Kong's combining rule.

In Sections 3.4.1 and 3.4.2, we will discuss phase equilibria of methane and H_2S with liquid hydrocarbons at high pressure, which are important when it comes to predicting phase behaviour in reservoir engineering, especially in enhanced recovery projects by gas reinjection. This subject will also provide an opportunity to illustrate the calculation of bubble points with the pseudo-ensemble introduced in Section 2.3.7, and the determination of critical coordinates for binary systems discussed in Section 2.4.8. The Section 3.4.3 will be devoted to phase equilibria involving CO_2 and alkanes. In addition to their application in reservoir engineering (for instance for CO_2 sequestration in deep reservoirs), these results open the way to the prediction of CO_2 solubility in polyethylene at high pressure. The solubility and permeability of gases in polyethylene – and other polymers – is indeed a problem faced by the industry when acid gases are being transported in flexible pipes in which polymers are used for their sealing properties. We will test the influence of the combining rule used to compute the unlike Lennard-Jones interaction parameters ($\varepsilon_{ij}, \sigma_{ij}$). Finally, Section 3.4.4 will focus on phase equilibria involving methanol, which is frequently encountered in gas treatment, either because it is added to avoid the formation of pipe-blocking gas hydrates during transport or because it is used as a selective solvent of acid gases at low temperature. As we have seen in the previous chapter, methanol is a hydrogen-bonded fluid. Modelling its phase equilibria with other compounds is an interesting challenge for molecular simulation.

3.4.1 Binary and Ternary Alkane Mixtures

When a supercritical component like methane is mixed with a liquid hydrocarbon, the liquid-vapour coexistence domain forms a loop in the (P, x) phase diagram, in which the maximum pressure may be considerably higher than the critical pressures of the pure components involved. This is illustrated by the methane-*n*-pentane mixture (Fig. 3.54) which has been investigated through two different algorithms.

The first algorithm is the Gibbs ensemble at imposed pressure and temperature (see Sections 2.1.1 and 2.3.4), in which care must be taken that the global composition of the system (liquid + vapour) is comprised in the central part of the liquid – vapour coexistence region. If the global composition is initialized too close to one of the boundaries of the coexistence domain, one simulation box is almost empty and the outcome of the simulation will not be reliable.

The second algorithm is the bubble point pseudo-ensemble discussed in Section 2.3.7, in which the temperature and the composition of the liquid phase are imposed. As *n*-pentane is a flexible molecule in which bending and torsion potentials must be accounted for, both algorithms use the configurational bias algorithm to sample correctly the various molecular conformations (Section 2.3.6) in the same way as done for pure alkanes. Methane interactions are described with the same potential as in previous sections [Möller *et al.*, 1992], and the Anisotropic United Atoms potential is used for *n*-pentane with the parameters discussed in Section 3.1. As can be seen from the phase diagram, both methods agree with experimental measurements of liquid and vapour mole fractions [Berry and Sage, 1970; Prodany and Williams, 1971; Reiff *et al.*, 1987] to within 7%. This may be considered as a reasonable degree of accuracy because the Lorentz-Berthelot combining rule has been used without fitting any interaction parameter. In fact, a significant proportion of the deviation from experimental values is due to the statistical uncertainties in the determination of phase compositions. These uncertainties are revealed by the fact that the methane content of the liquid phase is overestimated in some instances and underestimated in others.

These uncertainties are significantly higher in the bubble point pseudo-ensemble than in the Gibbs ensemble, which is not surprising because greater fluctuations would be expected in such a pseudo-ensemble. This indicates that the pseudo-ensemble is especially good at providing estimates of equilibrium conditions. When vapour and liquid compositions are not yet known on an *a priori* basis, it is difficult to find the a sufficiently accurate global composition to initialize Gibbs ensemble simulations, and a bubble point calculation may provide estimates for this purpose.

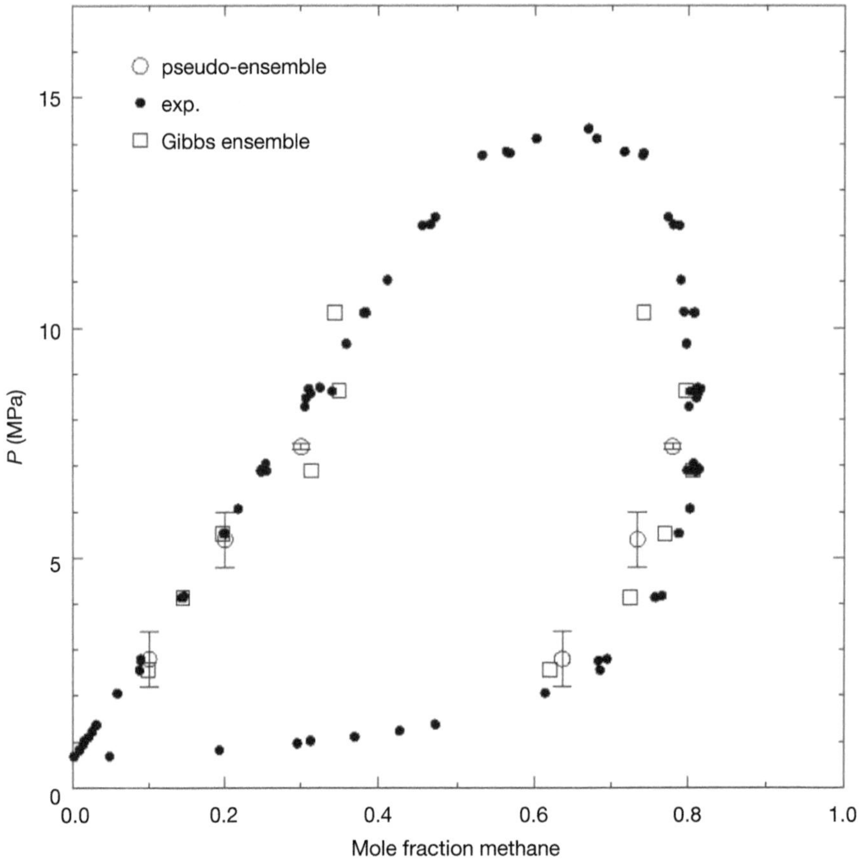

Figure 3.54 Phase diagram of the methane-*n*-pentane binary mixture at 377 K from simulations in the Gibbs ensemble and Bubble point pseudo-ensemble [Ungerer *et al.*, 2001] and from experimental measurements by Berry and Sage [1970]. Reprinted with permission from [Ungerer *et al.*, 2001] © Taylor & Francis Ltd (http://www.tandf.co.uk/journals).

If the methane-*n*-pentane study was repeated with today's computers (in which simulations ten times longer can be run), there is no doubt that the statistical uncertainty would be reduced by a factor of two or three. Nevertheless, these algorithms would still be subject to uncertainties in addressing the upper part of the phase diagram, where equilibrium pressures are greater than 11-12 MPa. Indeed, the area close to maximum equilibrium pressure is characterized by larger fluctuations, as this maximum corresponds to the critical point of the binary mixture where the liquid and vapour phases display the same composition. As with pure compounds, the cost in free energy of changing from the vapour to the liquid state tends toward zero when approaching the critical conditions. Compressivity is not infinite as with pure compounds but very large fluctuations of density and composition occur with small changes in free energy in the vicinity of the critical point. This is a fundamental limitation of either the Gibbs ensemble or the bubble point pseudo-ensemble. As will be illustrated by the example of the H_2S-cyclohexane system in Section 3.4.2, critical scaling can be used to extend molecular simulation predictions in the near-critical region of such binary systems.

Phase equilibrium calculations can also be applied to ternary systems, as illustrated by Figure 3.55. In this example of a ternary system methane-propane-*n*-decane, bubble point calculations were conducted in the same way as with the methane-*n*-pentane system (more details can be found in Ungerer *et al.* [2001]). The results are expressed as equilibrium constants $K_i = y_i / x_i$ where x_i and y_i are the molar fractions of component i in the liquid and vapour phases. The equilibrium constants are determined in bubble point conditions, so that they can be compared with interpolated experimental data [Sage and Berry, 1971] at various temperatures between 270 K and 470 K. Here again, no binary parameter has been calibrated

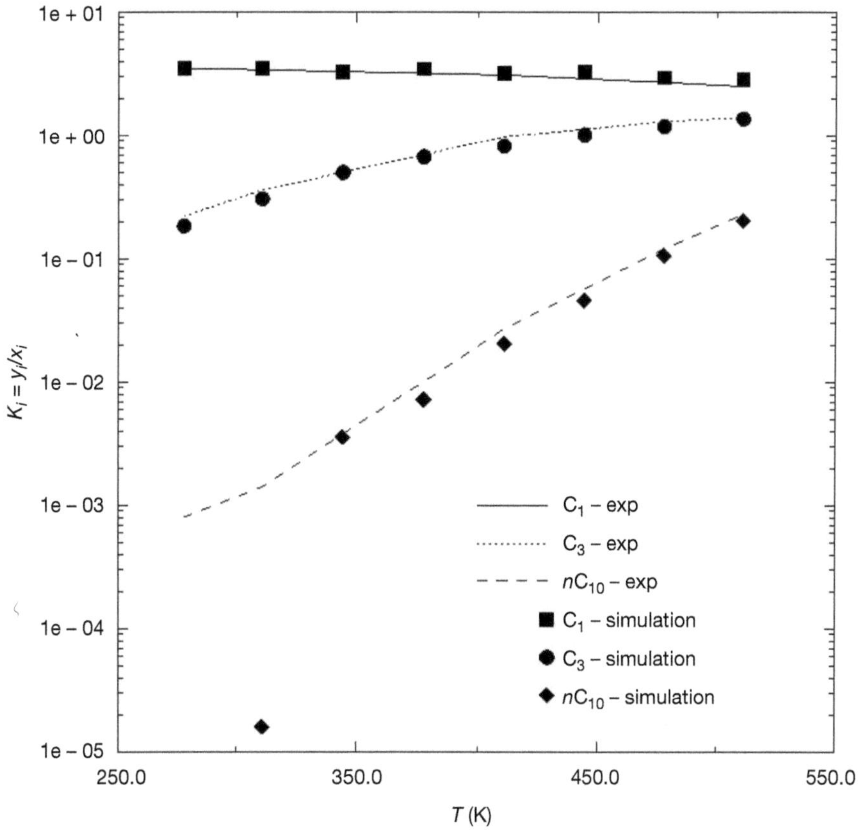

Figure 3.55 Vapour-liquid equilibrium constants in bubble point conditions for a ternary mixture containing 27.7% of methane, 13.1% of propane and 59.3% of *n*-decane. The lines have been obtained by interpolating the experimental measurements of Sage and Berry [1971], and the symbols are simulation results using the bubble point pseudo ensemble and Gibbs ensemble simulations [Ungerer *et al.*, 2001]. Reprinted with permission from [Ungerer *et al.*, 2001] © Taylor & Francis Ltd (http://www.tandf.co.uk/journals).

and nevertheless, the dependance of equilibrium constants with temperature is very well predicted for each component. Special mention must be made of the low temperature equilibrium constant of *n*-decane, which could not be computed at 270 K and for which an incorrect value is obtained at 310 K. The reason is that a minimum number of molecules in both phases is needed to evaluate an equilibrium constant, otherwise the assumptions underlined in defining the acceptance rates of the Monte Carlo transfers are not respected. It is indeed required that the number N of molecules in every simulation box is much larger than unity, because a simplified expression is used for $N!$, which is valid only for large N [McQuarrie, 1976; Rowley, 1994]. Nevertheless, it is found that a reasonable prediction of the equilibrium constant is achieved as soon as the average number of molecules in the vapour box is greater than unity, which is the case at temperatures over 320 K.

In addition to vapour composition, density data are available from the same source [Sage and Berry, 1971] to evaluate simulation results. While doing this comparison, two different parametrizations of the Lennard-Jones potential of methane were tested. In addition to the potential of Möller used to evaluate equilibrium constants, we considered a slighlty different set of parameters proposed by Errington [1998]. As shown by Figure 3.56, predictions are

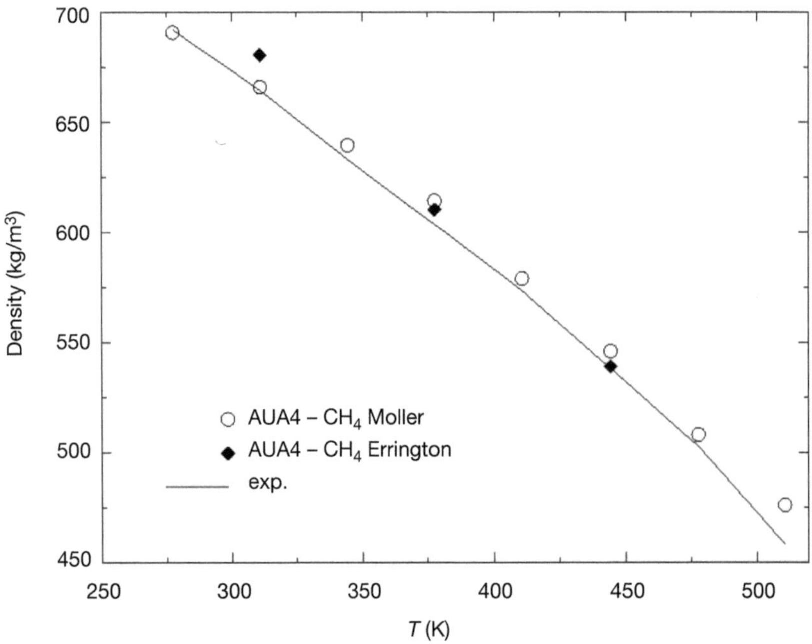

Figure 3.56 Liquid phase density in bubble point conditions for a ternary mixture containing 27.7% of methane, 13.1% of propane and 59.3% of *n*-decane. The lines are experimental measurements of Sage and Berry [1971], and the symbols are simulation results using the bubble point pseudo ensemble and Gibbs ensemble simulations [Ungerer *et al.*, 2001]. Two sets of Lennard-Jones parameters have been used for the methane molecule. Reprinted with permission from [Ungerer *et al.*, 2001] © Taylor & Francis Ltd (http://www.tandf.co.uk/journals).

correctly matching the decrease of density of the methane-propane-*n*-decane mixture when temperature is increased, whatever the intermolecular potential selected for methane.

3.4.2 Binary Mixtures of H_2S with Liquid Hydrocarbons

Phase equilibrium data on H_2S-*n*-alkane systems are sufficient to develop specific calibrations of binary interaction parameters in engineering equations of state, as investigated by Caroll and Mather [1995] for the Peng-Robinson equation. However, little information is available on binary systems of H_2S and other hydrocarbon families, and we will see here how molecular simulation can help.

When investigating the phase equilibria involving hydrogen sulphide (H_2S), a preliminary requisite is that a valid intermolecular potential model is available for this molecule. In this section, the potential proposed by Kristof and Liszi [1997] will be used, in the same way as a previous study of hydrogen sulphide-alkane mixtures [Delhommelle *et al.*, 1999]. Indeed, this potential appears to represent particularly well the equilibrium properties of H_2S (this point will be detailed in the Section 3.6 which is more specifically devoted to H_2S-rich gases). The model comprises one Lennard-Jones centre and four partial electrostatic charges, whose parameters are given in Section 3.6.

The determination of the (P, x) liquid-vapour phase equilibrium diagrams of H_2S mixtures with liquid hydrocarbons, such as the H_2S-*n*-pentane system (Fig. 3.57) is performed mainly with the Gibbs ensemble technique at constant pressure. The calculation of the potential energy must account for the electrostatic energy, but it is acting only between the H_2S molecules. Indeed, the electrostatic energy between H_2S molecules and *n*-pentane is zero, because the AUA potential has been used for alkanes without any electrostatic charges. The end points of the diagram, i.e. those corresponding to the pure compounds are treated in the Gibbs ensemble at constant volume, pressure being then an output of the simulation. Near the upper end of the diagram, i.e. for systems with more than 80% H_2S in the liquid phase, the Gibbs ensemble simulation at constant pressure is difficult to converge because one phase is often

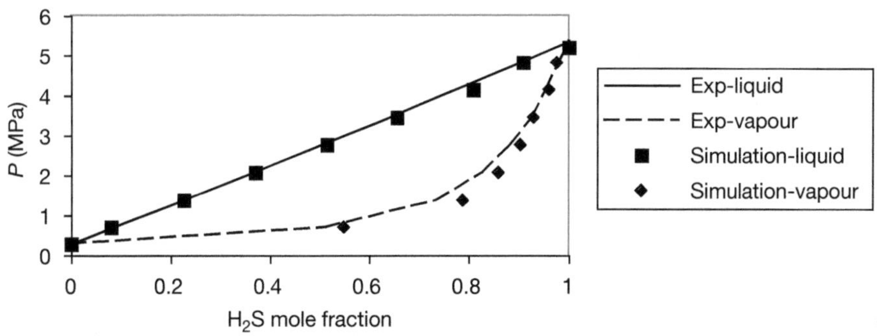

Figure 3.57 Phase diagram of the H_2S-*n*-pentane system at 344.3 K from Gibbs ensemble simulations (squares and diamonds) and from experimental measurements [Reamer *et al.*, 1953].

found to disappear. This may be attributed to the large fluctuations of density and composition that are characterizing this region. It is in fact more convenient to use Gibbs ensemble simulations at constant *volume* to simulate phase coexistence in this region, because this puts a constraint on the possible increase of one phase at the expense of the other. As can be seen from Figure 3.57, the prediction of the phase diagram is quite satisfactory, as the maximum deviation on liquid phase and vapour phase compositions does not exceed 5%.

Another phase diagram involving nC_{15} (*n*-pentadecane) instead of *n*-pentane is presented in Figure 3.58. Here, the temperature (422.6 K) is larger than the critical temperature of H_2S. As a result, the liquid-vapour coexistence domain does not merge to a single point on the right side of the diagram but forms a loop, in the same way as the methane-*n*-pentane system of Figure 3.54. However, the solubility of *n*-pentadecane in the vapour is so small that the right side boundary of the two-phase domain is almost confused to the vertical axis corresponding to $x_{H_2S} = 1$. This feature, as well as the composition of the liquid phase, is perfectly accounted for by simulation. However, simulations could not be extended up to the maximum coexistence pressure, for the same reasons as given in Section 3.4.1.

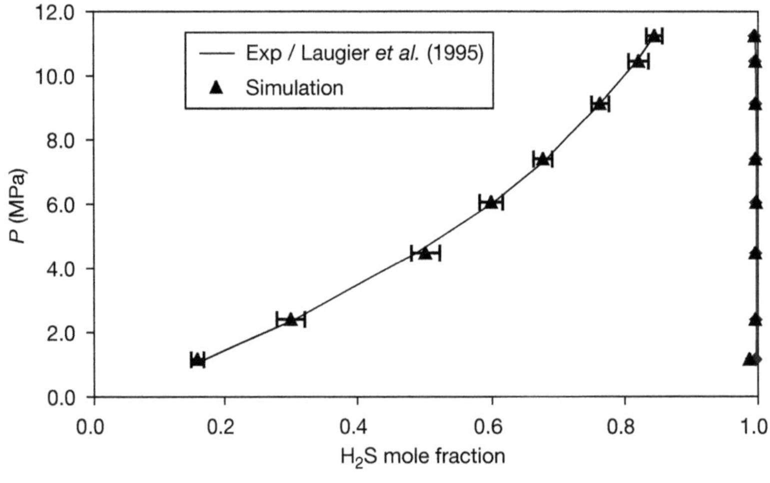

Figure 3.58 Phase diagram of the H_2S-*n*-pentadecane system at 422.6 K from Gibbs ensemble simulations (triangles) and compared with experimental measurements (lines) of Laugier and Richon [1995].

An interesting side aspect of phase equilibrium calculation is the prediction of densities, as illustrated in the case of the H_2S-*n*-pentane system (Fig. 3.59). The reasonable prediction of volumetric properties achieved in this system is an additional indication that molecular simulation is a valid technique to investigate mixtures of hydrogen sulphide and *n*-alkanes in large range of carbon number, temperature and pressure.

The agreement between simulation results and experimental data has been also found in the previous investigation of hydrogen sulphide-alkane mixtures [Delhommelle *et al.*, 1999] which used an earlier version of the AUA intermolecular potential. Among other findings, Delhommelle predicted correctly the phase diagrams involving pentane and propane.

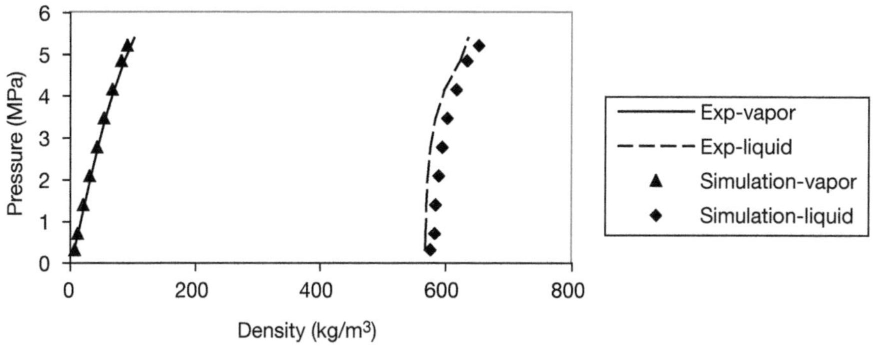

Figure 3.59 Density of coexisting liquid and vapour phases in the H_2S-*n*-pentane system at 344.3 K from Gibbs ensemble simulations (symbols) and from experimental measurements (lines) of Reamer *et al.* [1953].

In the case of the H_2S-benzene mixture, the same electrostatic potential model has been selected for hydrogen sulphide, while the AUA potential of Contreras-Camacho [2004a] has been used for benzene (Fig. 3.60). As a result, simulation predicts correctly the composition of the vapour phase, which is significantly richer in hydrogen sulphide than in the case of *n*-pentane at 344 K (Fig. 3.57). Qualitatively, this effect can be explained by Raoult's law ($Py_i = x_i P_{sat\,i}$): the lower the saturation pressure P_{sat} of the hydrocarbon, the lower its mole fraction y_i in the vapour. Although Raoult's law is too simple to represent fully this case, it has the merit of showing why an accurate representation of vapour pressures of the pure liquid component is needed if a good prediction of the vapour phase composition is expected. The content of hydrogen sulphide in the liquid phase seems to be underestimated by approximately 5%. It is not surprising that the model predicts a two-phase region slightly larger than actually observed. Indeed, the AUA potential of benzene does not include any electrostatic interactions of benzene with hydrogen sulphide. Although some of these interactions are implicitly taken into account in the Lennard-Jones potential for benzene alone, it is likely that they are slightly underestimated when benzene is surrounded by more polar molecules, as it is the case here. Regarding phase densities, it may be noticed that liquid densities decrease with increasing equilibrium pressure (Fig. 3.60). This effect, which is opposite to the mixture involving *n*-pentane, can be explained by the density of the pure compounds in the range of temperatures considered, i.e. 323 K to 344 K. In these conditions, the density of saturated liquid H_2S is indeed higher than pure *n*-pentane and lower than pure benzene. The density of the binary liquid mixture (which is intermediate between those of the pure compounds involved) changes simply according to pure component densities.

The simulation of the whole phase diagram between hydrogen sulphide and liquid hydrocarbons at temperatures larger than 373 K faces the difficulty of simulating the near-critical region, as discussed in Section 3.4.1 about the methane-pentane system. The H_2S-cyclohexane system at the temperature of 423 K is shown here as an example of this kind of diagram (Fig. 3.61). The AUA potential indicated in Section 3.1 is used for cyclohexane [Bourasseau *et al.*, 2002], and the model of Kristof and Liszi [1997] is selected for H_2S. Due to the larger

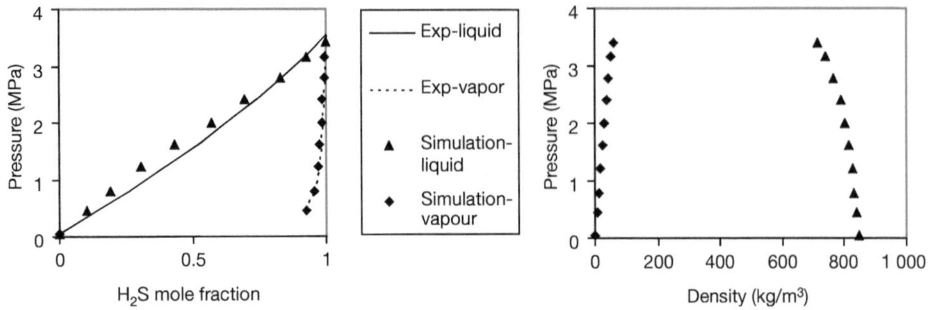

Figure 3.60 Phase diagram of the H_2S-benzene system at 323 K from Gibbs ensemble simulations (symbols), compared with experimental measurements (lines) of Laugier and Richon [1995].

and larger fluctuations with increasing pressure, the Gibbs ensemble simulations have been performed with larger box sizes at high pressure: the total number of molecules has been thus changed from 300 molecules at 1, 2 and 3 MPa to 500 molecules at 6 MPa and 1 500 molecules at 9 MPa. Investigating pressures significantly higher than 9 MPa would have required even larger systems, and thus several weeks of computation time would have been needed instead of one week for the higher pressure point. As shown by Figure (3.61a), simulation results agree very well with the experimental measurements of Laugier and Richon [1995]. The extrapolation technique presented in Section 2.4.8, based on the near-critical scaling laws (2.101) and (2.102), has been used to provide an estimation of the critical coordinates for this system at the temperature considered. As can be seen on the figure, the correlation of the composition coexistence curve (Fig. 2.61a) and of the density coexistence curve (Fig. 2.61b) is fully consistent with either experimental or simulation data. The critical pressure is thus estimated as $P_c = 11.1$ MPa, the critical density as $\rho_c = 353$ kg/m^3 and the critical composition as $x_c = 0.895$. Related uncertainties are estimated as 0.3 MPa, 10 kg/m^3 and 0.01.

In Figure 3.61c, the plot of the composition difference versus $(P_c - P)$ shows three different regimes with increasing distance from the critical pressure. Close to the critical pressure, for $(P_c - P) < 2$ MPa, the correlation is not linear. This is because the β-exponent term in Eq. (2.102) dominates for small values of $(P_c - P)$. This first regime occurs for conditions that are too close from the critical point to be investigated by simulation. At some distance from the critical point, for 2 MPa $< (P_c - P) <$ 8 MPa, a second regime is observed, in which the trend of Figure 2.61c is linear. This can be explained by the prevalence of the linear term in Eq. (2.102) for larger values of $(P_c - P)$. Far away from the critical point, i.e. for $(P_c - P)$ > 8 MPa, a third regime is characterized by a non-linear behaviour. The Taylor-like expansion of Eq. (2.102) is logically insufficient to reproduce the trend as it does not contain higher order terms. Care must be taken that the parameter regression is not based on coexistence points obeying the third regime. In the example shown, the regression is thus based on the coexistence points from 4 to 9 MPa only. On the contrary, the density difference between the phases is well reproduced by the single β-exponent term of Eq. (2.101), as illustrated by the linear trend of the density difference versus $(P_c - P)^\beta$ in Figure 2.61d.

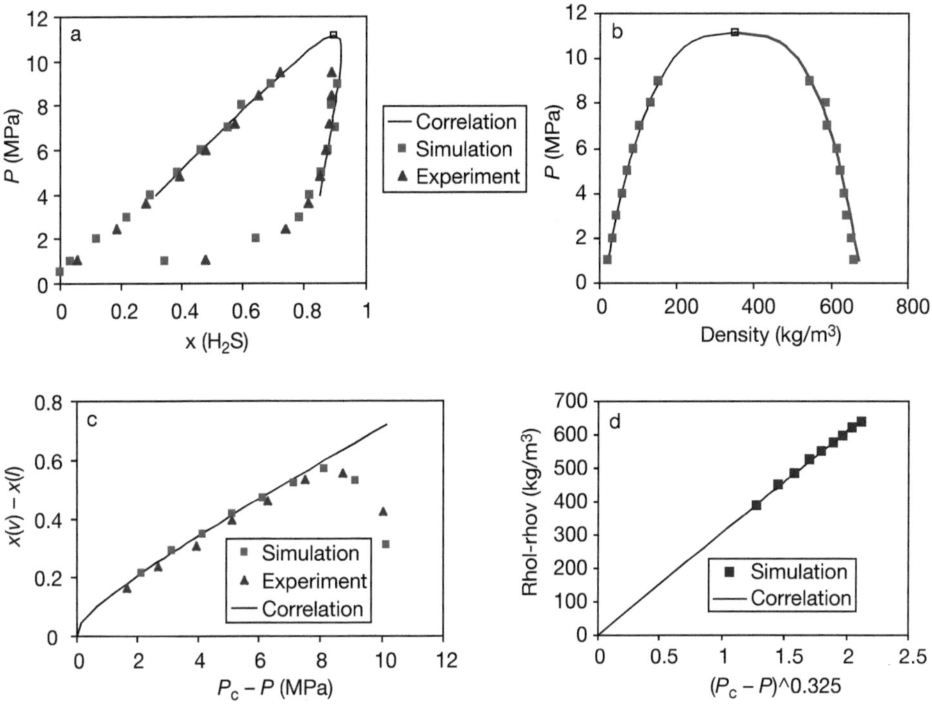

Figure 3.61 Phase diagram of the H_2S-cyclohexane system at 423 K from Gibbs ensemble simulations, compared with experimental measurements of Laugier and Richon [1995]. The continuous line is the correlation of the near-critical region with the Taylor-like expansions (2.101) and (2.102). (a) pressure-composition diagram (b) pressure-density diagram (c) composition difference between phases versus $(P_c - P)$ (d) density difference between phases versus $(P_c - P)^\beta$ with $\beta = 0.325$.

This example shows that it is now possible to locate with a reasonable accuracy the liquid-vapour critical locus of binary mixtures from molecular simulation, in a way that is fully consistent with critical scaling theory. It is worth noticing that such an extrapolation could not have been obtained with equations of state because the leading term in the near-critical region is $(P_c - P)^{0.5}$ instead of $(P_c - P)^\beta$ for *any* analytical equation of state [Levelt-Sengers, 1970; Barrat and Hansen, 2003].

As a concluding remark, it can be stated that the phase equilibria and densities of the four binary systems involving hydrogen sulphide and a liquid hydrocarbon are well predicted on average without binary parameter fitting. This justifies the use of simulation to produce predictions of full phase diagrams on other systems where the liquid hydrocarbon is a polyaromatic hydrocarbon, or a long chain alkylbenzene, or a naphtheoaromatic component, for which no experimental data is presently available. These results are also very encouraging for the prediction the solubility of hydrogen sulphide in polyethylene at high pressure, as this involves the same groups as pentane and pentadecane.

3.4.3 Phase Equilibria of CO_2 with Alkanes and Polyethylene

Simulating the phase behaviour of mixtures involving CO_2 and alkanes is possible using the same method as was used in the previous section for H_2S, with a suitable intermolecular potential for carbon dioxide. Several intermolecular potentials are available to describe the phase equilibrium properties of carbon dioxide. Some are using two Lennard-Jones force centres and a point quadrupole moment [Möller and Fischer, 1994; Vrabec and Fischer, 1996] while others use three atomic force centres and atomic electrostatic charges [Murthy *et al.*, 1983; Harris and Yung, 1995]. In this section, simulations are based on the intermolecular potential of Harris and Yung, in the same way as done by Delhommelle in his investigation of carbon dioxide-alkane mixtures [Delhommelle *et al.*, 1999]. However, it is likely that good results could be obtained as well with the other models, as they account also for the liquid-vapour equilibrium properties and for the strong quadrupole moment of carbon dioxide. The simulation of pure carbon dioxide will be discussed in more detailed in Section 3.6, in which potential parameters will be reviewed again.

When simulation in the Gibbs ensemble was applied to mixtures of carbon dioxide with *n*-alkanes, a very good prediction of the phase diagrams was obtained with *n*-pentane and a fair prediction with *n*-decane [Delhommelle *et al.*, 1999]. Here, results obtained with a slightly different intermolecular potential are presented. While Delhommelle *et al.* [1999] were using the original AUA potential [Toxvaerd, 1997] called AUA3, the results of the optimized AUA potential [Ungerer *et al.*, 2000] called AUA4 are also given. The (P, x) phase diagram of the CO_2-*n*-decane mixture, shown in Figure 3.62, illustrates this comparison. It may be noted that the composition of the vapour phase is better reproduced with the optimized potential. As the hydrocarbon content in the vapour is primarily influenced by the pure component vapour pressure, this may be related with the better account of saturated vapour pressures of heavy *n*-alkanes with the new AUA potential, which is specifically designed to match this property. The composition of the liquid phase is correctly reproduced, meaning that the solubility of carbon dioxide in alkanes is well predicted at high pressure.

It is tempting to extrapolate the prediction of carbon dioxide solubility from alkanes to polyethylene, since the CH_2 and CH_3 groups involved in both systems are identical. Determination of the solubility of carbon dioxide in very long chain alkanes is given here as a preliminary contribution to a discussion of the difficult problem of predicting gas solubility in polymers. In a general way, simulating polymer systems is difficult because the minimum size of a representative system is much larger than with liquids because of the high molecular weights of most common polymers (typically 10^3 to 10^6 monomers). Also, specific algorithms have to be used to generate initial configurations [Theodorou and Suter, 1985] and to sample all the possible configurations efficiently [Dodd *et al.*, 1993; Zervopoulou *et al.*, 2001]. Finally, many polymers are not amorphous but semi-crystalline, i.e. the chains organize locally in clusters with a regular crystal-like structure. Here, we will simply extend solubility calculations to alkanes up to 100 carbon atoms, at a temperature sufficiently high (433 K) that the polyethylene is amorphous.

As mentioned earlier in this chapter, the selection of the combining rule may influence phase equilibrium calculation involving groups of significantly different diameters, and this is the case of the systems formed by CO_2 and light alkanes [Potoff *et al.*, 1999]. These authors have shown that better predictions of fluid phase equilibria are achieved with Kong's combining rule.

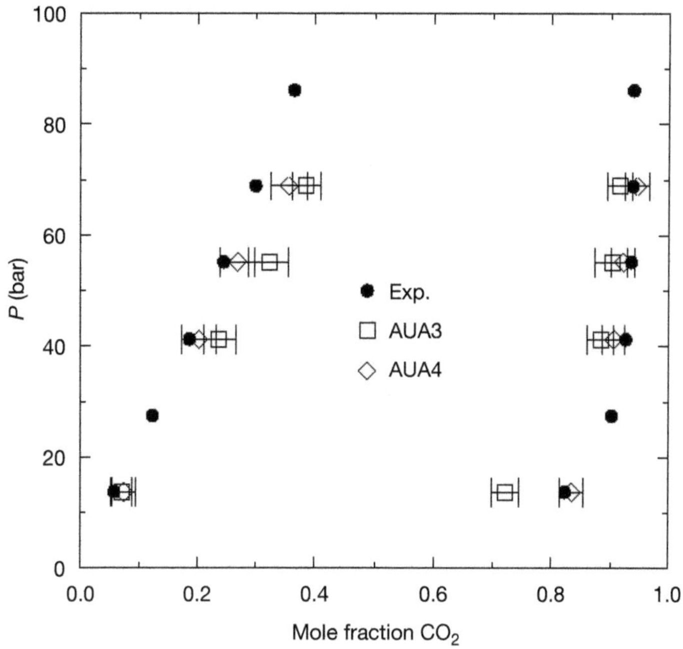

Figure 3.62 Phase diagram of the CO_2-*n*-decane system at 477 K from Gibbs ensemble simulations (open symbols), compared with experimental measurements (dots) of Reamer and Sage [1963]. AUA3 refers to the parametrization of Toxvaerd [1997] and AUA4 to Ungerer *et al.* [2000]. Reprinted with permission from [Ungerer *et al.*, 2000], © American Institute of Physics.

Using the AUA potential, we have found that fairly satisfactory predictions are obtained using the Lorentz Berthelot rule, but that Kong's rule is significantly better for predicting the solubility of CO_2 in alkanes. Also, we found that carbon dioxide solubility decreases with alkane chain length, but it can be expected to be more or less constant beyond 100 carbon atoms (Fig. 3.63). Interestingly, the predicted solubility in C_{100} corresponds fairly well with the experimental measurements of Sato *et al.* [1999] and of Chaudhary and Johns [1998] on polyethylene, especially when Kong's combining rule is used.

Another parameter that is key to characterising the system is its volumetric behaviour. It appears that the density of the polymer does not change significantly with the uptake of carbon dioxide (Table 3.11). From this information, it is possible to estimate the swelling of the polymer, i.e. the relative difference in volume between the saturated polymer and the empty polymer at constant pressure and temperature. For instance, the predicted swelling is 7.2% at 10 MPa. This significant difference shows that there is a substantial rearrangement of the polymer as it dissolves CO_2, and that the polymer-gas equilibrium under pressure cannot be considered as an adsorption process in which the adsorbent geometry is fixed.

The possibility of extending molecular simulation from long chain alkanes to polyethylene is extremely encouraging for future applications in which the interaction of polymers with

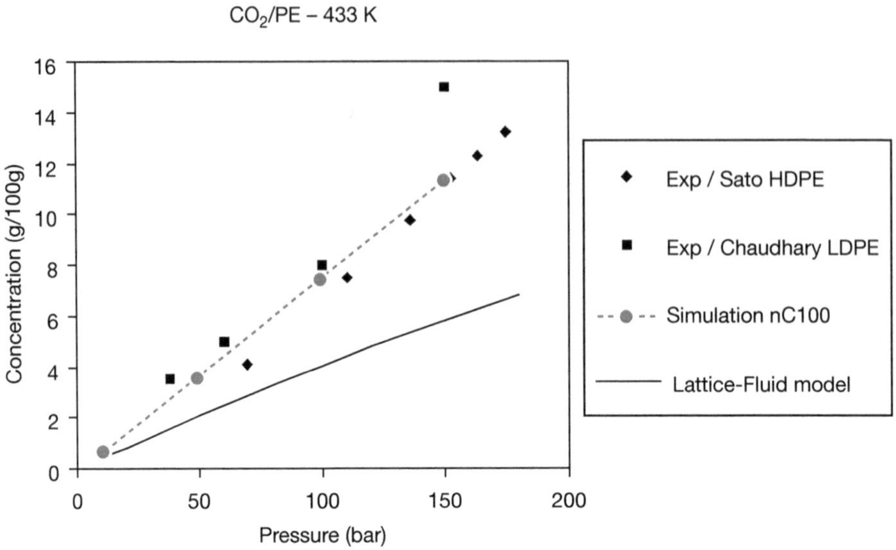

Figure 3.63 Phase diagram of the CO_2–nC_{100} system at 433 K from Gibbs ensemble simulations, compared with experimental measurements on high density (HDPE) and low density (LDPE) polyethylene and with a lattice fluid model.

Table 3.11 Simulation results obtained on the CO_2–nC_{100} system at 433 K. The liquid density corresponds to the polymer phase with dissolved carbon dioxide. Swelling is defined as the volume increase of the polymer upon CO_2 uptake with reference to the pure polymer at the same pressure and temperature.

P (MPa)	CO_2 concentration (g/100 g)	ρ_{liq} (kg/m³)	Swelling (%)
1	0.62 ± 0.01	785.3 ± 1.12	n.d.
5	3.47 ± 0.09	788.4 ± 1.6	n.d.
10	7.27 ± 0.12	788.7 ± 2.5	7.2
15	11.40 ± 0.13	784.9 ± 1.9	n.d.

high pressure natural gases in pipelines must be controlled. Indeed, there is a lack of reliable prediction method when complex mixtures comprising carbon dioxide and/or hydrogen sulphide are involved. Provided it can be adapted to the case of semi-crystalline polymers, molecular simulation can play an important role in this field.

3.4.4 Phase Equilibria Involving Methanol

Phase equilibria between methanol and hydrocarbons are challenging tests for thermodynamic models because these mixtures are so far from ideal, as illustrated by the presence of azeotropic behaviour in many instances. Yet, light alkanes do not display immiscibility with methanol at high pressure as they do with water. Is molecular simulation able to represent this behaviour? In order to answer this question, we have combined the use of van Leeuwen's and Chen's models for methanol and the AUA potential for *n*-alkanes. A system of 400 molecules has been used to apply the Gibbs ensemble calculation at imposed pressure. In this model, there is consequently no electrostatic interaction between alkanes and methanol. As illustrated by Figure 3.64, Gibbs ensemble simulation results are in surprisingly good agreement with the presence of an azeotrope with a composition $x_{\text{propane}} \approx 95\%$. It has been checked that there is no liquid-liquid phase split at pressures higher than the azeotrope equilibrium pressure. From this test, it can be concluded that intermolecular potentials account for the delicate balance between the tendency of both fluids to split, as a result of methanol's polarity, and to be miscible, as a result of the presence of an alkyl group in both. These results may be taken in parallel with the simulation results obtained by Lisal *et al.* [2001] on the ethane-methanol system at 298 K. Using the same intermolecular potential for methanol and another potential for ethane, these authors find the right qualitative behaviour, i.e. a liquid-liquid phase split at higher pressures than the saturation pressure of ethane (38 MPa). However, they do not obtain the right composition of the liquid phases which has been measured in the same conditions [Ishihara *et al.*, 1998]. As the intermolecular potential used by Lisal *et al.* overestimates by 25 K the critical temperature of ethane, it is conjectured that their prediction of the ethane-methanol phase diagram could be improved by using the same AUA potential as exposed here for the propane-methanol system.

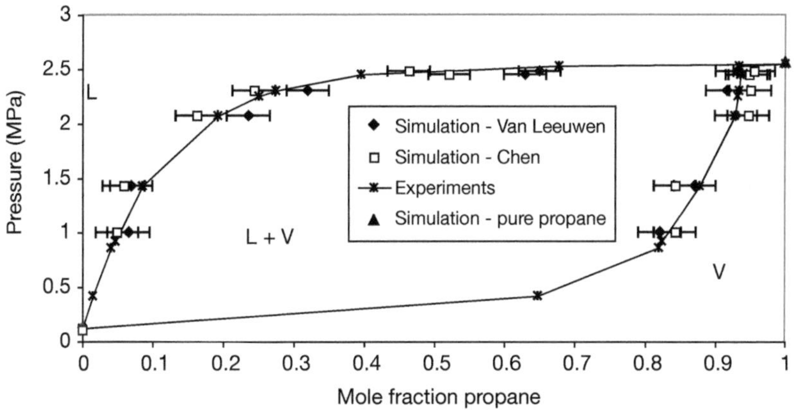

Figure 3.64 Phase diagram of the methanol-propane system at 343 K from Gibbs ensemble simulations (symbols) and experimental data (lines) from Galivel-Solastiouk *et al.* [1986].

In order to characterize the interactions in the liquid and gas phases, a cluster analysis can be implemented. The criterion that is generally used to determine if two molecules are hydrogen-bonded is that the distance between an oxygen atom and an hydrogen atom belonging to two different molecules is lower than some threshold [Nieto-Draghi *et al.*, 2003]. Applied to the methanol molecules with a threshold fixed to 2.09 Å, the cluster analysis based on this criterion provides the distribution of the clusters with size (Fig. 3.65). In polydisperse systems, it known that the number distribution x_n is less representative of the prevailing structures than the mass distribution $w_n = \frac{n x_n}{\sum_i i x_i}$. The comparison of both distributions (Fig. 3.65, left) in the liquid phase shows that the clusters of 5 molecules are those in which the largest number of methanol molecules can be found, while the most frequent clusters are those with one molecule only. The average number of methanol molecules per cluster is close to 4. In the vapour phase, the association of methanol molecules is much smaller but not negligible, as the cluster analysis reveals that only 20% of the methanol molecules belong to clusters of 2 to 5 molecules.

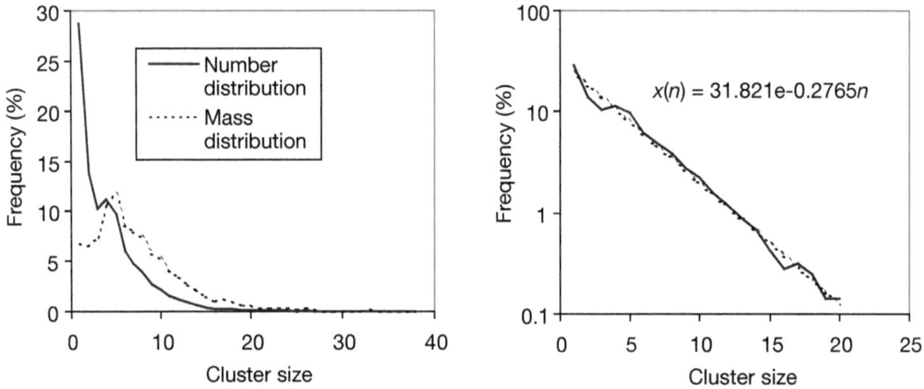

Figure 3.65 Cluster size distribution of methanol in the liquid phase of the methanol-propane system at 343 K and 2 MPa. Left: comparison between the distribution expressed on a number basis (probability of observing a cluster of given size) and on a mass basis (probability that a methanol molecule belongs to a cluster of given size). Right: logarithmic plot of the number-based distribution (continuous line) fitted with an exponential function (dotted line).

It is interesting to compare these results with the classical thermodynamic models used to represent associated mixtures, in which the following association reaction is supposed in equilibrium [Prausnitz *et al.*, 1986]:

$$A_n + A \rightarrow A_{n+1}$$

where A_n is a cluster of n molecules.

It is a common assumption to consider that the equilibrium constant ($k_n = x_{n+1}/x_n$) does not depend on n (this is equivalent to say that the free energy change of the above association reaction does not change with cluster size). According to this model, the number-based cluster size distribution should be exponentially decreasing.

The distribution resulting from the simulation of the methanol-propane system is indeed approximately exponential, as revealed by the logarithmic plot of Figure 3.65. The average distribution corresponds thus to an equilibrium constant $k_n = 0.758$ in the example shown (343 K and 2.07 MPa). The exponential distribution holds approximately between 1 and 20 carbon atoms, but some departures are observed. The clusters comprising 5 and 6 methanol molecules are over-represented, while clusters of 2 and 3 molecules are underrepresented. These variations are probably the result of entropic effects. However, the deviations from the average trend above 15 carbon atoms are likely to be meaningless, as the related cluster sizes are poorly abundant in the distribution.

Of course, the degree of association depends strongly on the exact conditions investigated. For instance, the association constant would be closer to unity in pure methanol, so that average clusters would be much larger.

The phase equilibrium of hydrogen sulphide with methanol is another interesting example, because simulation can account explicitly for its polarity through the same potential model as used in Sections 3.4.2 and 3.6 [Kristof and Liszi, 1997]. A remarkable degree of agreement with experimental results is found (Fig. 3.66). In contrast to propane, hydrogen sulphide develops a significant attractive electrostatic interaction with methanol as a result of its dipole moment so it is not surprising that the two-phase region is less developed with hydrogen sulphide than with propane.

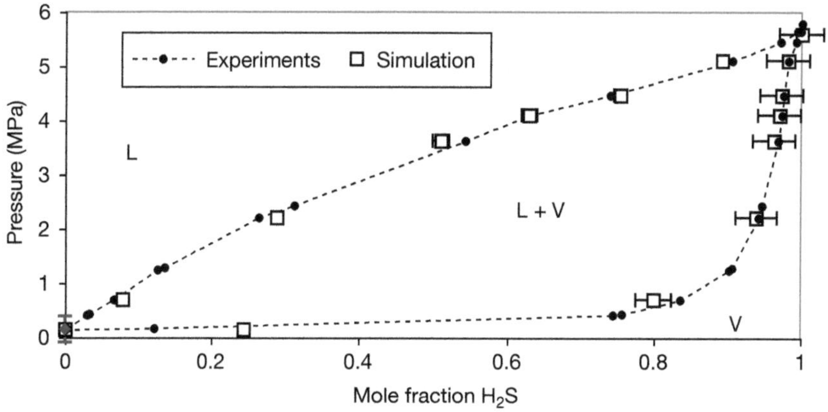

Figure 3.66 Phase diagram of the methanol-H_2S system at 348.15 K from Gibbs ensemble simulations (symbols) and experimental data (lines) from Leu *et al.* [1992].

Another interesting test of the simulation methods is the representation of methanol-water phase equilibrium, which is close to ideal. This was tested with the TIP4P model of water proposed by Jorgensen *et al.* [1983], as this model appears to reproduce reasonably well the liquid-vapour equilibrium of pure water (see Section 3.6 for more details). From Figure 3.67, it can be seen that the phase diagram of the water-methanol system at the temperature of 353 K is correctly reproduced. However, significant discrepancies are noticed with respect to phase composition and pure component vapour pressure. Among a range of possible causes, the possibility of an inaccurate intermolecular potential for water could be singled out. Nevertheless, the general shape of the diagram is well described, confirming the previous finding that activity coefficients in the methanol-water system could be qualitatively predicted by simulation with the same model for methanol and a slightly different potential for water [Slusher, 1999]. Compared with the mixtures of methanol with propane and hydrogen sulphide, it may be noted that the two-phase region is much diminished. Apart from the similarity in vapour pressure, this may be related to the strong attraction between methanol and water molecules, as they can from hydrogen bonds (Fig. 3.68). As a whole, the results are in good agreement with the methanol-water phase diagram at 373 K by Lisal *et al.* [2001] which was based on the same intermolecular potentials. These authors used a specific algorithm, the Reaction Ensemble Monte Carlo method (REMC), in which the acceptance probability of transfer moves are modified to correct the pure component vapour pressures. Using this method, they predicted the phase diagram in a quantitative way, which is very encouraging for future applications.

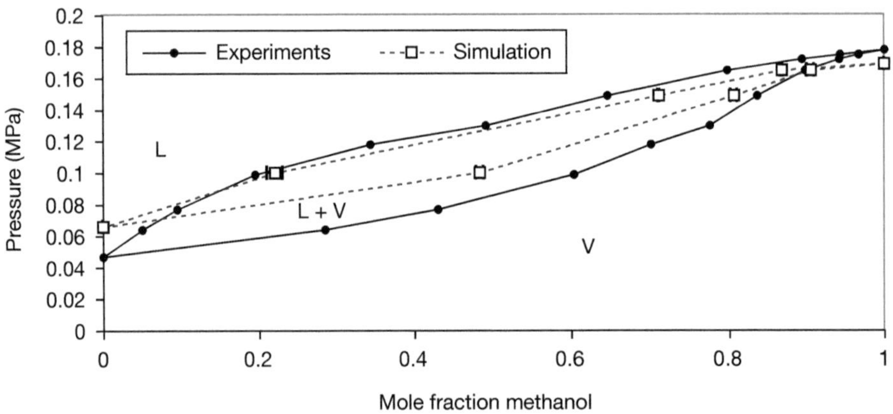

Figure 3.67 Phase diagram of the methanol-water system at 353 K from Gibbs ensemble simulations (symbols) and experimental data (lines) from Bao *et al.* [1995].

Similar investigations have been made on the ternary system methanol-methane-H_2S at different temperatures and pressures, for which comparisons with unpublished experimental data can be made. The intermolecular potentials used for the three compounds are identical to the simulations presented earlier in this section and to the simulations of Section 3.6 on H_2S-rich systems. A typical comparison of predictions with data is shown in Table 3.12 which is related to a system where H_2S has been contacted with a solvent made of 80% methanol and 20% water. It may be noticed that the equilibrium constants are qualitatively predicted. For instance, the equilibrium constant of H_2S is 15.9 at 0.831 MPa from experiments, while it is 18.9 from simulation. When analyzing these results, it appears that the overprediction of the equilibrium constant of water can be explained by the excessive vapour pressure

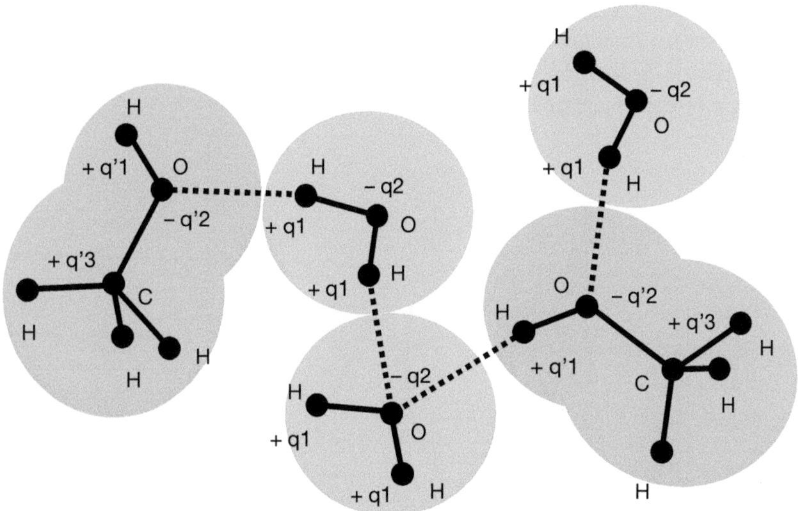

Figure 3.68 Schematic diagram of the hydrogen bonds occurring between in a water-methanol system.

at this temperature from the TIP4P model. This causes an overestimation of the mole fraction of water in the vapour phase, in the same way as it was found for the binary system methanol-water. These results could be significantly improved by using the reaction ensemble method with the same potentials.

Table 3.12 Equilibrium constants in a ternary system H_2S/water/methanol at 408 K, with a methanol/water ratio of 1/4.

	Equilibrium constant ($K_i = y_i/x_i$)					
	H_2S		Water		Methanol	
Pressure (MPa)	**Experiment**	**Simulation**	**Experiment**	**Simulation**	**Experiment**	**Simulation**
0.831	15.9	18.6	0.473	0.79	0.995	0.953
0.92	13.9	16.6	0.421	0.71	0.927	0.931

As predictions of the vapour pressure of water from the TIP4P model are more reliable at lower temperatures, it is likely that the same will be true of ternary mixtures. An example of a ternary phase diagram showing the composition of the coexisting phases at 343 K and 7 MPa is given in Figure 3.69. As the pressure is higher than the saturation pressure of H_2S and methanol at the temperature considered, the three pure components are in the liquid state in the conditions investigated. The binary system water-H_2S exhibits a liquid-liquid phase split unlike the methanol-H_2S and methanol-water binary systems which are fully miscible. The interesting feature of the system is that methanol acts as a cosolvent, which promotes the miscibility of water and H_2S. From the simulation results, it is expected that the ternary system is monophasic for methanol concentrations of over 25%, whatever the relative proportions of water and H_2S. This kind of information is particularly useful for the preliminary design of methanol-based gas treatment processes.

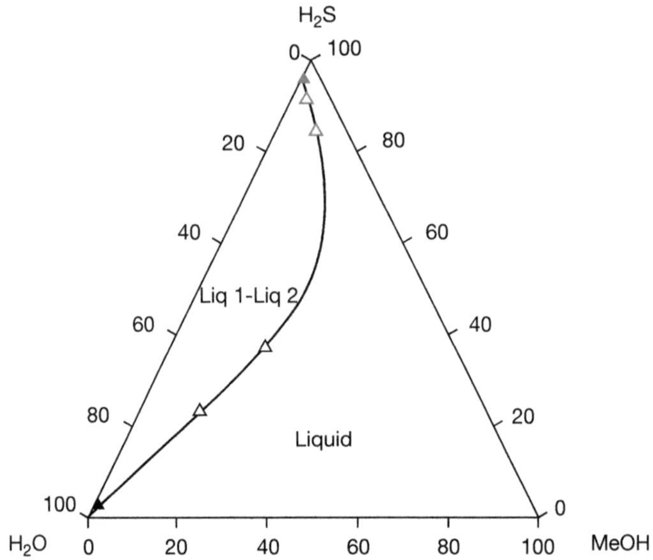

Figure 3.69 Ternary phase diagram of the methanol-water-H_2S system obtained by simulation in the Gibbs ensemble at 343 K and 7 MPa.

3.5 PROPERTIES OF NATURAL GASES AT HIGH PRESSURE

3.5.1 Possible Contribution of Molecular Simulation to Industrial Needs

The composition of the natural gases of industrial interest is generally dominated by methane (60-95%) which may be accompanied by numerous other hydrocarbons and variable amounts of non-hydrocarbon gases (CO_2, N_2, etc.). The heavier hydrocarbons follow a distribution that decreases more or less exponentially with carbon number [Pedersen *et al.*, 1989; Sportisse *et al.*, 1997] and the details of this distribution strongly influence the thermodynamic properties of the gas. When the distribution contains significant amounts of hydrocarbons with more than 7 carbon atoms, the gas often exhibits a retrograde dew point, i.e. phase separation when its pressure is decreased at constant reservoir temperature. When it displays this behaviour, it is referred to as a condensate gas.

In a general way, the production of a condensate gas entails problems because liquid phase condensation in the reservoir has to be avoided for economic reasons (the valuable liquid hydrocarbons are recovered less efficiently) and for fluid flow reasons (the condensed liquid phase hinders gas flow in the vicinity of the well). Therefore, it is always important that phase behaviour and volumetric behaviour are well characterised in condensate gas fields. The prediction of this behaviour is difficult because the exact composition across the entire range is seldom known, the analysis being generally limited to 10 carbon atoms and below (and often 7 carbon atoms and below). Moreover, the presence of this unknown heavy fraction is not the only difficulty because the lack of thermodynamic data about many hydrocarbon families would also prevent the application of engineering equations of state even if a full analysis were available. Also, equations of state require fine tuning of binary interaction coefficients between methane and heavy hydrocarbons to represent their phase equilibrium, whereas the necessary experimental information is unavailable for several families of heavy hydrocarbons. The prediction of condensate gas behaviour with equations of state in the industry is generally based on the use of pseudo-components (i.e. hydrocarbons of undefined molecular structure) whose thermodynamic properties are regressed to match the phase behaviour of the condensate gas under consideration. It is thus necessary to take large samples of the gas and perform systematic volumetric and phase behaviour measurements before field development can be planned (or abandoned).

Apart from phase behaviour, volumetric properties are very important for gas production, as illustrated by the systematic use of the compressibility factor ($Z = PV/RT$) variations versus pressure to evaluate reserves. More recently, the discovery of deep, high pressure gas wells has also triggered a need for a deeper understanding of thermal properties. In these fields, it has been observed that the gas tends to heat up in the production tubing as its pressure drops. This behaviour may be attributed to the Joule-Thomson effect (Fig. 3.70). In a very first approximation, the expansion of the gas in the production tubing may be assimilated to isenthalpic expansion when the production rate is high enough so that any heat exchange with the surroundings can be ignored. In this case, the thermal regime is governed by the Joule-Thomson coefficient which describes temperature changes in the course of isenthalpic expansion. Depending on temperature and pressure conditions, the Joule-Thomson effect may be positive or negative. In most gas reservoirs, the pressure is below 50 MPa and the gas tends to cool down as a result of the Joule-Thomson effect. Then the thermal regime is not a major industrial concern. In high pressure gas reservoirs, the Joule-Thomson effect tends to heat up the gas, and this makes additional demands on production equipment (tubes, valves, etc.). For this reason, the effect needs to be characterised although this is difficult using experimental methods: measuring the Joule-Thomson coefficient of gases in the laboratory requires both large amounts of sample and very good thermal insulation, both conditions which are difficult to fulfill at pressures which can reach 110 MPa. Predictions with classical equations of state are possible in principle [Kortekaas *et al.*, 1997] but they have not been validated by systematic comparisons with high pressure Joule-Thomson coefficient measurements, either for pure compounds or mixtures. In contrast, this type of validation was presented in Section 3.2, supporting the relevance of Monte Carlo simulation methods when it comes to predicting Joule-Thomson coefficients at high pressures.

The long-term challenge for molecular simulation in the natural gas industry is to be able to provide reasonably reliable predictions of the thermodynamic behaviour of a gas as long as a complete analysis of the mixture is available. The need for a complete analysis may seem unrealistic but much more information can be extracted from a chromatographic analysis than

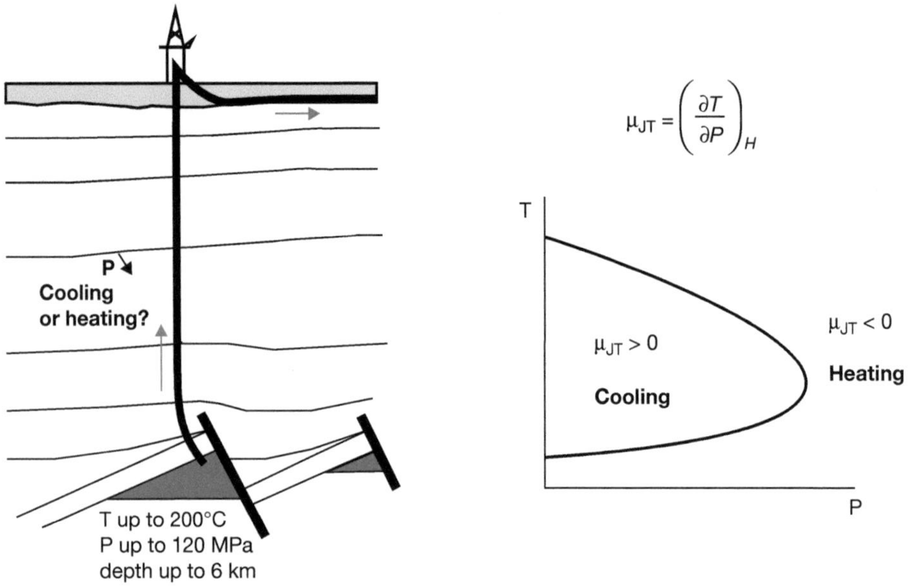

Figure 3.70 Influence of the Joule-Thomson effect in the thermal regime of deep gas wells.

is routinely extracted today, e.g. standard analytical techniques are available for molecules containing up to 20 carbon atoms [Durand *et al.*, 1989], and high temperature chromatography extends the analysis up to the highest molecular weights present in condensate gases, i.e. 35 carbon atoms. Mass spectrometry is also a technique that can be used to complement chromatography to provide a fuller evaluation of a hydrocarbon family. The advantage of coupling full analysis with molecular simulation is that a smaller amount of gas sample is needed, allowing a sooner prediction for new gas discoveries. The costly experiments conducted at reservoir pressure could be then focused on those natural gases that are most likely to be produced.

How far is molecular simulation from meeting this challenge? The answer differs depending on whether it is phase equilibria or monophasic properties which are under consideration. Monophasic properties are investigated by using the *NPT* or *NVT* statistical ensembles, and it is then possible to introduce a detailed account of the molecular composition of the gas, as it will be discussed in this section. Concerning phase equilibria, the problem is more complex because dew points cannot be determined directly from a Gibbs ensemble simulation, for the same reason as discussed in the context of bubble points in Section 2.3.7. Indeed, a dew point calculation requires that the *vapour* composition be imposed, while it is the *global* composition that is imposed in the Gibbs ensemble. Very encouraging dew point calculations have been reported in the literature [Escobedo, 1998; Escobedo, 1999] but, in these investigations, a full analysis of the natural gases under consideration was not available. Extension of these techniques to well-analysed gases is still open to further research.

As a consequence, only monophasic properties will be considered in the applications described in this section, including volumetric properties and the Joule-Thomson coefficient.

In the preliminary study of Neubauer *et al.* [1999b], it was shown – on the basis of natural gas examples taken from the literature – that the volumetric properties of natural gases can be simulated using Monte Carlo methods. The purpose of this section is to go a step further in this direction, based on the results of more recent research [Lagache *et al.*, 2004a]. We will look at, not only a classical gas example from the literature – a San Joaquin gas characterised by Sage *et al.* [1947], but also a condensate gas from a very deep North Sea reservoir characterised by Arnaud [1995]. This second fluid will be named "HP-HT fluid" in reference to the very high pressures (up to 110 MPa) and temperatures (up to 190 K) of the gas reservoirs found in this region [Ungerer *et al.*, 1995 and 1998]. A full analysis of the gas is provided to allow a representative modelling of the HP-HT gas, so that a large set of predictions are obtained about volumetric properties, its Joule-Thomson coefficient and thermodynamic derivative properties.

3.5.2 Representation of Natural Gas Composition in Monte Carlo Simulation

In applying molecular simulation to the property prediction of natural gases, it is obvious that there is a large advantage in benefiting from the availability of well-tested intermolecular potentials. The Lennard-Jones potential of Möller *et al.* [1993] which has been applied with success in Section 3.2 and the AUA potential of hydrocarbons tested in many instances (Sections 3.1 and 3.2) provides a firm basis for this purpose. In Figure 3.71, the compressibility factor predicted for methane at high pressure is shown for two temperatures, confirming that very good predictions can be obtained. These intermolecular potentials will be used in this section for natural gases.

It is worth pointing out that such intermolecular potentials can only be used if every component of the simulated system has a well-defined molecular structure. It is thus uneasy to introduce pseudo-components, i.e. components of undefined molecular structure, as it is commonly done when modelling the heavy fraction of natural gases with equations of state.

Non-hydrocarbon gases like CO_2 and N_2 are also taken into account when modelling the simulation of gas. Generally, the intermolecular potentials used for these molecules comprise electrostatic charges [Harris and Yung, 1995; Delhommelle, 2000], and electrostatic interactions are an important part of the interaction energy. However, the contribution of electrostatic energy to the whole interaction energy is negligible in this case, because CO_2 and N_2 are mostly surrounded by non-polar molecules, e.g. the contribution of electrostatic energy to the interaction energy of the HP-HT gas discussed in this section was found to be only 0.02%. In these circumstances, ignoring the electrostatic energy did not have any significant influence on any property derived by molecular simulation. This is an interesting example of a more general rule, i.e. that the important interaction terms for a given molecule depend on the polarity of its environment. In this case, it could be used to speed up the simulations of the natural gases shown.

The way to represent the composition of a gas is illustrated by the example of the San Joaquin gas investigated by Sage *et al.* [1947]. The composition is given as fractions by carbon number up to seven carbon atoms. The distinction between *n*-alkanes and branched alkanes is provided for the C_4 and C_5 fractions, but not for C_6 or C_7 as shown in Figure 3.72.

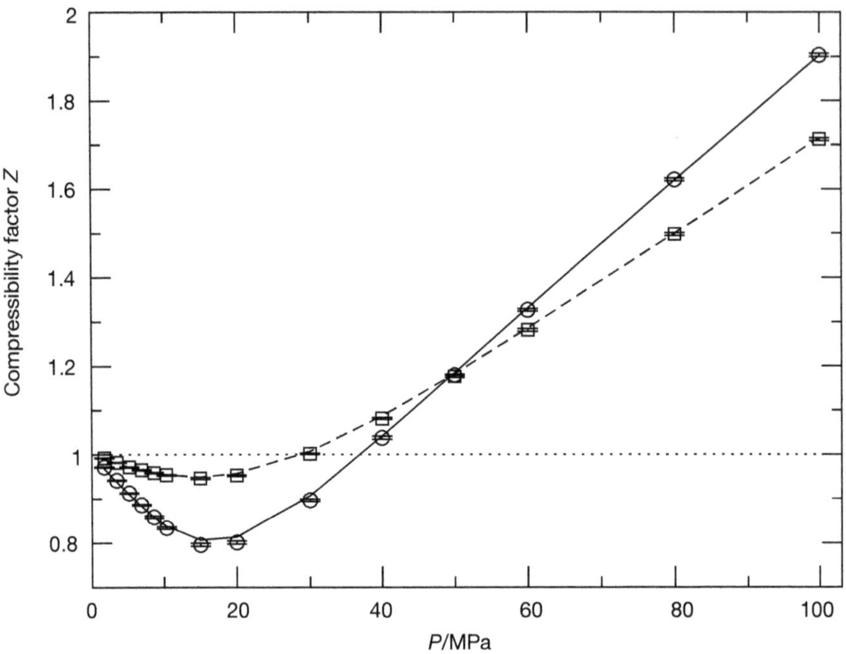

Figure 3.71 Compressibility factor of methane determined from *NPT* simulations at 294.25K (o) and 377.55 K (❑). The lines represent experimental observations [Lagache *et al.*, 2001], reproduced by permission of the PCCP Owner Societies.

No indication is given about the chemical family of the heavy fraction (C_8+). For the purpose of our simulations, the fluid is represented by 1 000 molecules, as indicated in Table 3.13. As a consequence of the total number of molecules, the molar fractions are reproduced within 0.1%, which is probably less than the experimental uncertainty on the hydrocarbon distribution. In the absence of specific information, the C_6 and C_7 fractions are represented by *n*-hexane and *n*-heptane respectively, although it is suspected that they contain a significant amount of cyclic alkanes and aromatics. The heavy fraction C_8+ is also represented by normal alkanes. In order to match the experimentally determined average molecular weight of the fluid, the C_8+ fraction is simulated as 50% *n*-nonane and 50% *n*-decane.

When a detailed analysis of the gas is available, the use of well-defined components raises two problems. First, the number of identified individual components in a natural gas exceeds 200 with recent chromatographic methods, so that it is practically impossible to account for all the different hydrocarbons in simulation. This problem is illustrated by the composition of the HP-HT fluid shown in Figure 3.73 [Lagache *et al.*, 2004a]. The detailed composition of the HP-HT fluid is in fact even more complex than shown in this figure because some fractions like C_8 isoalkanes or C_9 aromatic compounds contain several isomers which are resolved from one another in the chromatographic analysis. Second, the analysis is not available at the molecular level for heavy fractions, but it is expressed in fractions by molecular

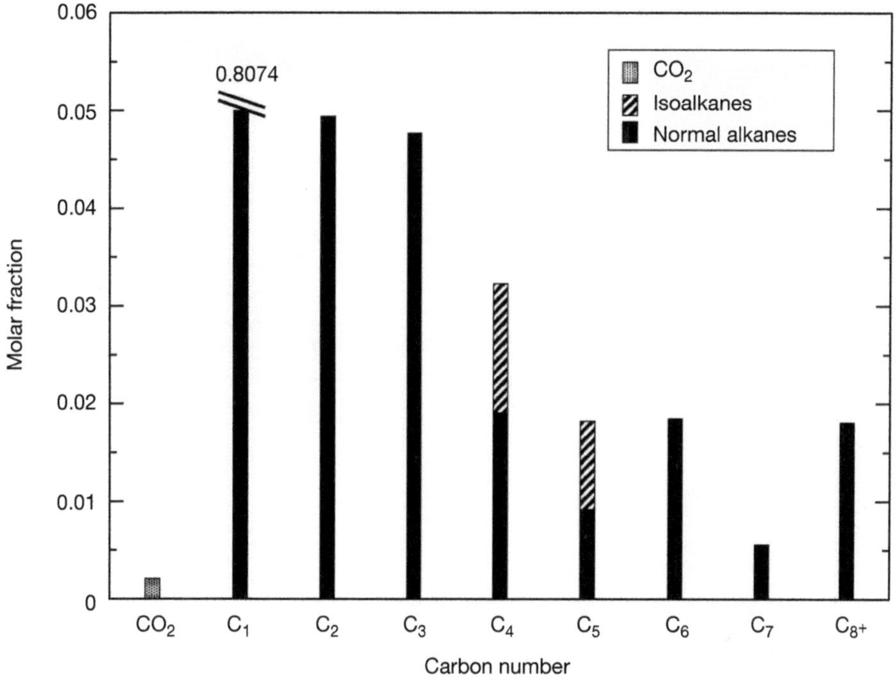

Figure 3.72 Molar composition of the San Joaquin gas [fluid C of Sage and Olds, 1947]. Methane is not drawn to scale. Reprinted from [Lagache *et al.*, 2004], © 2004 with permission from Elsevier.

Table 3.13 Composition of the San Joaquin Valley natural gas [fluid C of Sage and Olds, 1947] and representation by a discrete number of molecules [Lagache *et al.*, 2004a].

Component	Experimental mole fraction	*NPT* simulation Number of molecules	Molar fraction
Carbon dioxide	0.0021	2	0.002
Methane	0.8074	807	0.807
Ethane	0.0494	49	0.049
Propane	0.0477	48	0.048
Isobutane	0.0132	13	0.013
n-butane	0.0191	19	0.019
Isopentane	0.0097	10	0.010
n-pentane	0.0092	9	0.009
Hexanes	0.0185	19 nC_6	0.019
Heptanes	0.0056	6 nC_7	0.006
Octane+	0.0181	9 nC_9 + 9 nC_{10}	0.018
Total	1	1 000	1
Average molar mass (g/mol)	24.38		24.47

number and chemical family (for instance "C_{14} aromatics"). The precise nature of the molecules making up the fraction is then unknown. In order to circumvent these problems, it is proposed to proceed in two steps.

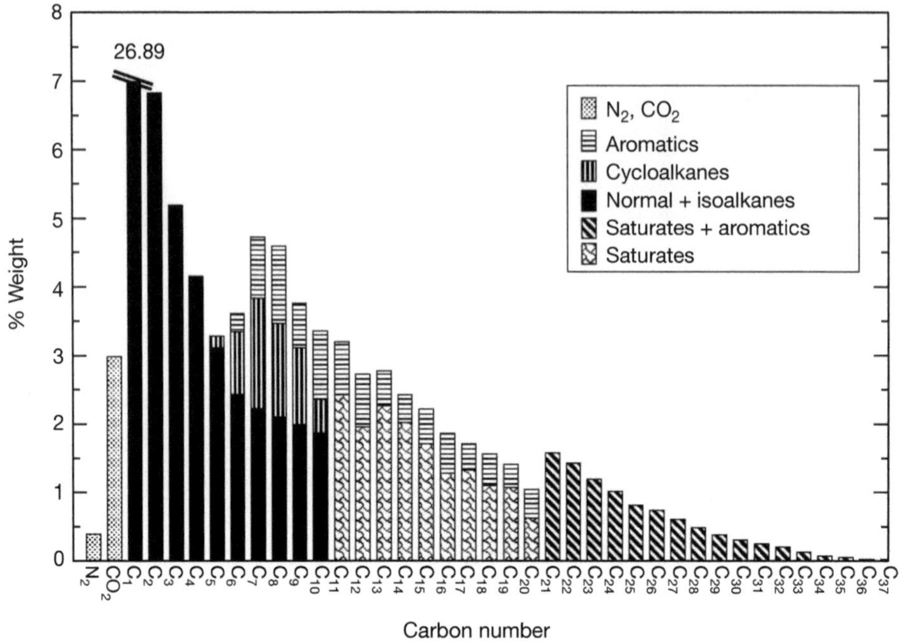

Figure 3.73 Molar composition of the HP-HT fluid after Arnaud [1995] and Lagache *et al.* [2004a], © 2004 with permission from Elsevier. Methane is not drawn to scale.

In a first step, the detailed analysis is simplified by division into in a manageable number of fractions (typically 10 to 20 fractions) keeping paraffins, cyclic alkanes and aromatic compounds in separate fractions when the breakdown is known (note that paraffins refers to *n*-alkanes and branched alkanes). The hydrocarbons belonging to a given family (either paraffin, cyclic or aromatic) are divided into fractions grouping together hydrocarbons of similar carbon numbers (for instance C_{11} to C_{15} aromatics).

In a second step, a representative component is selected for each fraction. Paraffins are represented either by *n*-alkanes or by branched alkanes. Cyclic alkanes lower than 10 carbon atoms are represented by alkylated cyclohexanes, the most abundant type of hydrocarbons in this family. Similarly, aromatics lower than 10 carbon atoms are simulated by alkylated benzenes. As condensed polyaromatics are known to be major aromatic components above 10 carbon atoms [Petrov, 1984; Tissot and Welte, 1984], it is more realistic to use molecules like alkyl-naphthalenes (2 cycles) or alkyphenanthrene (3 cycles) rather than *n*-alkylbenzenes to represent heavy aromatic fractions.

The corresponding procedure is illustrated by Table 3.14, in which 19 fractions have been used to represent the gas. These fractions comprise not only *n*-alkanes, but also branched alkanes, cyclic alkanes and aromatics. A total of 500 molecules is used to define the simulated system. As a consequence, the simulated molar fractions are multiples of 1/500 i.e. 0.2%. The molar fractions are thus represented in an approximate way, especially for the heavy fractions.

Table 3.14 Composition of the HP-HT fluid [Arnaud, 1995; Lagache *et al.*, 2004a] and representation by a discrete number of molecules.

Lumped fraction	Mole fraction (%)	Representative molecule	Number of molecules	Simulated mole fraction (%)
N_2	0.543	N_2	3	0.6
CO_2	2.642	CO_2	13	2.6
C_1	65.568	C_1	328	65.6
C_2	8.882	C_2	44	8.8
C_3	4.602	C_3	23	4.6
$n + iC_4$	2.792	nC_4	14	2.8
$n + iC_5$	1.686	nC_5	8	1.6
n + iso C_6-C_7	1.972	iC_7	10	2.0
n + iso C_8-C_{10}	1.839	nC_9	9	1.8
Cyclic alkanes C_5-C_{10}	2.120	methyl-cyclohexane	11	2.2
Aroatics C_6-C_{10}	1.422	*p*-xylene	7	1.4
Saturates C_{11}-C_{14}	2.281	nC_{12}	11	2.2
Aromatics C_{11}-C_{15}	0.805	dimethyl-naphtalene	4	0.8
Saturates C_{15}-C_{18}	1.083	nC_{16}	6	1.2
Aromatics C_{16}-C_{20}	0.418	dimethyl-phenanthrene	2	0.4
C_{19}-C_{21}	0.495	nC_{20}	3	0.6
C_{22}-C_{27}	0.673	nC_{25}	3	0.6
C_{28}-C_{37}	0.178	nC_{30}	1	0.2
Total C_1-C_{37}	100	Total C_1-C_{37}	500	100

In addition to the simplifications introduced by the discretization of the composition, care must be taken that the analysis of the gas may contain some errors. In Figure 3.73, it may be noticed that the distribution exhibits two maxima at 7 and 21 carbon atoms in a globally decreasing trend. The first of these maxima is likely to be real, since it corresponds to the appearance of cyclic alkanes and aromatics, which are absent below 6 carbon atoms. On the opposite, the second maximum is probably an artifact of the analysis, as the molar fractions for molecules containing more than 20 carbon atoms have been obtained from a more approximate mass balance than the lighter part of the distribution.

The pattern for the simulated HP-HT fluid (Fig. 3.74) reveals that the gaseous hydrocarbons (C_1 to C_4) are not dominating the composition as strongly as might be expected from the distribution of Figure 3.73. The heavier hydrocarbons (C_5 and above) are minor in terms of the number of moles present, but they are approximately equivalent in terms of mass and absolute numbers of carbon atoms. This provides a qualitative explanation of why this fluid displays liquid-like properties at high pressure, as will be seen later.

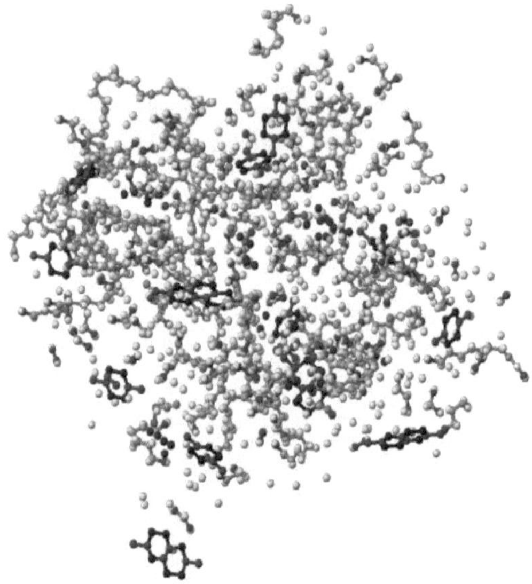

Figure 3.74 Example of a configuration of the HP-HT fluid under reservoir conditions (463 K and 110 MPa), simulated with 500 molecules belonging to 19 different molecular types.

3.5.3 Volumetric Properties

On the basis of the 1 000 molecule representation shown in Table 3.13, simulations of the San Joaquin gas have been conducted in the *NPT* ensemble at three different temperatures for which experimental measurements of the compressibility factor $Z = PV/RT$ were available [Sage and Olds, 1947]. These simulations have been conducted only at pressures greater than the retrograde dew point pressure, which is 20.7 MPa. The results (Fig. 3.75) indicate that the influence of pressure and temperature on Z is correctly reproduced, but the compressibility factor appears systematically underestimated compared with measurements. Nevertheless, the discrepancy between the result obtained by the Monte Carlo simulation method and the experimental measurement does not exceed 3%. This discrepancy is significantly greater than the estimated uncertainty for a mixture of known composition, which is inferred to be lower than 2% on the basis of the tests performed on pure components (see Figure 3.71). The deviation may be explained – at least partly – by the approximate analysis on which the modelled composition is based. Another possible cause of error might be the representation of the heavy fraction. In order to test this, simulations have been performed in which the heavy fraction has been represented by *n*-decane alone instead of *n*-decane and *n*-nonane. The compressibility factor is not significantly changed, indicating that the representation of the heavy fraction was not a major issue for this particular gas.

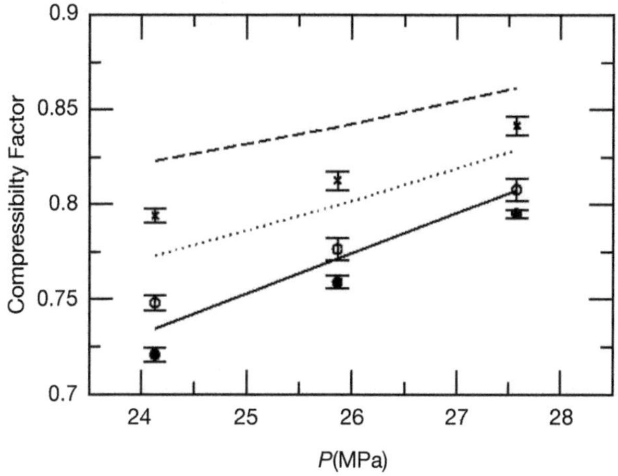

Figure 3.75 Compressibility factor of the San Joaquin gas from *NPT* simulations at 277.57K (•), at 310.93K (o), at 344.26K (×) after Lagache *et al.* [2004a]. The lines represent the experimental measurements of Sage and Olds [1947]. Reprinted from [Lagache *et al.*, 2004], © 2004 with permission from Elsevier.

In the case of the HP-HT gas, volumetric measurements in reservoir conditions have been made by Arnaud [1995]. The Monte Carlo simulation of this gas was carried out using the *NPT* ensemble at the reservoir temperature with the composition shown in Table 3.14. Although the large number of components might be an obstacle to satisfactory convergence, it is observed that density converges in a reasonable number of Monte Carlo steps: after 20 millions of iterations, the limiting value is obtained within 2% (Fig. 3.76). The density of the HP-HT fluid is particularly high as it reaches 500 kg/m^3 at 110 MPa (Fig. 3.77), i.e. greater than the densities of methane and ethane in the same conditions. This high density can be explained by the predominance of heavy hydrocarbons in the gas, as mentioned above when discussing the image of a configuration box (Fig. 3.74). Experimental gas density measurements agree closely with simulation predictionswhich is not surprising when considering the good representation of the densities of methane and ethane, which are the most abundant compounds in the fluid. There is however a difference in temperature with respect to the measurements of Arnaud [1995] which were made at the reservoir temperature as estimated at that time, i.e. 453 K, whereas the simulations were made at a more recently estimated reservoir temperature, i.e. 463 K. If the simulation had been conducted at 453 K, a slight overestimation of the density would have been found. Nevertheless, the results are extremely encouraging because the HP-HT fluid is a particularly challenging example of a gas condensate.

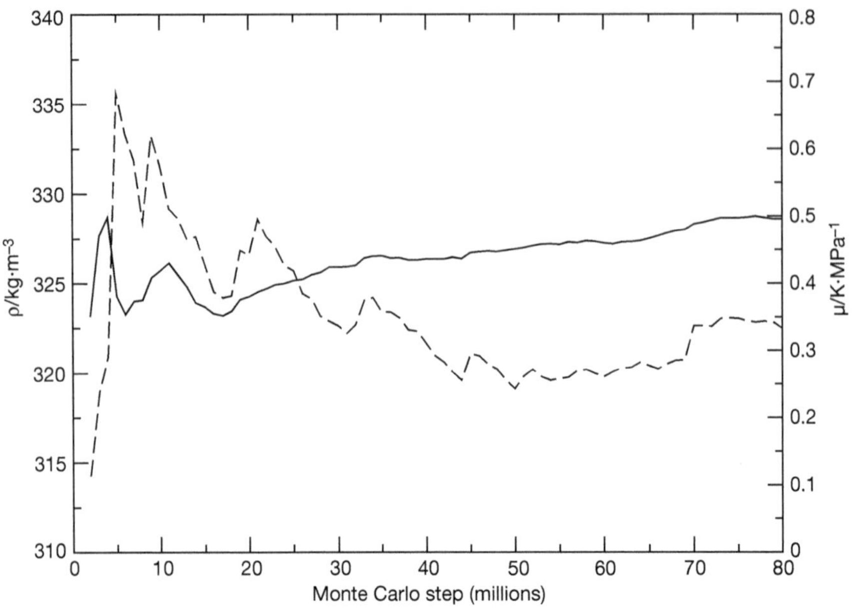

Figure 3.76 Convergence of the running density average (continuous line) and of the Joule-Thomson coefficient (dotted line) of the HP-HT fluid in a typical simulation in the *NPT* ensemble [Lagache *et al.* 2004a] at $T = 463.15$ K and $P = 30$ MPa.

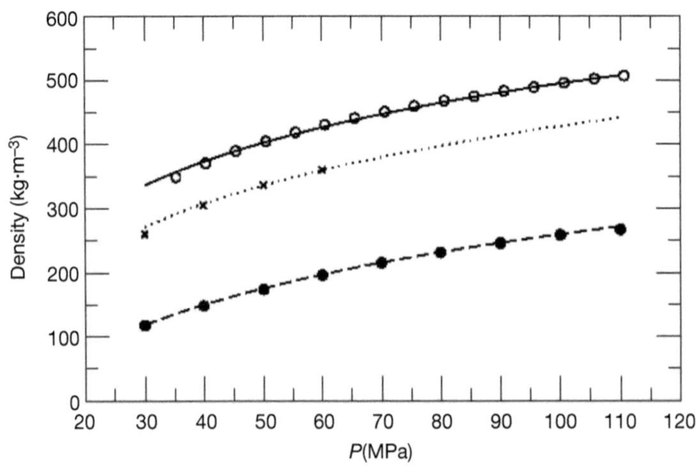

Figure 3.77 Density of methane (---), ethane (·····) and of the HP-HT fluid (——) obtained by *NPT* simulation at 463.15 K, compared with the recommended values of IUPAC for methane (•) at 463.15 K, with the values of Friend *et al.* [1991] for ethane (×) at 463.15 K, and with the values of Arnaud [1995] for the HP-HT fluid (o) at 452.65 K [Lagache *et al.*, 2004a]. Reprinted from [Lagache *et al.*, 2004], © 2004 with permission from Elsevier.

3.5.4 Joule-Thomson Coefficient and Derivative Properties

As seen in Sections 2.4.8 and 3.2, the derivative properties and the Joule-Thomson coefficient can be obtained by combining simulation results in the *NPT* ensemble with the ideal heat capacities of pure components. In this study, the ideal heat capacities have been determined from the correlations available in the Dortmund Data Bank. In Figure 3.76, the convergence of the Joule-Thomson coefficient is shown to require longer simulations than the density, and the statistical uncertainty on this property is approximately 0.05 $K \cdot MPa^{-1}$. The computed Joule-Thomson coefficients have been plotted as continuous curves which represent the raw simulation results to within this degree of accuracy.

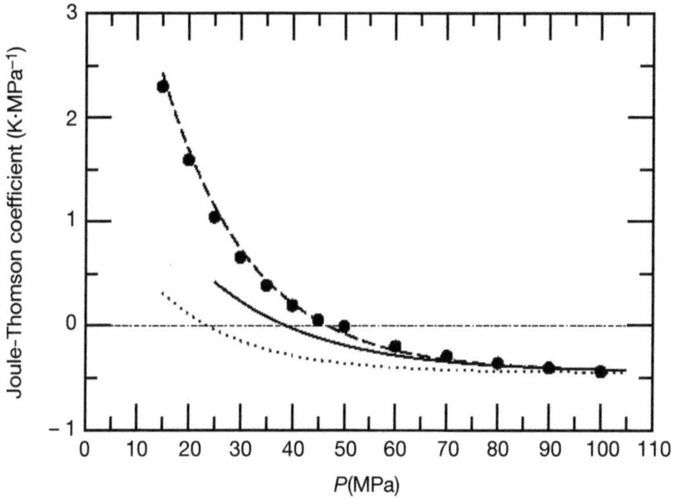

Figure 3.78 Joule-Thomson coefficient of the San Joaquin gas (——), of methane (---) and of ethane (·····) at 310.93 K, compared with the recommended values of IUPAC [Wagner, 1996] for methane (•) at the same temperature [Lagache *et al.*, 2004a]. Reprinted from [Lagache *et al.*, 2004], © 2004 with permission from Elsevier.

In the case of the San Joaquin gas, the Joule-Thomson has been computed by this method in a large range of pressures at 310.9 K, i.e. one of the temperatures at which the volumetric behaviour has been tested in Section 3.5.3. This allows to determine the Joule-Thomson inversion pressure at this temperature. The same determination is also shown for methane and ethane (Fig. 3.78). These results show that the behaviour of the natural gas is intermediate between methane and ethane. This is not surprising, as methane and ethane account for more than 85% of the gas. The Joule-Thomson inversion pressure is approximately 39 MPa, while it is 47 MPa for methane and 24 MPa for ethane at the same temperature. At high pressure, the Joule-Thomson coefficient tends to the same limiting value of – 0.4 K/MPa for the San Joaquin gas and the two pure hydrocarbons.

In the case of the HP-HT gas, the same type of calculation has been performed at the reservoir temperature, i.e. 463 K (Fig. 3.79). In the same way as for the density of the same fluid, the Joule-Thomson coefficient of the gas no longer falls between those of methane and ethane, e.g. the Joule-Thomson inversion pressure for the gas as obtained by simulation is 42 MPa, while it is 51 MPa for ethane and 58 MPa for methane. Although no direct measurements of the Joule-Thomson coefficient have been made on the HP-HT fluid, the inversion conditions are known from its volumetric behaviour. Indeed, the density has been measured at several pressures and temperatures, allowing a good estimate of the thermal expansivity and of the molar volume. From these informations, it is possible to estimate the conditions for which the product $(\alpha T - 1)$ is zero. This indirect determination of the Joule-Thomson inversion pressure is 42.6 MPa, a value in excellent agreement with simulation results. Here again, it is likely that such an agreement is partly fortuitous, as the deviation is lower than the estimated statistical uncertainty (approximately 2 MPa), but it is a likely indication that no major bias is involved in the predictions of this natural gas.

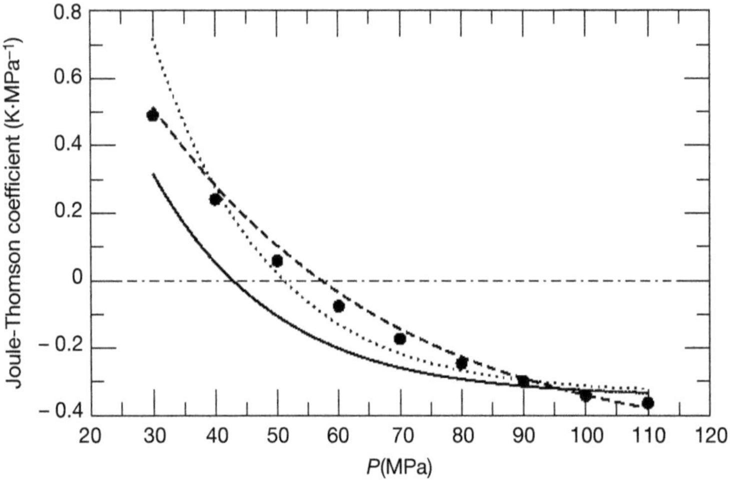

Figure 3.79 Joule-Thomson coefficient of the HP-HT fluid (——), of methane (---) and of ethane (·····) obtained by *NPT* simulation at 463.15 K, compared with the recommended values of IUPAC [Wagner, 1996] for methane (•) at the same temperature [Lagache *et al.*, 2004a]. Reprinted from [Lagache *et al.*, 2004], © 2004 with permission from Elsevier.

The trend of isothermal compressibility is given in Figure 3.80 for the HP-HT fluid. There is no experimental validation for this property, so the computed values are pure predictions which are indirectly supported by the correct prediction of the other properties. With increasing pressure, the isothermal compressibility decreases significantly, confirming that the fluid tends to behave like a liquid in the upper part of the pressure range.

As a concluding remark, it may be said that the detailed composition of condensate gases may now be addressed by molecular simulation and that the prediction of various thermodynamic properties appears feasible. More detailed investigations are needed to define how sensitive the outputs are to the way in which fluid composition is modelled. Nevertheless, the results are encouraging enough to envisage extending this type of prediction to phase equilibrium behaviour with a reasonable expectation of success in the years to come.

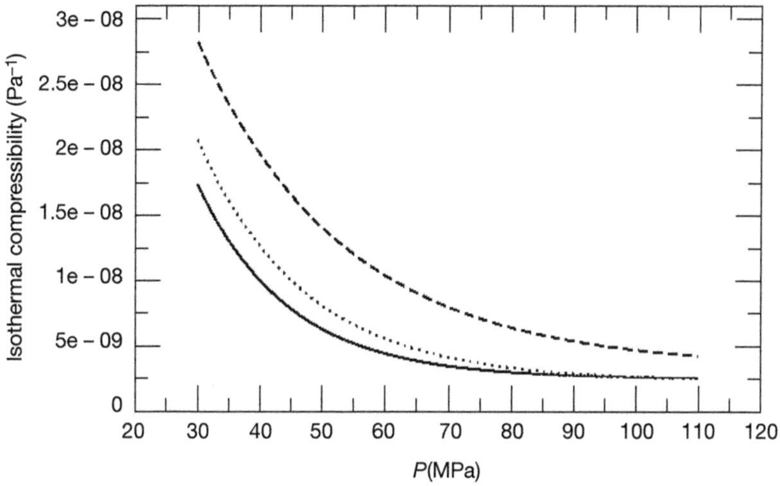

Figure 3.80 Isothermal compressibility of the HP-HT fluid (——), of methane (---) and of ethane (·····) obtained by *NPT* simulation at 463.15 K. Reprinted from [Lagache *et al.*, 2004], © 2004 with permission from Elsevier.

3.6 THERMODYNAMIC PROPERTIES OF ACID GASES AT HIGH PRESSURE

The production of acid gases (i.e. natural gases containing carbon dioxide or hydrogen sulphide) raises particular operational problems for H_2S concentrations above 10%. These problems correspond to the cost of the amine separation unit which is necessary to reduce the H_2S concentration to the specified level (which is very low) and of the Claus unit in which the separated H_2S must be converted to elementary sulphur. The overall cost of this treatment makes exploiting several important natural gas deposits unfeasible.

In order to reduce production costs, a preliminary simple separation step may be considered so that most of the H_2S in the gas is recovered in the liquid state and can be reinjected into a deep reservoir. This reduces considerably the size of the amine separation unit (which is still necessary to comply with the H_2S specification) and of the associated Claus unit. The lower investment and operational costs of these units compensates largely for the additional cost of the proposed preliminary separation. This preliminary separation step is achieved using a stripper in which the natural gas stream is brought into contact with the refrigerated recycled stream which contains mostly H_2S and water separated in the cold tank. The whole preliminary treatment operation proceeds at a pressure which is high enough (typically 7 MPa) to ensure that the H_2S-rich effluent is in liquid form so that, if necessary, it can be pumped for the purposes of reinjection (Fig. 3.81). In addition to H_2S and hydrocarbon gases, attention must be paid to the fate of water and CO_2 in the process. As much as possible of these undesirable components of the natural gas should be pumped into the reservoir together with the H_2S.

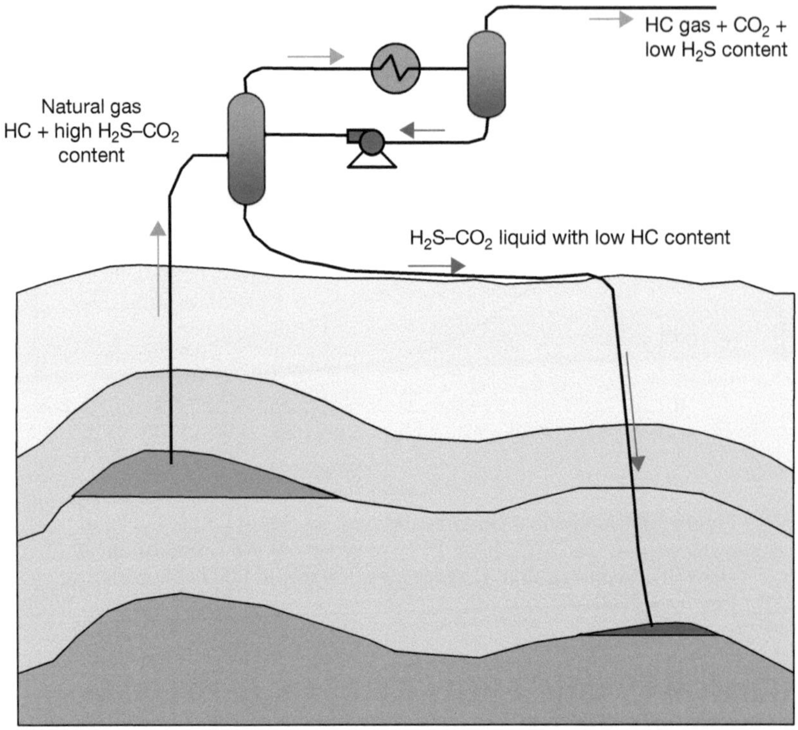

Figure 3.81 General scheme of the simplified acid gas treatment and reinjection. The typical depth of the reservoirs considered is 2 to 6 km. Reprinted with permission from [Ungerer *et al.*, 2004], © Taylor & Francis Ltd (http://www.tandf.co.uk/journals).

There is not much data to help with the design of such a process because the toxicity and corrosiveness of H_2S makes experimental measurements difficult and costly, particularly at high pressure.

Firstly, there is little information on phase equilibria between methane, H_2S and water at high pressure. The general behaviour is known for the binary systems methane-water [Culberson, 1951; Gillespie and Wilson, 1982], H_2S-water [Gillespie and Wilson, 1982; Carroll and Mather, 1989] and methane-H_2S [Kohn and Kurata, 1959a-b]. However it is suspected that liquid-liquid phase splitting could occur at a low temperature in the H_2S-methane system at high pressure, although there are no published data covering such conditions. There are no published experimental results on the ternary system methane-water-H_2S, and little unpublished information. Secondly, there is little information on the enthalpies of mixing, which may infuence the energy balance in separation units. Finally, there is a gap in our knowledge of the volumetric behaviour of H_2S-rich mixtures at very high pressure (up to 80 MPa); this is a problem because such pressure levels are encountered in the deep reservoir rocks where reinjection is considered.

Predicting the desired properties solely from equations of state or from other classical thermodynamic models (such as group-contribution-based excess Gibbs energy models) would be risky for the types of mixture and condition involved, because these models are known to be poorly predictive for mixtures involving such polar components as water and H_2S. The purpose of this chapter is to expose the contribution of molecular simulation to these problems, using the material published in a recent article [Ungerer *et al.*, 2004].

Among the properties investigated here, we will mainly concentrate on phase diagrams, densities, excess volumes, excess enthalpies and derivative properties (such as the isothermal compressivity or the Joule-Thomson coefficient). The reason why high pressure volumetric properties (density, compressivity) are important is that the necessary reinjection pressure depends strongly on liquid density for deep wells. If the H_2S-rich fluid is reinjected into an oil reservoir, its density determines whether it will channel on top of the oil, push it forwards or sink to the bottom. The reason for investigating excess enthalpy is the need to be able to predict thermal regimes in separation units. Finally, knowing the Joule-Thomson coefficient is important when it comes to understanding the temperature regime in reinjection wells.

3.6.1 Intermolecular Potential for CH_4, Water, CO_2 and H_2S

A. Methane

This study focuses on a temperature range where methane is supercritical. Therefore, the intermolecular potential must not only account for the liquid-vapour equilibrium properties of pure methane, but also for its supercritical behavior. As seen in Sections 3.2 and 3.5, the Lennard-Jones 6-12 potential of Möller *et al.* [1992] accurately describes the density and the Joule-Thomson coefficient of methane at pressures up to 100 MPa and temperatures higher than ambient. The exp-6 potential developed recently by Errington and coworkers [Errington and Panagiotopoulos, 1998] would have been ideal for the investigation of water solubilities [Errington *et al.*, 1998] but the absence of similar exp-6 potential for H_2S would have made investigating related mixtures difficult. As a result, Möller's potential is used for methane in this section.

B. H_2S

Two potentials have been considered for H_2S in this study: the first is the potential proposed by Kristof and Liszi [1997] which involves a single Lennard-Jones 6-12 site and four electrostatic point charges, but does not take polarisation energy into account. The second potential investigated [Delhommelle *et al.*, 2000a] also involves a single Lennard-Jones site and four point charges with slightly different parameters, and accounts for the polarisation energy. The parameters of both models are indicated in Table 3.15. The polarizability of the H_2S molecule is taken as 3.78 $Å^3$ (see Section 2.2). A third potential involving three atomic charges and a single Lennard-Jones site [Nath, 2003] was advanced after completion of the study presented here.

The simulations are made with numerical conditions similar to those used by Kristof and Liszi. A total number of 512 molecules is used, and electrostatic interactions are cut off at half of the box length without Ewald summation. A typical simulation length is 25 000 Monte Carlo cycles. The simulation results obtained in the Gibbs ensemble have been compared with experimental correlations of vapour pressure, vaporization enthalpy and saturated liquid density (Fig. 3.82). As reported by Kristof and Liszi, excellent agreement is found with H_2S coexistence properties. The results are also in very good agreement with results obtained more

Table 3.15 Parameters of the intermolecular potential models investigated in the present study for the various components. The force centers and charges are either located on atomic nuclei (C, S, H, O) or on intermediate points (M).

	Force center or charge	Position (Å)			Lennard Jones parameters		Charge q/e	References
		X	Y	Z	ϵ/k (K)	σ (Å)		
CH_4	C	0	0	0	149.92	3.7327	0	Möller *et al.*, 1992
H_2S non polarisable	S	0	0	0	250	3.73	0.40	Kristof and Liszi, 1997
	H1	0.9639	0.9308	0			0.25	
	H2	– 0.9639	0.9308	0			0.25	
	M	0	0.1862	0			– 0.9	
H_2S polarisable	S	0	0	0	230	3.74	1.393	Delhommelle *et al.*, 2000
	H1	0.9591	0.9246	0			0.323	
	H2	– 0.9591	0.9246	0			0.323	
	M	0	0.1933	0			– 2.039	
H_2O (TIP4P)	O	0	0	0	78.03	3.1536	0	Jorgensen, 1983
	M	0.15	0	0			– 1.04	
	H1	0.5859	0.757	0			0.52	
	H2	0.5859	– 0.757	0			0.52	
CO_2 (HY-EPM2)	C	0	0	0	28.129	2.757	0.6512	Harris and Yung, 1995
	O1	1.149	0	0	80.507	3.033	– 0.3256	
	O2	– 1.149	0	0	80.507	3.033	– 0.3256	
CO_2 (Vrabec)	M1	1.2088	0	0	133.22	2.9847	–	Vrabec *et al.*, 2001
	M2	– 1.2088	0	0	133.22	2.9847	–	
	C	0	0	0	–	–	3.1597	
	M3	– 0.5	0	0	–	–	– 1.57985	
	M4	0.5	0	0	–	–	– 1.57985	
CO_2 (EPM-W)	C	0	0	0	28.999	2.785	0.6645	Ungerer *et al.*, 2004
	O1	1.16	0	0	81.23	3.0356	– 0.33225	
	O2	– 1.16	0	0	81.23	3.0356	– 0.33225	

recently using Ewald summation [Vorholz *et al.*, 2002], e.g. vapour pressures are only slightly underestimated (Fig. 3.82a) and vaporization enthalpy above 300 K is slightly over-estimated (Fig. 3.82b), in agreement with Vorholz *et al.* However, the simulation results of Delhommelle are not reproduced with the polarizable model, and there is a significant discrepancy *vis-à-vis* the experimental measurements. As this application is not the focus of this work, no attempt has been made to probe the reason for this disagreement. The potential of Kristof and Liszi is thus used in subsequent simulations. This intermolecular potential is further tested against high pressure density data taken from Goodwin [1983], as shown in Figure 3.83. The maximum deviation on high pressure density is found to be 0.8%, which is quite satisfactory.

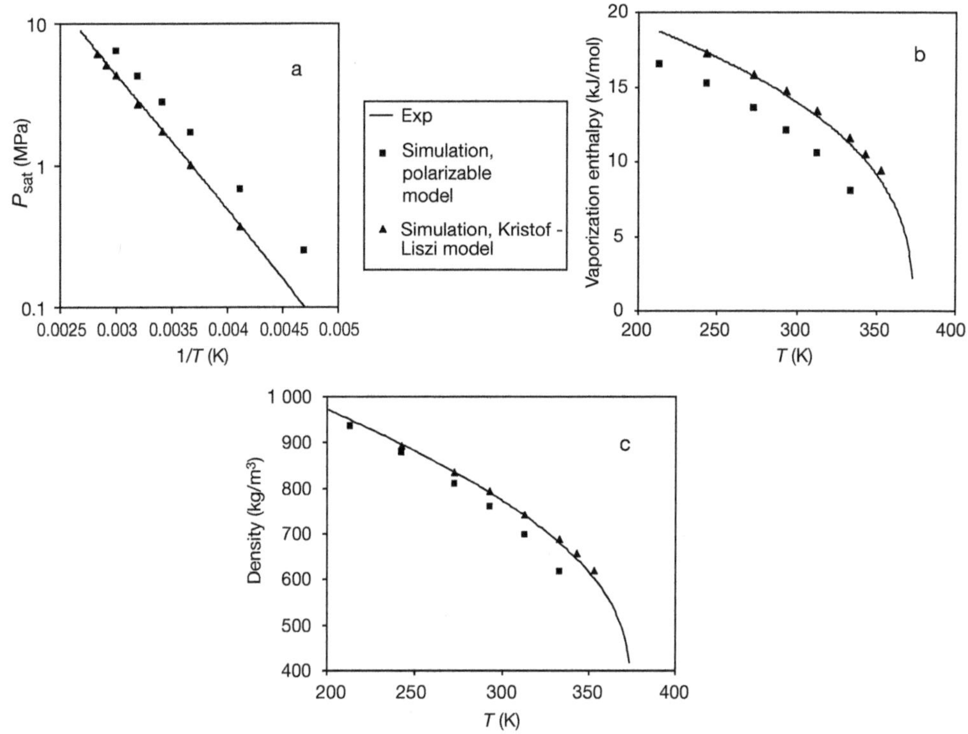

Figure 3.82 Gibbs ensemble simulation results for pure H_2S. Simulations based on the potential of Kristof and Liszi [1997] and the polarizable potential of Delhommelle [2000] are compared with experimental correlations from the Dortmund Data Bank (DDB). (a) Saturated vapor pressure (b) Vaporization enthalpy (c) Saturated liquid density.

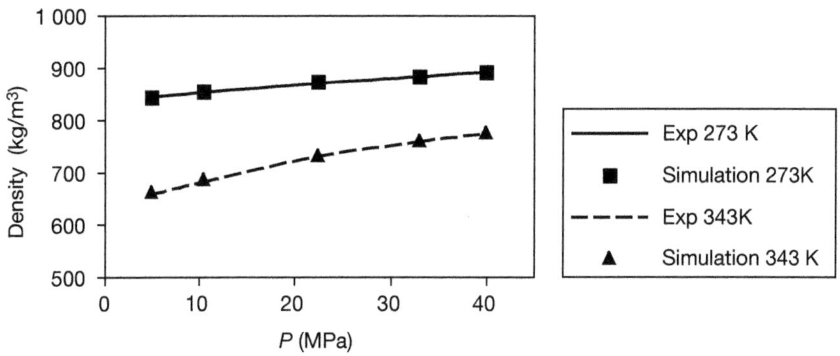

Figure 3.83 High pressure density of H_2S obtained by simulation in the *NPT* ensemble at 273 K and 343 K, compared with experimental results from Goodwin [1983].

C. Water

The literature presents several intermolecular potentials to model water. The simplest model is the SPC potential in which the three electrostatic charges are located on the atomic centres of the oxygen and hydrogen atoms. In order to get a more realistic representation of the properties of liquid water, it was proposed that the negative charge be shifted from the oxygen nucleus to an intermediate axial position, yielding the TIP4P potential [Jorgensen *et al.*, 1983]. It is known that the TIP4P model represents the coexistence curve of water more accurately than the SPC potential [Panagiotopoulos, 1994], particularly at temperatures lower than 500 K [de Pablo and Prausnitz, 1989]. The SPC model has been reparametrized to take better account of water's properties over a large temperature range [Boulougouris *et al.*, 1998]. Numerous alternative models have been proposed since: an exp-6 potential [Errington and Panagiotopoulos, 1998], polarisable models [Jedlovszky and Vallauri, 2001; Kiyohara *et al.*, 1998] or distributed electrostatic charges [Guillot and Guissani, 2001]. However, many would have been incompatible with the non-polarizable models selected for the other pure components. In addition, considering polarisability would not seem sufficient to afford significant improvement *vis-à-vis* coexistence properties. In the temperature range required for the study (ambient to 100°C), the TIP4P potential has been found to reproduce the properties of water reasonably well [Vorholz *et al.*, 2000] so this potential model was used in our tests.

The TIP4P potential has been tested with two ways of evaluating the electrostatic energy: Ewald summation and a cut-off at 10 Å. When Ewald summation is used, a preliminary test is conducted to select suitable range parameters, i.e. $\alpha = 2\pi/L$ Å^{-1} and $k_{max} = 7$, where L is the size of the simulation box. The coexistence properties are obtained by Gibbs ensemble simulations, using 300 molecules, with simulation lengths of at least 150 000 Monte Carlo cycles. These long simulations are indeed necessary to obtain reasonably consistent results, as the temperatures considered are lower than the normal boiling point. As shown in Table 3.16, the saturated vapour pressure, the vaporization enthalpy and the liquid density computed with a cut-off of 10 Å agree with Ewald summation almost within the estimated statistical uncertainties. Larger deviations are logically found with cut-off values of 7.5 and 5 Å, which overestimate density and cohesive energy. The trend of saturated vapour pressure of water between 273 K and 373 K appears similar with a 10 Å cut-off and with Ewald summation, indicating an overestimation of this property (Fig. 3.84). These results are in very good agreement with previous results obtained with the TIP4P potential [Vorholz *et al.*,

Table 3.16 Comparison of Ewald summation and a spherical cut-off of electrostatic interactions on the basis of the equilibrium properties of pure water with the TIP4P potential Ewald [Ungerer *et al.*, 2004].

Property	Ewald	Cut-off		
		5 Å	7.5 Å	10 Å
Saturation pressure (kPa)	64_5	58_6	65_3	61_3
Liquid density (kg/m^3)	957_4	1006_4	971_4	955_3
Vaporization enthalpy (kJ/mol)	40.9_2	43.3_2	41.5_1	40.9_1

2000; Lisal *et al.*, 2001]. As the TIP4P potential was optimized on the basis of simulations performed with a cut-off of electrostatic interactions at 7.5 Å [Jorgensen *et al.*, 1983] and did not consider vapour pressure, it is not surprising that such systematic differences are found. Similar comparisons have been made for vaporization enthalpy, indicating deviations between + 1.4 kJ/mol at 273 K and – 1.0 kJ/mol at 373 K, while the statistical uncertainty is approximately 0.2 kJ/mol. The deviations on liquid density were found to range between + 10 kg/m^3 at 273 K and – 24 kg/m^3 at 373 K. Both properties are well described, but their variation with temperature is exaggerated compared with experimental observations. Here again, these results are identical to previous simulation results [Vorholz *et al.*, 2000].

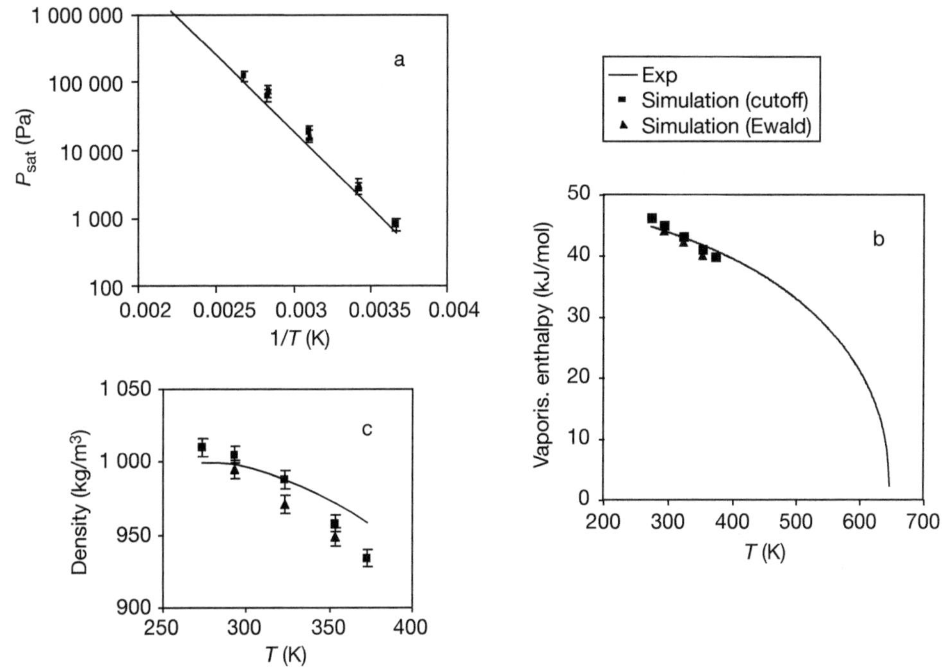

Figure 3.84 Gibbs ensemble simulation results for pure water. based on the TIP4P potential. Simulations performed with Ewald summation and with a 10 Å cut-off of the electrostatic interactions are compared with experimental correlations from the Dortmund Data Bank (DDB). (a) Saturated vapor pressure (b) Vaporization enthalpy (c) Saturated liquid density. Reprinted with permission from [Ungerer *et al.*, 2004], © Taylor & Francis Ltd (http://www.tandf.co.uk/journals).

Considering the acceptable results obtained by the cut-off method for the example of water (which is the most polar fluid studied), the method is used to evaluate the electrostatic energy in the following simulations. An advantage of this option is that it requires approximately one-quarter as much computer time as Ewald summation. For a given computing time, the use of a 10 Å cut-off allows much longer simulations and thereby provides much better convergence of thermodynamic averages. However, care has to be taken that the results may depend somewhat on the size of the system, as illustrated by a recent study of water with a slightly different intermolecular potential [Lisal *et al.*, 2002]. In order to limit the possible effect of numerical parameters, most subsequent simulations are performed with comparable box sizes, i.e. approximately 300 water molecules in the aqueous liquid box.

D. Carbon dioxide

Carbon dioxide is simulated first with the rigid version of the EPM2 intermolecular potential of Harris and Yung [Harris and Yung, 1995] and with the two-centres + quadrupole model of Vrabec *et al.* [Vrabec *et al.*, 2001]. While testing these models, the cut-off method is used to evaluate electrostatic energy in order to maintain consistency with the previously defined conditions for H_2S and water. The simulations are conducted with 400 molecules over 75 000 cycles. In the case of the Vrabec model, the point quadrupole is simulated as a set of three point charges placed on the molecular axis at +/– 0.5 Å from the carbon nucleus (Table 3.15) in order to match the quadrupole moment of the molecule, which is $4.3\ 10^{-40}\ C \cdot m^2$ [Graham *et al.*, 1998].

These intermolecular potentials have been tested at four different state points. The first two state points correspond to the vapour-liquid equilibrium at 216.6 K, i.e. at the triple point, and at 270 K, i.e. close to the critical temperature which is 304.1 K. These conditions allow checking of the representation of the coexistence properties with the Gibbs ensemble. The other state points are selected in the supercritical region ($T = 370$ K) at two pressure levels (7 and 80 MPa). Their role is to test the volumetric behavior of CO_2 in representative conditions of possible CO_2 reinjection, using monophasic *NPT* simulations.

The results of these tests are shown in Table 3.17. The Harris and Yung potential predicts densities and vaporization enthalpies within 3%. High densities, either in the liquid state at 216 K and 270 K or in the supercritical region at high pressure, are predicted with an accuracy better than 2%. This encouraging result confirms previous findings with the EPM2 model [Delhommelle *et al.*, 1999; Vorholz *et al.*, 2000; Lisal *et al.*, 2001]. However, saturated vapour pressures are overestimated by 14% which is consistent with the slight overestimation of vapour pressures observed in previous work based on the EPM2 model of Harris and Yung [Vorholz *et al.*, 2000; Lisal *et al.*, 2001].

The Vrabec potential predicts saturated vapour pressures with a better precision (10%), but the density discrepancies are disappointing (reaching 4% at high temperature and high pressure).

In order to improve these results, it is possible to further optimize the parameters of the Harris-Yung potential, using the optimization method outlined in Section 2.4.4. The least-squares error criterion is defined on the basis of the experimental measurements indicated in Table 3.17, including the two supercritical fluid densities as additional reference data. This error criterion is minimized to get the optimum Lennard-Jones parameters, using the EPM model of Harris and Yung as a starting point, but with the C–O distance set at the recommended value of 1.16 Å [Lide, 1992]. Two parameters may be optimized per atom, i.e. four altogether. It is found that optimizing three or four parameters results in important variations from the EPM parameters, indicating that the optimization is unstable. This may be attributed to an insufficient amount of independent information in the error criterion

Table 3.17 Comparison of pure CO_2 simulation results with the EPM2 rigid CO_2 model of Harris and Yung [1995], the potential of Vrabec *et al.* [2001] and the potential slightly reparametrized by Ungerer *et al.* [2004] at five state points.

			EPM2 [Haris and Yung, 1995]		Vrabec *et al.*, 2001		EPM-W [Ungerer *et al.*, 2004]	
		Exp.	MC simul.	Deviation (%)	MC simul.	Deviation (%)	MC simul.	Deviation (%)
216.6 K	P_{sat} (MPa)	0.518	0.592	14	0.569	9.8	0.485	6.4
	$\Delta_{vap}h$ (kJ/mol)	15.15	15.05	0.7	15.49	2.2	15.96	5.3
	ρ (kg/m^3)	1178	1165	1.2	1158	1.8	1184	0.4
270 K	P_{sat}(MPa)	3.203	3.65	14	3.38	5.5	3.04	5
	$\Delta_{vap}h$ (kJ/mol)	10.56	10.28	2.7	10.64	0.7	11.37	7.6
	ρ (kg/m^3)	946	936	1	919	3	962	1.6
370 K	ρ at $P = 7$ MPa	122	118	3	118	3.3	120	1.7
	ρ at $P = 40$ MPa	767	/	/	/	/	764	0.4
	ρ at $P = 80$ MPa	940	924	1.8	903	4	935	0.5

compared with the number of optimized parameters. When optimization is restricted to the two Lennard-Jones parameters of the oxygen, small variations are found. This is satisfactory for refining the potential, as attempted here. Related parameters are given in Table 3.15. As shown in Table 3.17, densities are predicted to an accuracy of 1.7% and saturated vapour pressures within 7% for the reference state points. However, vaporization enthalpies are less accurately predicted with errors as great as 7.6%. Nevertheless, the optimized EPM-W model would seem to be a better compromise, density prediction being important for our purpose.

3.6.2 Phase Behaviour of the H_2S–CH_4–H_2O System

A. Methane-H_2S System

The methane-H_2S system was investigated in the Gibbs ensemble at imposed pressure and at three temperatures, namely 223, 273 and 343 K. The total number of molecules is 600 to 800, and the length of the simulations is 10^7 to 3 10^7 iterations. As a general feature, convergence is more difficult for higher pressures in a given phase diagram, and density exchange between phases is observed towards the critical point. At 273 K and 343 K, the phase diagram is reasonably predicted (Fig. 3.85a and 3.85b).

At the lower temperature of 223 K, bubble pressures appear once more to be somewhat underestimated, but dew points are well represented (Fig. 3.85c). Higher pressures than those shown in Figure 3.84c were also investigated but convergence is very slow and results remain uncertain, thus leaving the prediction of the liquid-liquid splitting unsolved.

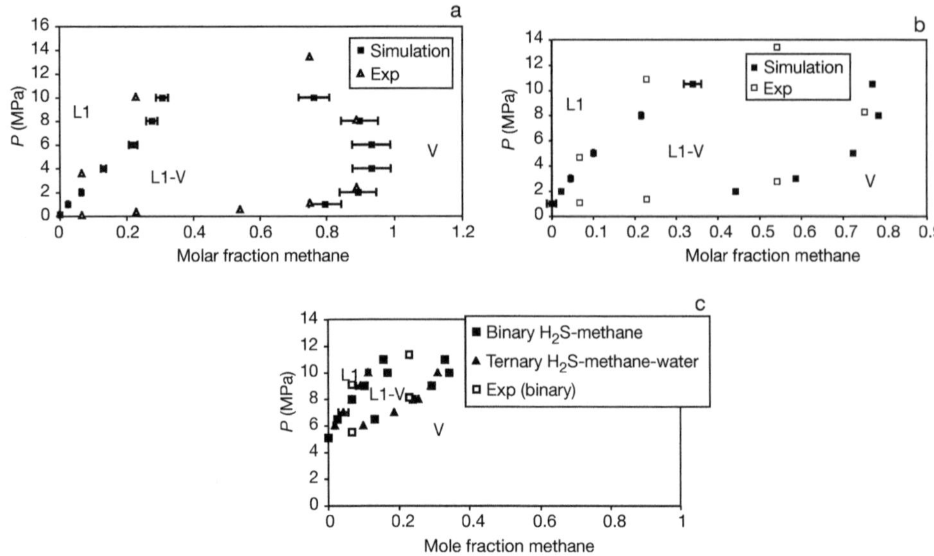

Figure 3.85 Phase equilibrium simulation of the methane-H_2S binary system, compared with interpolated data points of Kohn and Kurata [1959a and 1959b]. (a) 223 K (b) 273 K (c) 343 K. Reprinted with permission from [Ungerer *et al.*, 2004], © Taylor & Francis Ltd (http://www.tandf.co.uk/journals).

B. Methane-Water system

The phase behavior of the methane-water system is investigated in the two-phase Gibbs ensemble at imposed pressure, as three-phase equilibrium is not suspected. A total number of 500 water molecules and 200 methane molecules is used in these simulations. The purpose of having a higher number of water molecules is to avoid having too few molecules of methane in the aqueous phase, due the low solubility. The cut-off distance, which is common to Lennard-Jones and electrostatic interactions, is set to 10 Å in order to keep approximately the same way of evaluating energy as in the study of pure water. The simulated solubility of methane in the water phase, shown in Figure 3.86a, shows that the solubility of methane is underestimated by approximately 30% at high pressure. This may seem quite high but the deviation is in fact similar to the deviations found with other potentials when computing Henry's constants [Errington *et al.*, 1998; Economou, 2001]. We can notice that the apparent Henry's constant (P/x) is increasing with increasing pressure, as observed previously [Culberson, 1951; Dhima *et al.*, 1998]. Decreased water solubility in the methane-rich vapour phase with increasing pressure is well reproduced, as shown by Figure 3.86b.

C. Water-H_2S System

In the 273 K-373 K range, the H_2S-water system is known to exhibit liquid-vapour equilibrium at low pressure and liquid-liquid equilibrium at high pressure [Gillespie and Wilson, 1982],

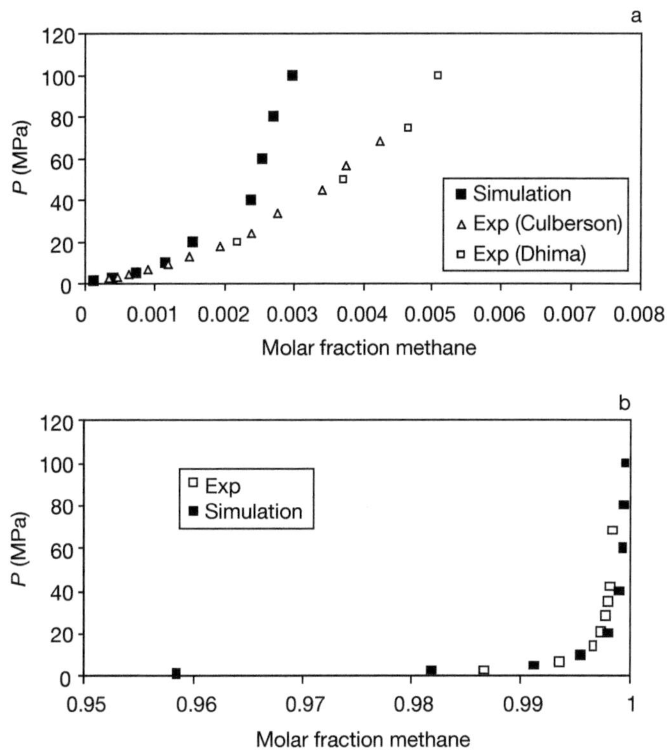

Figure 3.86 (P, x) phase diagram of the methane-water binary system at 343 K (a) solubility of methane in the aqueous phase (b) composition of the vapour phase. Reprinted with permission from [Ungerer *et al.*, 2004], © Taylor & Francis Ltd (http://www.tandf.co.uk/journals).

both domains being connected by a three-phase equilibrium line. This system was investigated at 343 K (Fig. 3.87) using either the Gibbs ensemble at imposed pressure for two-phase conditions or the Gibbs ensemble at imposed volume for three-phase calculations. The simulated system contained approximately 300 water molecules and 500 H_2S molecules. 40 000 to 70 000 MC cycles proved sufficient for satisfactory convergence of simulation averages. When analyzing typical configurations, it is observed that the water molecules are organized in clusters in the H_2S-rich liquid phase (Fig. 3.88) above the three-phase equilibrium pressure, while such a self-association of water is not found in the vapour phase. In order to test the sensitivity of this association to system size, the simulations were repeated with 900 H_2S molecules instead of 500, so that the H_2S-rich liquid contains a more representative number of clusters. Water association is then slightly increased and water solubility in liquid H_2S increases accordingly. This test shows that the association does not depend strongly on box size.

The whole predicted phase diagram is satisfactory, since the liquid-liquid and liquid vapour domains are found as expected. The three-phase equilibrium pressure determined from the three-phase Gibbs ensemble calculation at imposed volume is quite consistent with

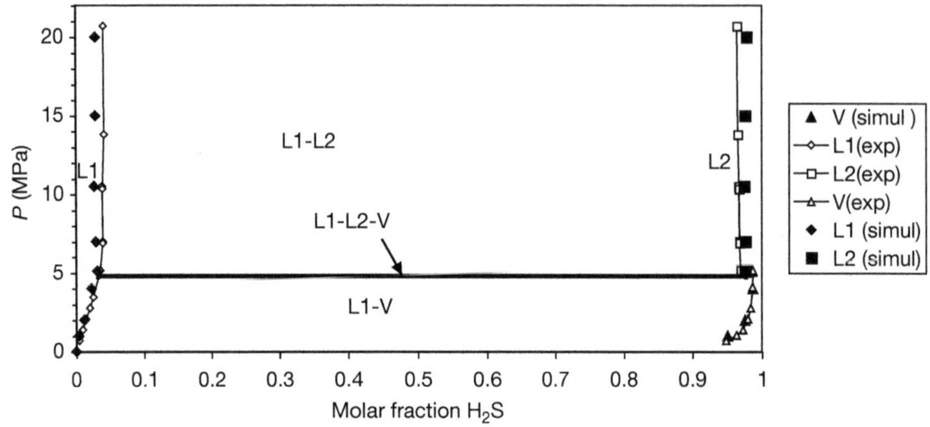

Figure 3.87 (P, x) phase diagram of the water-H_2S binary system at 343 K compared with experimental phase equilibrium data [Gillespie and Wilson, 1982]. Reprinted with permission from [Ungerer *et al.*, 2004], © Taylor & Francis Ltd (http://www.tandf.co.uk/journals).

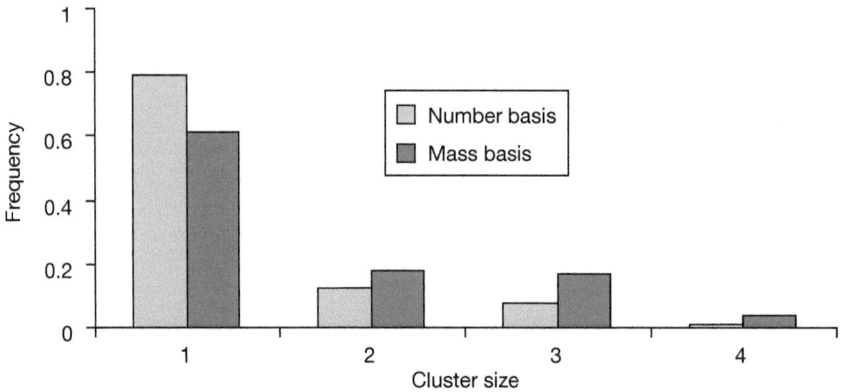

Figure 3.88 Size distribution of the water clusters in the H_2S-rich liquid phase in equilibrium with a vapor phase and an aqueous phase at 343 K. Reprinted with permission from [Ungerer *et al.*, 2004], © Taylor & Francis Ltd (http://www.tandf.co.uk/journals).

the observed behavior. In qualitative terms, simulation results predict the higher solubility of water in the H_2S-rich liquid phase (approximately 3%) than in the vapour phase (approximately 1.3%). These results are in general agreement with a previous investigation of the water-H_2S system using molecular simulation with the same potential by Vorholz *et al.* [2002] who also observed liquid-liquid phase splitting at high pressure but did not determine explicitly the L-L-V three-phase equilibrium line. The solubility of H_2S in the aqueous phase below the three-phase equilibrium pressure is in good agreement with Vorholz *et al.* although the smaller size of the H_2S-rich liquid box in their study explains why they found a smaller water content in the H_2S-rich liquid phase at high pressure.

As highly polar fluids are a challenge for classical thermodynamic models, it is interesting to test a classical equation of state on the water-H_2S system. The tests performed with the Soave-Redlich-Kwong equation of state [Soave, 1972] indicate that the water-H_2S phase diagram cannot be reproduced, even qualitatively, unless a special binary parameter is fitted. When fitting a binary parameter onto the H_2S-water equilibrium data, the solubility of water in H_2S can be reproduced correctly, but not the solubility of H_2S in water which is predicted to be lower than 1%. In order to represent the phase diagram with a common model for both phases, a more advanced equation of state using a mixing rule based on an excess Gibbs energy model [Huron and Vidal, 1979] must be used.

D. Water-methane-H_2S Ternary System

The behaviour of the ternary system has been investigated with three-phase and two-phase Gibbs ensemble calculations at imposed temperature and pressure. At imposed pressure and temperature, the compositions of the three coexisting phases do not depend on the global composition of the system. Selected results obtained at 343 K, 6 and 10 MPa, are shown in the ternary diagrams of Figure 3.89a and 3.89b. From these graphs, it can be seen that the composition of the vapour phase and of the non-aqueous liquid phases are little affected by the presence of water. Indeed, the water content is small in these phases. When the H_2S-rich liquid phase is diluted with methane, the solubility of water drops rapidly, e.g. the solubility of water in a mixture of 80% H_2S and 20% methane is only one-third to one-half of the solubility of water in pure H_2S (Fig. 3.90a).

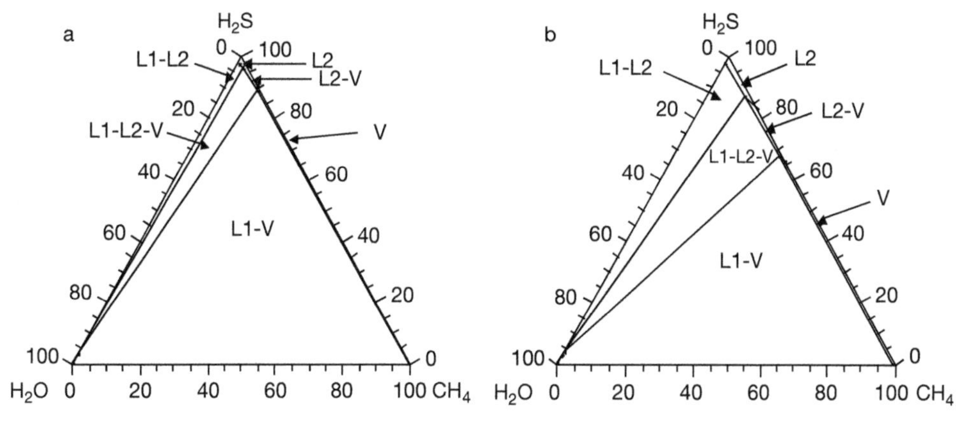

Figure 3.89 Ternary phase diagram of the water-H_2S-methane system at 343 K at pressures of 6 MPa (a) and 10 MPa (b). Part a is reprinted with permission from [Ungerer *et al.*, 2004] © Taylor & Francis Ltd (http://www.tandf.co.uk/journals).

3.6.3 Volumetric Properties

A. Methane-water-H_2S System

In the conditions of our study, the density of the non-aqueous phase at high pressures changes strongly with composition, as shown in Figure 3.90b. It may be noticed that the addition of 20% methane to H_2S decreases the density by more than 20%. This effect is likely to decrease the solvent capacity of the H_2S-rich phase with respect to water, providing therefore the explanation for the rapid decrease in water solubility when H_2S is diluted with methane.

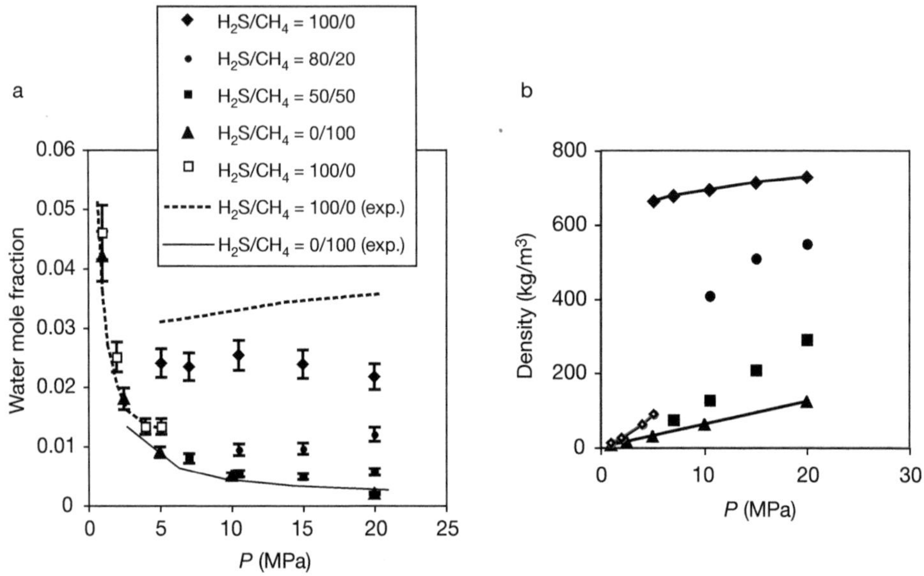

Figure 3.90 Properties of the non-aqueous phase in equilibrium with the water phase in the ternary system H_2S-methane-water, with various H_2S/methane ratios. (a) Solubility of water in the non-aqueous phases at 343 K (b) density. The full lines represent the experimental data on the water-methane system, and dotted lines in the H_2S-water system [Gillespie and Wilson, 1982]. Open symbols refer to the vapor phase in the H_2S-water binary system, below the three-phase equilibrium pressure of 5.1 MPa. Full symbols refer to dense phases at higher pressures. Part a is reprinted with permission from [Ungerer *et al.*, 2004] © Taylor & Francis Ltd (http://www.tandf.co.uk/journals).

B. H_2S–CO_2 System

In order to test the capacity of molecular simulation to provide accurate density predictions for H_2S–CO_2 mixtures, comparisons have been made with the experimental measurements provided by the GPA [Kellermann *et al.*, 1995]. These measurements have been performed for H_2S concentrations up to 50% and pressures up to 23 MPa. The tests have been conducted along three isotherms (273 K, 350 K and 400 K) within the pressure and composition range over which data were available for these isotherms.

The first tests were performed with the Lorentz-Berthelot combining rule. At 273 K, we recorded discrepancies of up to 10.1% in the liquid density of the equimolar H_2S–CO_2 mixture. At 350 K, a maximum discrepancy of 126% was found in the density of the equimolar mixture at 10 MPa. The prediction at 400 K showed deviations from experimental measurements of less than 1.2%. Maximum deviations at 273 K and 350 K are extremely disappointing, because they are much worse than the 2-3% deviations observed in the densities of pure components. However, it may be noticed that the greatest discrepancies at 350 K are limited to a specific part of the supercritical region where system density is extremely sensitive to small changes of pressure, temperature or composition.

In order to improve predictions without having to fit additional model parameters, alternative combining rules have been tested. It has been reported that alternative combining rules (Kong or Waldmann-Hagler) yield better predictions of liquid-vapour phase equilibria [Potoff *et al.*, 1999; Delhommelle and Millié, 2001]. This is particularly the case when the atoms (or united atoms) involved in the mixture differ significantly in diameter. Indeed, Kong and Waldmann-Hagler combining rules are based on a geometric mean value in the attraction term $\varepsilon\sigma^6$ rather than in the Lennard-Jones parameter ε. In the present case, the main interaction between H_2S and CO_2 originates from the oxygen-H_2S interaction involving groups of 3.0356 and 3.73 Å. This significant size difference is suggestive of a combining rule effect.

Indeed, replacement of the Lorentz-Berthelot by the Kong or Waldmann-Hagler combining rules produces very significant changes in density predictions. As illustrated by Figures 3.91a and 3.91b (which correspond to the most problematic conditions with the Lorentz-Berthelot rule), alternative combining rules provide a much more satisfactory prediction of mixture densities in this region. The alternative combining rules also give better predictions of liquid density at 273 K (Table 3.18), since the maximum discrepancies are only 1.3% (Kong) and 1.9% (Waldmann-Hagler). At 400 K, densities are predicted to within 1.2% by the Lorentz-Berthelot combining rules, while higher maximum discrepancies are found with Kong rules and with Waldmann-Hagler rules (Table 3.19). Nevertheless, the Kong combining rules appear as the best possible choice for the H_2S–CO_2 system when all state points are considered.

Table 3.18 Influence of the combining rule on the density of H_2S-CO_2 mixtures in the liquid state at 273 K, compared with experimental measurements [Kellermann *et al.*, 1995].

Molar fraction of H_2S	Pressure (MPa)	ρ_{exp} (kg/m^3)	Lorentz-Berthelot		Kong		Waldmann-Hagler	
			ρ_{simul}	%Δ	ρ_{simul}	%Δ	ρ_{simul}	%Δ
0.0955	15	951.9	1003.1	+ 5.4	964.5	+ 1.3	956.0	+ 0.4
0.2933	15	913.8	987.6	+ 8.1	926.0	+ 1.3	902.7	– 1.2
0.500	2.7	846	931.4	+ 10.1	853.8	+ 0.9	829.7	– 1.9

Table 3.19 Influence of the combining rule on the density of H_2S-CO_2 mixtures at 400 K. compared with experimental measurements [Kellermann *et al.*, 1995].

Molar fraction of H_2S	Pressure (MPa)	ρ_{exp} (kg/m^3)	Lorentz-Berthelot		Kong		Waldmann-Hagler	
			ρ_{simul}	%Δ	ρ_{simul}	%Δ	ρ_{simul}	%Δ
0.0955	5	71.9	71.3	− 0.8	71	− 1.5	70.7	− 1.6
	7.5	113.9	113	− 1.1	111.2	− 2.3	110.8	− 2.7
0.2933	5	69.6	69	− 0.7	68.2	− 1.9	67.5	− 3
	7.5	111.0	110	− 0.8	107.5	− 3.1	107	− 3.5
0.500	5	67.5	67	− 1.1	65.5	− 2.9	65.2	− 3.4
	7.5	107.9	109	+ 1.2	105	− 2.8	103.4	− 4.2

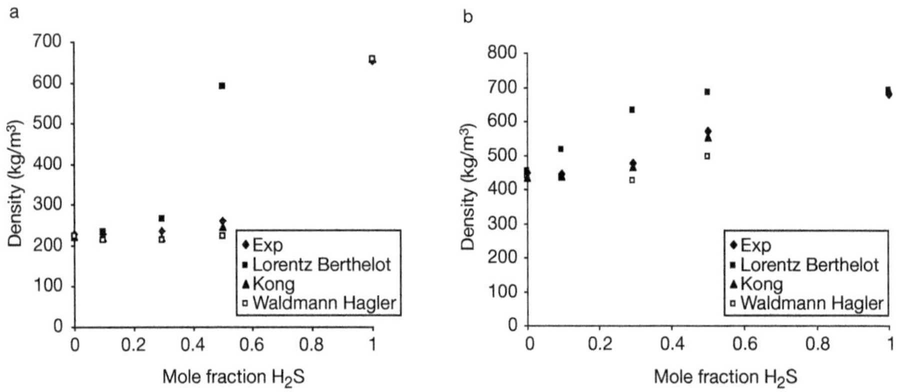

Figure 3.91 Prediction of the density of the H_2S–CO_2 system with various combining rules at 350 K as a function of composition (a) 10 MPa (b) 15 MPa. The experimental data are taken from Kellermann *et al.* [1995]. Reprinted with permission from [Ungerer *et al.*, 2004], © Taylor & Francis Ltd (http://www.tandf.co.uk/journals).

A potential problem when changing the combining rules is that they may change the properties of pure CO_2 since carbon-oxygen interactions play a role in pure component properties and the combining rule influences their parameters. This point has been tested by re-computing the coexistence curve and supercritical densities of CO_2 with the EPM-W potential parameters and the Kong combining rule. Significantly smaller maximum discrepancies in saturated liquid densities, vaporization enthalpies and saturation pressures are found with the Kong rules than with the Lorentz-Berthelot rules. Supercritical CO_2 densities are predicted with a maximum discrepancy of 3.3%, a slightly worse prediction than the 2.5% maximum which is observed with the Lorentz-Berthelot combining rule. As a whole, it may be concluded from this test that the properties of pure fluid CO_2 are still accurately represented when the Kong combining rule is used with the optimized EPM-W parameter set.

C. Prediction of Densities and Excess Volumes at High Pressure

As established above, Kong combining rules offer the best prediction of mixture densities in the H_2S–CO_2 system for a large range of test conditions, either in the vapour, liquid or supercritical state. Molecular simulation can now be used to provide reference predictions for the same system at high pressure, where no experimental data are currently available. These reference predictions aim at capturing the influence of pressure, temperature and composition. We investigate all the possible combinations involving four temperatures (273, 323, 373 and 423 K), four pressures (20, 40, 60 and 80 MPa) and five mole fractions (0, 25, 50, 75 and 100% H_2S). Thus, 80 simulations are performed.

The densities obtained at 323 K and 423 K are shown in Figure 3.92a and 3.92b for pressures ranging from 20 to 80 MPa. These graphs show that the influence of composition changes significantly in the high pressure range, e.g. an increase of H_2S molar fraction at 373 K or 423 K causes an increase in density at 20 MPa while it causes a decrease of density at 80 MPa.

The volumetric properties of the H_2S–CO_2 system have been also investigated for the same state points, using SBWR, a modified version of the Benedict-Webb-Rubin equation of state [Soave, 1995]. The densities predicted by this route are in fair agreement with molecular simulation. This equation of state seems to predict slightly lower densities for intermediate molar fractions (25 to 75%) however.

From the simulation results, it is also possible to compute excess volumes. The statistical uncertainty in the determination of excess volumes is estimated at 0.5 to 1 cm^3/mol (depending on conditions). As a general feature, excess volumes are positive in the investigated range of conditions and they change with H_2S molar fraction in an approximately parabolic fashion, as illustrated by Figure 3.93a. At 273 and 323 K, the pure components are both displaying liquid-like densities across the whole pressure range and the observed excess volumes are small (typically 2 cm^3/mol or less). Higher excess volumes are found at 373 K and 423 K, and especially at 20 MPa, where they exceed 10 cm^3/mol.

3.6.4 Prediction of Excess Enthalpies

It is possible to derive excess enthalpy in the same way as excess volume, using the method outlined in Section 2.4.8. The statistical uncertainty is approximately 100 J/mol. Thus excess enthalpies are readily obtained from the same set of simulations as excess volumes (pressures of 20 to 80 MPa, temperatures 273 to 423 K and compositions of 0, 25, 50, 75 and 100% H_2S). The excess enthalpies in the H_2S–CO_2 system are found to parallel the behavior of the excess volumes. Indeed, excess enthalpy also showed a parabolic dependence with composition. The amplitude of excess enthalpies is also found higher at the lower pressure (20 MPa) and the higher temperature (423 K) of the investigated range (Fig. 3.93b).

a

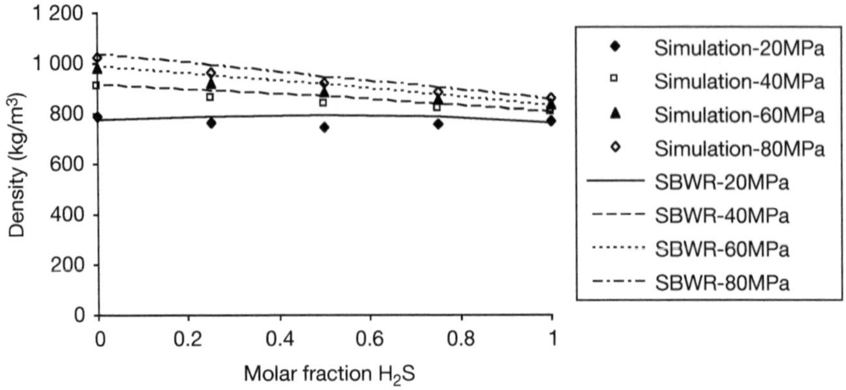

b

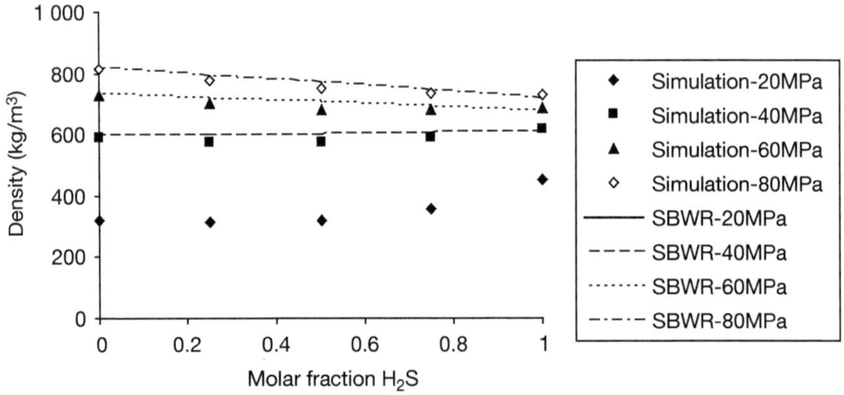

Figure 3.92 Density prediction of H_2S–CO_2 system with the Kong combining rule as a function of composition and pressure at 323 K (a) and 423 K (b). Reprinted with permission from [Ungerer *et al.*, 2004], © Taylor & Francis Ltd (http://www.tandf.co.uk/journals).

3.6.5 Prediction of Derivative Properties

In Table 3.20, three derivative properties (heat capacity, isobaric thermal expansion coefficient and isothermal compressivity) and the Joule-Thomson coefficient are given for pure CO_2 and pure H_2S in test conditions (350 K) where CO_2 is supercritical while H_2S is subcritical. It may be seen that the simulation predicts the first three properties with a maximum discrepancy of 13%. The predicted Joule-Thomson coefficient appears to be less close to the true value with discrepancies of up to 30% but this is due to the proximity of the Joule-Thomson inversion. In fact, the statistical uncertainty in the Joule-Thomson coefficient is estimated to be approximately 0.1 $K \cdot MPa^{-1}$ so the predicted values are in reasonable agreement with experimentally-based values. The inversion of the Joule-Thomson coefficient when changing from H_2S to CO_2 is well described.

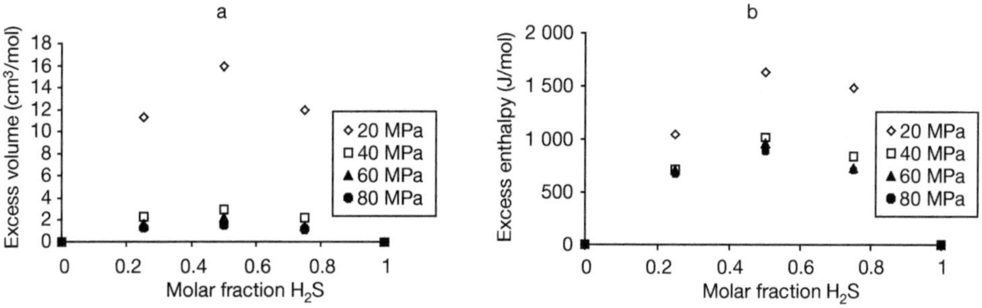

Figure 3.93 Excess properties in the H_2S–CO_2 system as a function of composition and pressure at 423 K, predicted with the Kong combining rule (a) Excess volume (b) Excess enthalpy. Part b is reprinted from [Ungerer *et al.*, 2004] © Taylor & Francis Ltd (http://www.tandf.co.uk/journals).

As a conclusion to this section, we can say that Monte Carlo simulation has provided a unified prediction of phase behaviour and the volumetric properties of systems comprising acid gases. It has also provided a detailed understanding of fluid structure, as illustrated by the clustering of water molecules in liquid H_2S which explains why water solubility is increased in this phase. Choosing the appropriate combining rule is however a delicate issue when there are large differences in size between different groups, and this point merits further investigation. Nevertheless, it is striking to observe that, if no experimental data had been available, all phase diagrams would have been predicted reasonably. This is highly encouraging when it comes to applying molecular simulation methods to early process design when poorly known systems are under consideration.

Table 3.20 Comparison of thermodynamic derivative properties of pure H_2S and pure CO_2 at 350 K and 40 MPa. Experimental data are from the IUPAC correlation [Angus *et al.*, 1973] in the case of CO_2 and from Goodwin [1983] in the case of H_2S.

	H_2S			CO_2		
	Experiment	**Simulation**	**Deviation**	**Experiment**	**Simulation**	**Deviation**
Heat capacity ($J \cdot mol^{-1} \cdot K^{-1}$)	61.62	66.7	+ 8.3 %	81.3	79.2	– 3.8 %
Isobaric thermal expansion coefficient ($K^{-1} \times 10^{-3}$)	2.43	2.34	– 3.7 %	4.03	3.66	– 9.2 %
Isothermal compressibility (kPa^{-1})	2.88	2.77	– 3.8 %	7.27	6.30	– 13.3 %
Joule-Thomson coefficient ($K \cdot MPa^{-1}$)	– 0.109	– 0.148	– 36 %	0.266	0.191	– 28 %

4

Adsorption

Of the variety of practical problems which can be addressed by Monte Carlo molecular simulation, predicting the adsorptive properties of microporous solids is amongst the most challenging and promising.

The industry has a need for predictive tools. In order to develop a new industrial separation process by adsorption, one needs to examine the properties of a large variety of microporous solids, differing in terms of crystalline structures, chemical composition, etc. Of course, there is no way to study them all, and no reason to synthesise them in the absence of some evidence that they might constitute suitable candidates for the intended industrial application.

Classical models describing the thermodynamics of adsorption have their limitations. The most commonly used, the Langmuir model, has serious limitations when applied to the prediction of multicomponent adsorption and selectivity. It is also not very effective when applied in a predictive manner. For instance, with xylene isomers adsorption on Faujasite type zeolites, pure compounds yield almost identical isotherms, but when the parameters for pure compounds are fitted to these data and used to predict the adsorption selectivity of a mixture, the generalised Langmuir model fails [Tournier *et al.* 2000]. This is especially true when the pure compounds' isotherms are very similar, as in the example given here.

Grand canonical Monte Carlo simulation appears as a serious alternative and an increasingly powerful tool to help select microporous solids as potential candidates for a given application. Enhanced computer performance makes the study of realistic systems possible and makes it possible to sample the phase space of such systems in a meaningful way.

Thus:

1) Different pieces of information can be ascertained from a single simulation, including adsorbed quantities and selectivities of adsorption, heats of adsorption, Henry constants, etc.
2) One can gain insight into the microscopic details of the processes involved, through meticulous investigation of the microscopic configurations of the adsorbates in the cavities of the solids. For instance, it is easy to separate the different components of total adsorption energy – the adsorbate-adsorbate and adsorbate-adsorbent parts – and thereby assess the relative importance of the different physical effects which influence adsorption selectivity.

3) Experimentally, the study of multicomponent systems can be – and usually is – more difficult than that of pure components. With molecular simulation, once a method of investigating the pure compounds has been developed (representation of the pure components' isotherms with appropriate optimisation of the model's parameters), tackling more complex systems is quite straightforward (e.g. to predict adsorption selectivities in a mixture with no further adjustment of the model's parameters).

Thus, one of the principal aims of molecular simulation of adsorption in zeolites is to predict one or a few candidate microporous solids for a given industrial application, prior to synthesis of said candidates with a view to experimentation and testing.

The adsorptive properties of zeolites are very sensitive to a number of factors. For zeolites of almost identical structure, slight variations of the Si/Al ratio, the nature of the counter-charge cations, or the presence of strongly adsorbed species such as water, can dramatically change the properties of the adsorbent, even as far as changing the selectivity from one component in a binary mixture to the other component.

This chapter presents a few results of practical interest obtained by Monte Carlo molecular simulation. This chapter begins with a description of how to use the technique – no attempt is made at generality, the purpose being merely to present one way of addressing the problem.

4.1 A PRACTICAL EXAMPLE OF GRAND CANONICAL MONTE CARLO SIMULATION OF ADSORPTION

As already stated in Section 2.4.2, simulation of adsorption in microporous solids is usually performed in the Grand Canonical ensemble. The atoms or pseudo-atoms taken into account for the relevant Monte Carlo movements are the ones constituting the adsorbed species, whereas the atoms of the solid itself are considered only through the interaction potential that they generate, and to which the adsorbed species are subject. This is possible because the atoms of the solid are considered as being fixed during the course of the simulation. Of course, when one wishes to study the position or mobility of counter-charge ions, then these species must be treated in a different way from the atoms constituting the skeleton of the microporous solid (see Section 4.2.1.).

4.1.1 Construction of the System: the Solid

Crystalline species such as zeolites have a periodic structure and are described in relevant space groups. Space groups are defined by a choice of an origin for the atom coordinates and a set of translations and symmetries which make it possible to construct the entire unit cell of the solid from the knowledge of the positions of a minimum group of atoms, called the asymmetric unit. Thus, the asymmetric unit is the smallest set of atoms for which the entire solid can be constructed by applying all the translations and symmetries of the space group, and this set is not reducible by any of these symmetries or translations. The multiplicity of a crystallographic site is defined as the number of distinct replicates which are obtained in one unit cell of the solid when all the translations and symmetries of the space group are applied to the site's coordinates.

The systems modelled always represent a discrete number of zeolite unit cells. Some of the most commonly used and studied zeolites are represented in Figures 4.1 and 4.2: silicalite (described in the orthorhombic space group Pnma, with cell dimensions a = 20.1 Å, b = 19.9 Å, c = 13.4 Å); type A zeolites (cell dimensions a, b, c = 12.3 or 24.6 Å depending on the space group employed to describe the crystal, respectively Pm-3m and Fm-3c).

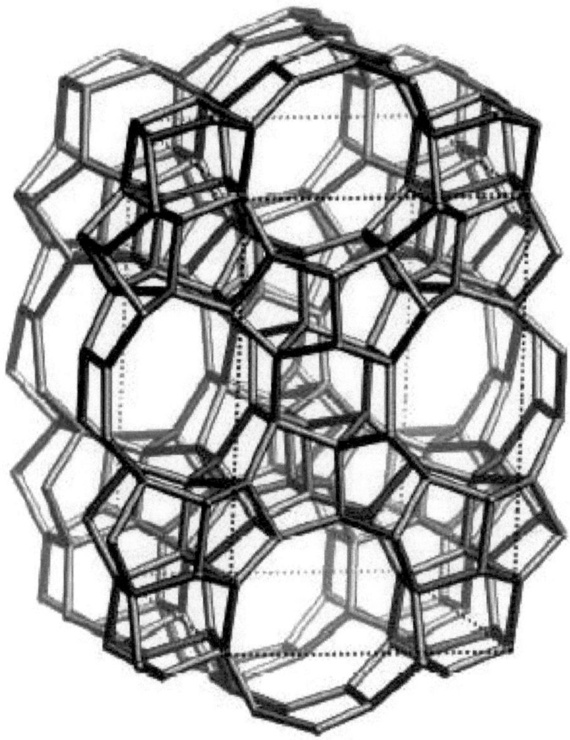

Figure 4.1 Unit cell of silicalite-1. Framework viewed along axis [010].

The Faujasite type zeolites are probably the most widely used zeolites in the industry. Their structure is based on the linking of sodalite cages through hexagonal prisms, to form a tridimensional structure of interlinked supercages of about 12.5 Å in diameter (Fig. 4.3). These supercages are interconnected through 12-ring windows of about 7.4 Å in diameter. Cations located around crystallographic sites counter-balance the negative charges induced by the presence of aluminium atoms in the framework.

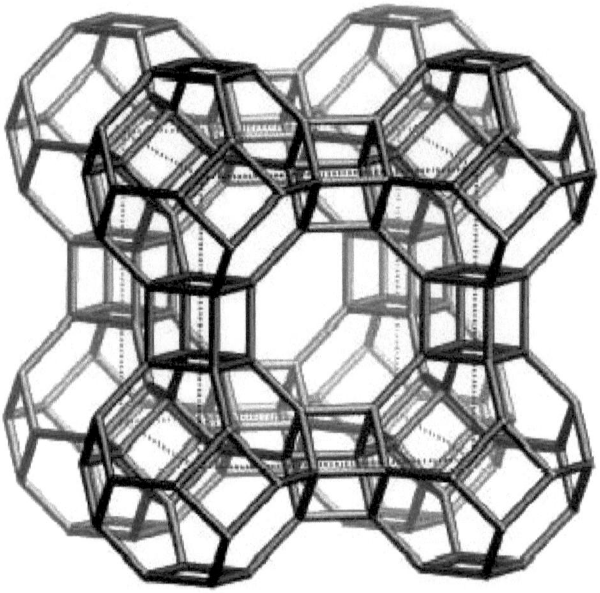

Figure 4.2 Unit cell of LTA zeolite. Framework viewed along axis [001].

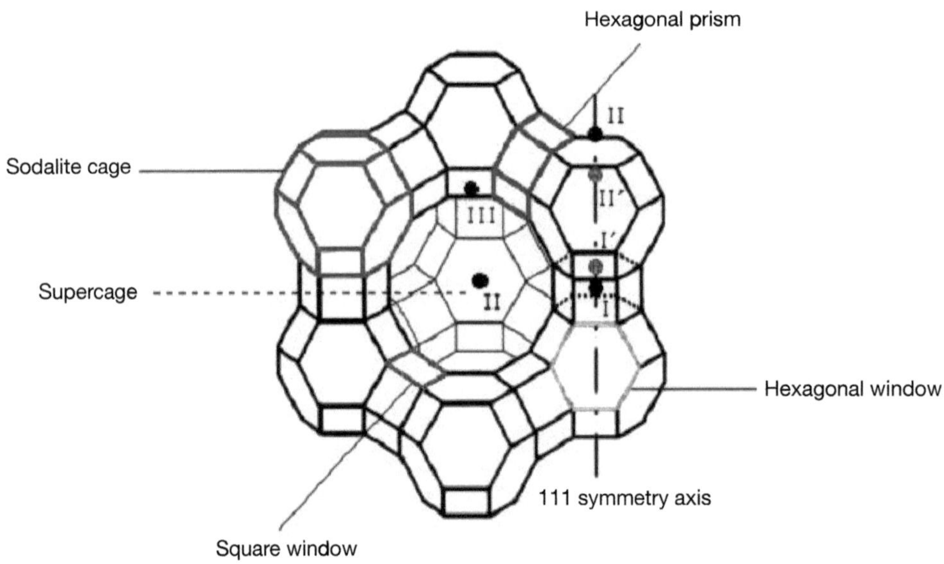

Figure 4.3 Partial view of one unit cell of faujasite. Some characteristic features are shown, as well as the most commonly occupied cationic sites.

Simulations are performed on systems periodically reproduced in the three space dimensions, to mimic a perfect and infinite crystal. Thus, the simulation box is always commensurate to the unit cell of the crystal: it corresponds to a whole number of unit cells.

The main considerations taken into account in the definition of the simulation box are:

- Choice of the cut-off distance for calculation of the interactions. The dispersion-repulsion term, often a Lennard-Jones expression, typically requires a cut-off of 10 to 15 Å, and long distance corrections to account for interactions with atoms located further than the cut-off. The use of periodic boundary conditions with the minimum image convention implies that the smallest dimension of the simulation box should be at least twice this distance. Thus, for silicalite, one needs to take at least two unit cells in the z direction for the system. In this case, the dimensions of the simulation box being a, b, c = 20.1, 19.9, 26.8 Å respectively, the maximum cut-off is then b/2 = 9.95 Å. For Faujasite, simulation of one single unit cell makes possible a cut-off greater than 12 Å, which is suitable. The same prevails for zeolite A described in the space group Fm-3c with a = b = c = 24.6 Å, whereas it is advised to choose at least $2 \times 2 \times 2 = 8$ unit cells if the zeolite is described in the space group Pm-3m with a = b = c = 12.3 Å, to ensure a sufficient cut-off.
- The electrostatic contribution – if taken into consideration – is generally computed using the Ewald summation technique (see Section 2.2.2). One generally chooses the same cut-off as above for the computation of the contribution in real space. Were the Ewald summation technique not used, then of course the cut-off would have to be much larger, since electrostatic interactions are relatively long-range. A much larger cut-off would greatly increase the number of interactions to be computed, which is impractical.
- The number of atoms in the system: with the choices above, the number of zeolite atoms in the simulation box can be around 500-1 000. But in the case of simulations carried on at low coverages of the solid (low pressure, calculation of Henry constants or heats of adsorption at low coverage), one might wish to extend the size of the system in order to obtain better statistics on the number of molecules adsorbed.

To construct the solid, one needs to know the positions of all the atoms it contains. Crystallographic data (mainly from X-ray diffraction analysis) are usually the source for this information: it precises the space group, the dimensions of the unit cell, the positions of atoms in the asymmetric unit and the occupancy factors. Application of the space group symmetries makes it possible to generate all the atoms of the unit cell, and by translation, of the simulation box. The occupancy factors may be less than unity for a number of the atoms of the asymmetric unit, which means that the effective number of atoms in the unit cell is less than the "multiplicity" of the site (i.e. the number of sites obtained when applying all the symmetries of the space group to the coordinates of the atom of the asymmetric unit). In this case, there is some liberty as to which crystallographic sites are populated and which are not (this can especially be the case for the charge-compensating cations in aluminosilicate zeolites: see for instance the study described in Section 4.4.7).

Instead of differentiating the Si and Al atoms of the zeolitic framework, an approximation – more accurately, a simplification involves considering a "mean" atom (called the "T-atom") with interaction parameters and a charge intermediate between those of Si and Al. This is often a necessary approximation, because, in the first place, there is not usually any evidence or experimental data about the location of the Si and Al atoms in the zeolitic skeleton, and, in the second place, any attempt to differentiate them in the model would lead to major artefacts because of the periodicity of the system.

Commercial tools are available to construct zeolite unit cells on the basis of the positions of the asymmetric unit atoms, the space group and unit cell parameters, all data which may be found in the literature. One such program was used in most of the cases presented in this chapter, namely Cerius2 software from Accelrys.

The solids addressed in this book are zeolites, more precisely orthorhombic aluminosilicates. Orthorhombic solids are crystals in which the unit cell has orthogonal axis – cubic crystals, in particular, are orthorhombic. No other adsorbent (e.g. carbons, gels or glasses) is presented. Many tools and procedures are specific to the type of problem addressed, of course, as discussed in a good review on the subject of molecular modelling of adsorption in zeolites [Fuchs and Cheetham, 2001].

4.1.2 Calculation of the Energy Grids

To save computing time during grand canonical Monte Carlo simulation, calculation at each step of the adsorbate-zeolite interaction energy is not performed by summing the contributions of every zeolite atom located in the cut-off radius, but rather by pre-computing these interactions into an "energy grid" (i.e. at a discrete number of points inside the solid) and then simply interpolating between those points in the course of simulation (see Section 2.4.2). To ensure sufficient precision, typical grid meshes should not exceed 0.2-0.3 Å. On a Faujasite zeolite with a cubic configuration and a cell parameter of about 25 Å, a grid mesh of 0.2 Å leads to an energy grid containing close to 2 million data points.

One energy grid per type of atom or pseudo-atom adsorbed is needed for GCMC calculation. If the dispersion-repulsion term is a Lennard-Jones expression, then it is possible to store only two grids per atom type in the zeolite, namely the sums of $1/r^6$ and $1/r^{12}$. If electrostatic energy is taken into account, then another grid corresponding to the interaction energy between one unit charge and the solid will be required.

Grids must be commensurate to one unit cell: there is no need to extend the computation of the grid over the whole simulation box if it contains more than one unit cell because of the system's periodicity.

4.1.3 Running a Grand Canonical Simulation

A. Determination of the Chemical Potentials

The chemical potentials of every individual compound adsorbed in the fluid phase in equilibrium with the adsorbed phase constitute input data for grand canonical (μVT) Monte Carlo simulation. They appear in the acceptance criteria for Monte Carlo creation and deletion moves, and thus have to be calculated prior to simulation in the *NPT* statistical ensemble, the

Gibbs ensemble or the canonical (*NVT*) ensemble at the desired (P, T, x) conditions (the same than those for the subsequent GCMC calculation and obtained with the same force field).

When the gas phase can be considered as ideal, the chemical potential can be calculated analytically from the fugacity $y_i P$ of the vapour phase by using the ideal gas relation:

$$\frac{\mu_i}{k} = T \ln \frac{y_i P}{P_o} \tag{4.1}$$

where P_o is the pressure of the reference state, corresponding to a fictitious ideal gas at one molecule per cube Angström. If the gas cannot be considered as ideal, then a simulation is performed in the *NPT* statistical ensemble, with Widom test insertions to calculate the chemical potentials of the adsorbates (see Section 2.3.5).

B. Types of Move in GCMC

Aside from the usual translation and rotation Monte Carlo moves, and moves specifically designed for one type of molecule (such as "pivot" or "reptation" moves for hydrocarbon chains) which one encounters in any Monte Carlo simulation in a statistical ensemble, two steps specific to the Grand Canonical ensemble must be implemented in the simulation program. These are tentative deletion of an adsorbed molecule and tentative creation of an adsorbate molecule inside the solid. Depending on the type of adsorbate molecule, different insertion and deletion bias can be implemented to accelerate convergence (see Section 2.3.3).

C. Computation of Adsorbed Quantities and Selectivities

This is quite immediate, the number of adsorbed molecules of each species being the extensive variables associated with the chemical potentials given as input, and thus the adsorbed quantites are direct results from the grand canonical Monte Carlo simulation. Adsorption selectivities are obtained according to their definition, i.e. for adsorption of a binary mixture:

$$\sigma_{AB} = \frac{x_A / x_B}{y_A / y_B} = \frac{\langle N_A \rangle / \langle N_B \rangle}{y_A / y_B} \tag{4.2}$$

where x_A and x_B are the molar fractions of compounds A and B in the adsorbed phase, $\langle N_A \rangle$ and $\langle N_B \rangle$ are the average number of adsorbed molecules per unit mass of the solid, y_A and y_B the molar fractions of A and B in the fluid phase, and σ_{AB} is the adsorbent selectivity for A relatively to B.

4.1.4 Computation of Heats of Adsorption

There are several ways in which the heat of adsorption can be defined. The isosteric heat of adsorption, which is the one we are interested in here, is the difference between the molar enthalpy of the adsorbate in the fluid phase and the partial molar enthalpy in the adsorbed phase:

$$-\Delta H^\circ = H_g - \overline{H}_s \tag{4.3}$$

The enthalpy is a function of the internal energy and the product PV. In the case of a vapour phase, PV is assumed to be equal to RT, and the molecular volume of the adsorbed phase is neglected. The isosteric heat of adsorption can thus be expressed as a function of the total molar energy in the vapour phase $\overline{E}_{\text{tot}}^{g}$ and in the adsorbed phase $\overline{E}_{\text{tot}}^{s}$:

$$-\Delta H^\circ = RT - \overline{E}_{\text{tot}}^{s} + \overline{E}_{\text{tot}}^{g} \tag{4.4}$$

In GCMC simulations it is equivalent to calculate $-\Delta H^\circ$ using the partial derivative of the average total energy with respect to the average number of adsorbed molecules, in both gas and adsorbed phases:

$$-\Delta H^\circ = RT - \frac{\partial < E_{\text{tot}}^{s} >}{\partial < N_s >} + \frac{\partial < E_{\text{tot}}^{g} >}{\partial < N_g >} \tag{4.5}$$

where $<N_s>$ and $<N_g>$ are the average number of molecules in the adsorbed and gas phases, respectively. The heat of adsorption Q_{st} can thus be calculated by the fluctuations method:

$$Q_{st} = RT - \frac{< E_{\text{tot}}^{s} N > - < E_{\text{tot}}^{s} >< N >}{< N_s^2 > - < N_s >^2} + \frac{< E_{\text{tot}}^{g} N > - < E_{\text{tot}}^{g} >< N >}{< N_g^2 > - < N_g >^2} \tag{4.6}$$

The heat of adsorption Q_{st} calculation requires thus a Monte Carlo simulation of the vapour phase in the Grand Canonical ensemble to determine the third part of Eq. (4.6). These simulations are usually performed so that the number of molecules fluctuates reasonably around five or six, which often implies simulation box dimensions of about 1 000 Å. But usually, the calculation of Q_{st} is achieved without this second simulation of the vapour phase.

Indeed, if the vapour phase is assumed to be ideal, the third part of the equation above is equivalent to the molar intramolecular energy of molecules in the vapour phase. Moreover, if internal degrees of freedom are considered not to be affected by adsorption, the molar intramolecular energies of the vapour phase and the adsorbed phase are equal. These approximations lead to the following expression for the heat of adsorption, proposed by Nicholson and Parsonage [1982]:

$$Q_{st} = RT - \frac{< E_{\text{ext}}^{s} N > - < E_{\text{ext}}^{s} >< N >}{< N^2 > - < N >^2} \tag{4.7}$$

where E_{ext}^{s} represents intermolecular interactions in the adsorbed phase. This expression, commonly used to determine the isosteric heat of adsorption, is based on severe approximations: ideal gas for the fluid phase, and the assumption that intramolecular energies in the vapour phase and the adsorbed phase are equal. This last hypothesis is not always fulfilled: for instance, it has been shown recently [Pascual and Boutin, 2004] that *n*-hexane adsorbing in the ferrierite zeolite experiences a *trans* to *cis* conformational change, which of course impacts on the intramolecular energy of the alkane molecule. Nevertheless, at low coverage, when sorbate/sorbate interactions are very weak, the isosteric heat Q_{st} can be determined through Eq. (4.7) by considering only the adsorbate/zeolite interactions in the intermolecular potential energy.

4.2 ADSORPTION OF C_8 AROMATICS AND WATER IN FAUJASITE TYPE ZEOLITES

One system of great interest is composed by the Faujasite zeolites, which are used on an industrial scale to separate the isomers of xylene, an aromatic hydrocarbon with 8 carbon atoms. The four isomers are paraxylene (1,4-dimethylbenzene), metaxylene (1,3-dimethylbenzene), orthoxylene (1,2-dimethylbenzene), and ethylbenzene. Paraxylene is widely used in the petrochemical industry; it is also very difficult to separate it from metaxylene (the two isomers are impossible to separate by any distillation process). The fine composition and structure of the adsorbent determine the separative properties. In order to develop and optimise a separation process, one looks for a solid with the highest possible selectivity towards one or the other of the components to be separated. The Si/Al ratio is an important parameter: for instance, NaY zeolite is selective towards metaxylene, though NaX zeolite exhibits no selectivity at all. The nature of the charge-compensating cation is another determining parameter: NaY is selective for metaxylene but KY exhibits a strong selectivity towards paraxylene, at similar temperatures [Lachet *et al.*, 1998].

4.2.1 Cation Distribution vs Si/Al Ratio

Si/Al ratio and the number of charge-compensating cations are related factors. When one wishes to perform simulations of adsorption on zeolite models, it is important to know where to place the extra-framework cations in the model. It has been emphasised above that, although they are in fact mobile species in the zeolite framework (sharing no chemical bonds with other atoms of the solid), they usually are considered to be fixed during the course of the adsorption process. This is of course an approximation, and the detailed effect of cation mobility is currently being investigated, by molecular simulation techniques among others [Buttefey, 2002; Beauvais *et al.*, 2004a and b].

Cations occupy crystallographic sites of the zeolite structures, and thus have been studied experimentally (mainly by neutron diffraction) as well as at the theoretical level, e.g. the *Electronegativity Equalising Method* [Uytterhoeven *et al.*, 1992]. Depending on the authors, several crystallographic sites have been described (Table 4.1). The most important are (Fig. 4.3): site I (located at the centre of the hexagonal prism linking two sodalite cages), I' (located on the 111 symmetry axis of the crystal, slightly into a sodalite cage), II (located in the supercage, on the 111 symmetry axis of the solid and facing a hexagonal window of a sodalite unit), III (located in the supercage, close to a dodecagonal window linking two supercages). Other sites, listed in Table 4.1, are referred to less often in crystallographic models and are not shown in Figure 4.3. The precise location of the various sites differ slightly from one source to the other and, for site III for instance, even the multiplicity may vary, depending notably on the space group used to describe the zeolite.

Table 4.1 Location and description of the most current cationic sites in Faujasites.

Cationic site	Multiplicity	Location	Description
I	16	Hexagonal prism	Centre of the hexagonal prism
I'	32	Sodalite cage	On the 111 symmetry axis, close to a hexagonal window leading into a prism
II'	32		On the 111 symmetry axis, close to a hexagonal window leading into the supercage
U	8		Centre of the sodalite cage
II	32	Supercage	On the 111 symmetry axis, in the centre of a hexagonal window of a sodalite cage. Often identified to site II*
II*	32		On the 111 symmetry axis, into the supercage. Sites II and II* are usually described simply as site II
III	48 or 96		Close to three square windows, the middle one belonging to a hexagonal prism linking two sodalite cages
IV	8		Centre of the supercage
V	16		Centre of the dodecagonal window linking two supercages

The number of cationic crystallographic sites greatly exceeds the number of cations necessary to neutralise the charge defects induced by the presence of aluminium atoms, even in the case of low Si/Al ratios (X Faujasites). For Y zeolites, sites III are free: only sites I, I' and II are occupied.

N.B. As a result of the use of periodic boundary conditions, the system actually considered is periodic, which is of course not the case in the real material. But it is the only way to proceed as molecular modelling cannot take into account systems which are too large (a few thousands of atoms at most): simulations should be performed on configurations of minimal energies, and the dependency of the adsorption results on the specific cation distributions should always be evaluated. This has been done below in the specific case of $Na_{56}Y$ (Section 4.2.1.2), as well as for the adsorptive separation of monobranched and dibranched alkane isomers (see Section 4.4.7).

Table 4.2 summarises a few crystallographic results concerning the cation distribution in various zeolites.

A. Si/Al Between 2 and 3 (Number of Na Cations Between 48 and 64)

Experiments (see Table 4.2) indicate that all sites II (multiplicity 32) are occupied. The remainder of the cations occupy sites I or I'. Cations tend to occupy the available crystallographic sites in such a manner as to minimise the electrostatic repulsive energy. Figure 4.3

Table 4.2 Experimental distribution of cations for various dehydrated Faujasites.

Zeolite	Number of Al atoms per unit cell	Unit cell parameter (Å)	Number of cations per unit cell: I (multiplicity 16)	I' (multiplicity 32)	II (multiplicity 32)	III (multiplicity 96)	Reference
NaX	88	25.10	2.1	29.1	31.0	29.8	Olson *et al.* [1981]
NaY	57	24.71	7.8	20.2	31.2	–	Eulenberger *et al.* [1967]
KY	57	24.80	12.0	14.6	31.0	–	Eulenberger *et al.* [1967]
BaX	84.5	25.18	12.0	2.8	20.4	–	Mellot *et al.* [1994]
BaY	52	24.85	7.0	4.7	11.4	–	Pluth [1971]

shows that the minimum distance with a site II is larger for a site I than a site I'. Thus, one would suppose that, the 32 sites II being occupied, the first 16 extra cations occupy the sites I. The following cations (from the 49th and above) will tend to occupy a site I'. In a Faujasite unit cell, there are 16 "trimers" I'-I-I' (the two I' sites belonging to two sodalite cages linked through a hexagonal prism, in the centre of which lies the site I). When two cations must occupy one trimer, the electrostatic repulsion will enforce a configuration where both sites I' are occupied, the site I remaining empty: the distance of 2.18 Å between site I and a neighbouring site I' is indeed too short for a simultaneous occupancy of both sites.

Thus, a theoretical model of occupancy of cationic sites I, I' and II for Faujasites with a Si/Al ratio between 2 and 3 would be that summarised in Table 4.3, in which the results are compared to available experimental data [Marra *et al.*, 1997; van Dun and Mortier, 1988; Jirak *et al.*, 1980; Eulenberger *et al.*, 1967; Fitch *et al.*, 1986]. The agreement with experimental data is quite good, the only notable discrepancies being for the zeolites with 52 and 54 cations. For these, sites I' have a higher rate of occupancy than predicted, sites I a lower, and the authors [Marra *et al.*, 1997; van Dun and Mortier, 1988] describe another kind of site, with a filling of less than 4 molecules per unit cell.

B. *Specific Study of a Single Case: $Na_{56}Y$*

The above scheme for the cation distribution suggests an occupancy of 8, 16 and 32 cations, respectively, for sites I, I' and II, with half the I'-I-I' trimers filled with two cations in position I', and the other half with only one cation in position I. If one draws the configurational energy histogram of all 12 870 possible combinations, it is possible to determine which configurations are most stable (Fig. 4.4), and then to choose the lowest energy configuration, or a few of the lowest energy configurations, to perform the simulations. It might not be sufficient to choose only the one configuration corresponding to the global minimum energy, since this configuration would be reproduced periodically in the simulation model, whereas the real material is likely to exhibit a statistical distribution of the lowest energy configurations.

Table 4.3 Cationic crystallographic sites occupancies for Na Faujasites with Si/Al ratios comprised between 2 and 3. Results of the simulations and experimental data.

Si/Al ratio	Number of cations	Simulation			Experiment				Reference
		I	I'	II	I	I'	II	Other sites	
3	48	16	0	32					
2.92	49	15	2	32					
2.84	50	14	4	32					
2.76	51	13	6	32					
2.69	52	12	8	32	9.3	13.7	25.3	3.5	Marra *et al.* [1997]
2.62	53	11	10	32					
2.56	54	10	12	32	7.04	13.76	29.44	3.76	Van Dun & Mortier [1988]
2.49	55	9	14	32	4.0	17.6	32.0	1.4	Jirak *et al.* [1980]
2.43	56	8	16	32					
2.37	57	7	18	32	8.0	18.88	30.08	0.04	Eulenberger *et al.* [1967]
2.31	58	6	20	32	7.1	18.6	32.2	–	Fitch *et al.* [1986]
2.25	59	5	22	32					
2.20	60	4	24	32					
2.15	61	3	26	32					
2.10	62	2	28	32					
2.05	63	1	30	32					
2.00	64	0	32	32					

The question is now to check if molecular simulation can lead to cationic configurations in agreement both with the above theoretical model, and with available experimental data. The magnitude of the electrostatic energies suggests that cation moves are highly cooperative. Classical ways of exploring configurational space fail to converge in a reasonable amount of computer time.

The Simulated Annealing technique involves simulating the system in the canonical ensemble at high temperature (1 500 K), then progressively lowering the temperature. It leads to a great number of configurations, not necessarily those that are lowest in energy but only local minima. A site-to-site hopping technique leads to the configuration predicted by the model (8 cations is site I, 16 in site I' and 32 in site II), but this sort of algorithm is restricted to moves of cations on pre-determined sites. The choice method to tackle this particular problem is that of Parallel Tempering, which is described in Section 2.3.9. It is the only method which makes it possible to explore efficiently such rough energy landscapes.

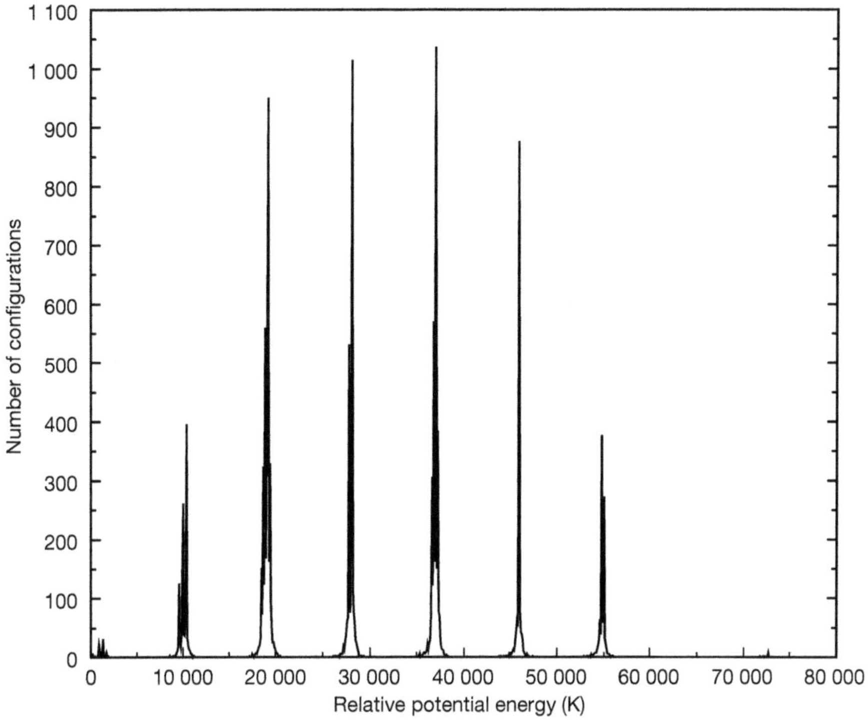

Figure 4.4 Potential energy histogram of the 12 870 possible cationic configurations of zeolite $Na_{56}Y$. Zero energy corresponds to the most stable configuration. Reprinted with permission from [Buttefey *et al.*, 2001], © 2001 American Chemical Society.

In this particular study, the Parallel Tempering method will also make it possible to make predictions on the cationic configuration for Si/Al ratios above 3, which cannot be predicted with the model described above, and for X Faujasites [Buttefey, 2002; Beauvais *et al.*, 2004b].

Monte Carlo calculations in the *NVT* statistical ensemble have thus been performed on sodium Faujasites containing 1 to 96 cations per unit cell (i.e. Si/Al ratios ranging from 1 to 191). Sodium cations are considered as "hosts", the remainder of the zeolitic structure being the "guest". Cation/framework interactions are computed using a dispersion-repulsion term between sodium and oxygen species only, and a coulombic term with partial charges on all atoms, silicium and aluminium atoms being modeled as "T-atoms" with a partial charge adjusted so as to ensure the electrostatic neutrality of the system.

Eight temperatures were used for the parallel tempering simulations, ranging from 300 K to above 2 300 K [Beauvais *et al.*, 2004b]. As necessary with the parallel tempering simulation technique, the energy histograms corresponding to these temperatures partially overlap each other (Fig. 4.5a), to ensure that that 1° exchange of configurations between two neighbouring temperatures can be accepted with a sufficient frequency (Fig. 4.5b), and 2° any specific configuration effectively "visits" all the specified temperatures, so that exploration and sampling of the phase space is correct.

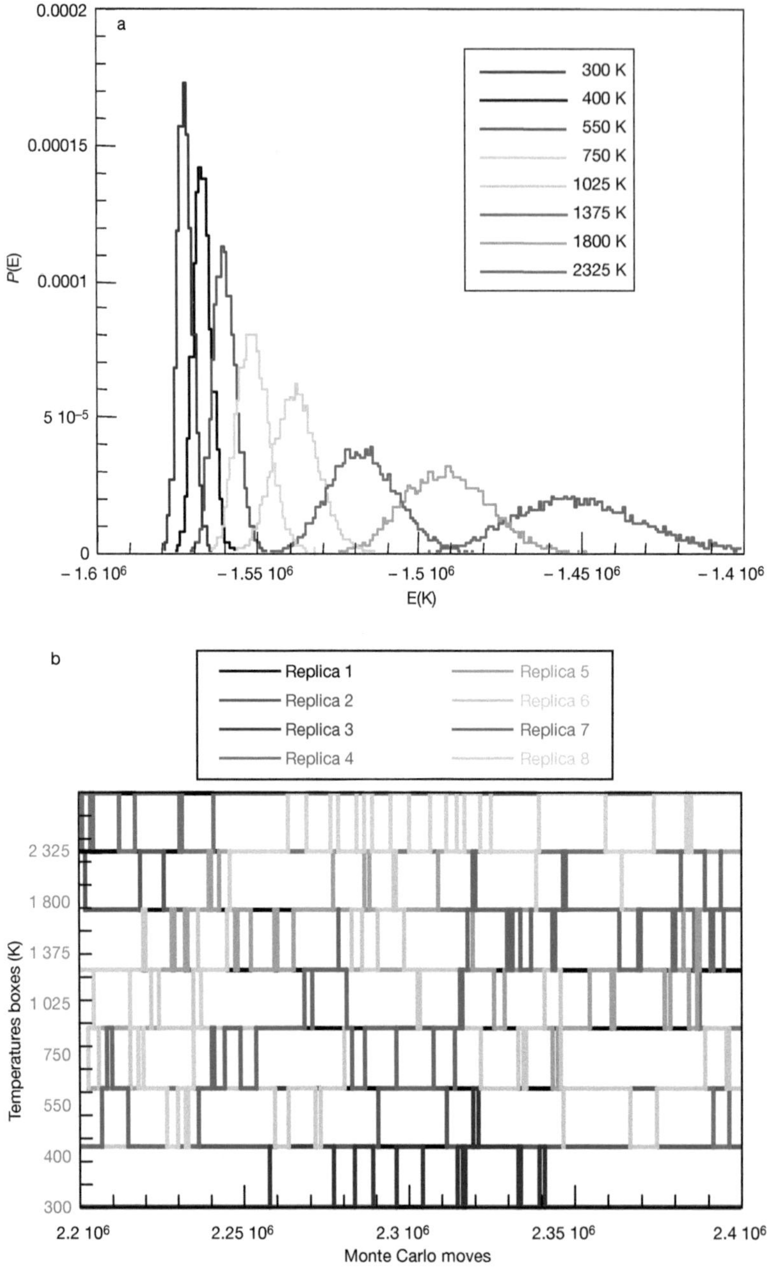

Figure 4.5 a) Energy histograms of the 8 simulation boxes used in the parallel tempering technique. x-axis: energy, y-axis: occurrence probability. Reprinted with permission from [Beauvais *et al.*, 2004b], © 2004 American Chemical Society. b) Exchange events between the different temperature boxes.

The results are presented in Figure 4.6, and compared with available crystallographic data found in the literature [Fitch *et al.*, 1986; Jirak *et al.*, 1980; van Dun *et al.*, 1988; Eulenberger *et al.*, 1967; Olson, 1995; Vitale *et al.*, 1997; Gallezot *et al.*, 1975; Porcher *et al.*, 1999; Lievens *et al.*, 1992a, 1992b, 1992c, 1992d; Hseu, 1972;, Smolin *et al.*, 1983]. The agreement is particularly remarkable wherever experimental data exist, and it should be stressed that molecular simulation is able to make predictions on systems which have not been studied experimentally.

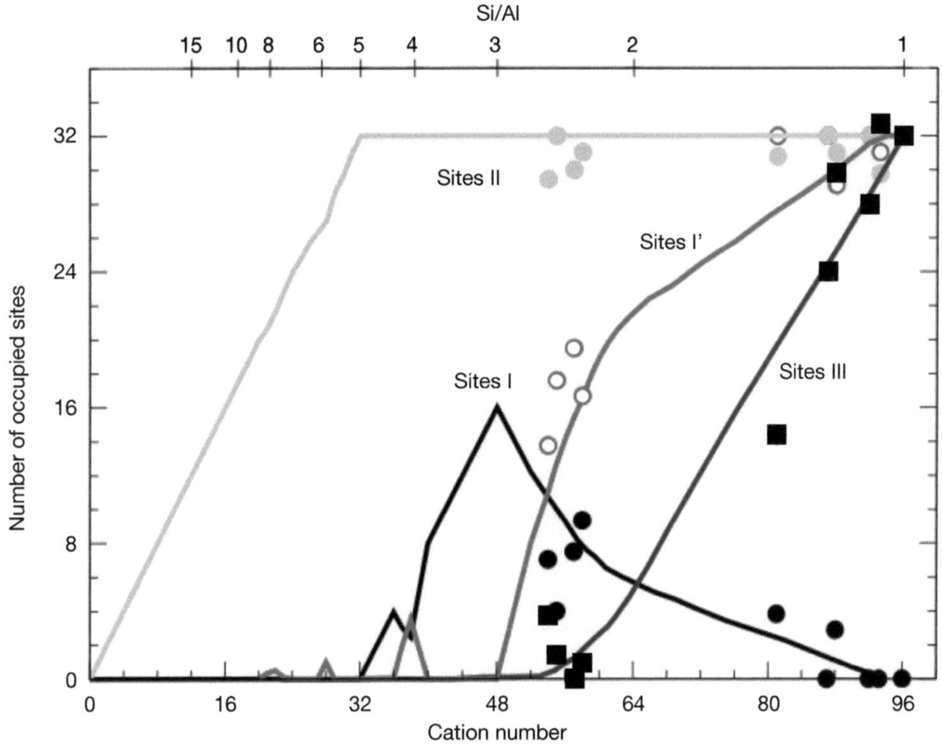

Figure 4.6 Cationic crystallographic sites filling as a function of the number of cations per unit cell for Na-faujasites. Points: experimental data. Lines: parallel tempering Monte Carlo simulation results.

The above results can also be compared to the theoretical predictions given by a statistical thermodynamic method developed by Mortier and co-workers [Uytterhoeven *et al.*, 1992] to study the cation distributions in zeolites. This model is based on the chemical potentials equalisation method, which derives from the density functional theory and is described in Mortier *et al.* [1986].

The degree of agreement between the two predictive methods – that of Mortier and molecular simulation – is remarkable. This strengthens the cation-framework interaction potential developed, as well as the parallel tempering method. The most noticeable discrepancies concern the preferential occupation of sites I' or II for Faujasites with very few cations, and this is most probably due to the fact that Si and Al atoms are not differentiated in the molecular model, and also because of the interaction potentials used, that of Mortier leading to a lower energy for a cation in site II than one in site I', whereas the potential used in molecular simulation leads to the opposite result.

C. Case of NaX Zeolites

When the number of cations reaches 64 (Si/Al = 2), the above-mentioned Mortier model as well as simulation both predict total occupation of sites I' and II (32 cations in either type of sites). For lower Si/Al ratios, i. e. higher numbers of cations, the cations in excess of 64 occupy the sites III [Vitale *et al.,* 1989], near a square window of a sodalite cage belonging to a 12-membered ring or "window" connecting two supercages. Several authors describe multiple sites III, which some call sites III' [Olson, 1995; Zhu and Seff, 1999], located more or less in the vicinity, somewhat translated towards a square window of one hexagonal prism connecting two sodalite cages, but also near a 12-membered ring between two supercages. Moreover, the positions found for these sites III or III' differ from one author to another. These discrepancies may result from differences in the zeolites studied, mainly with respect to the degree of hydration or dehydration of the materials employed, temperature, or even the quality of the diffraction spectra.

Localisation of Na cations for Faujasites with Si/Al ratios lesser than 2 has been studied by Auerbach and co-workers by molecular simulation [Jaramillo and Auerbach, 1999]. In this work, the Si and Al atoms are differentiated. We will present simulation results obtained with our interaction potential model, which considers only mean T-atoms, on NaX zeolites with 96 and 86 cations (Si/Al = 1.00 and 1.23 respectively).

Starting configurations were characterised, for $Na_{96}X$, by an occupation of 32 cations in site I', 32 in site II and 32 in site III. For $Na_{86}X$, only 22 cations were placed in sites III. Several configurations were chosen, differing only by the distribution of the cations in site III.

Figures 4.7 and 4.8 represent the distance histograms obtained between the cations and the sites I' (4.7a and 4.8a), II (4.7b and 4.8b), and III (4.7c and 4.8c). One can see that:

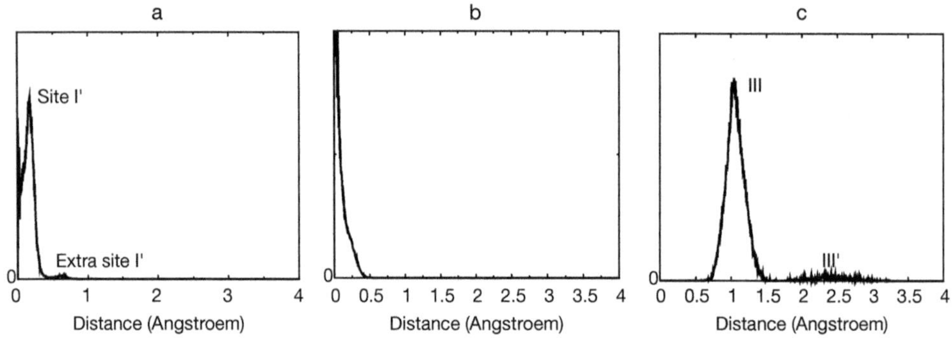

Figure 4.7 Distance histograms between the cations and the crystallographic sites in $Na_{86}X$. a) site I'; b) site II; c) site III. Arbitrary units on the vertical axis.

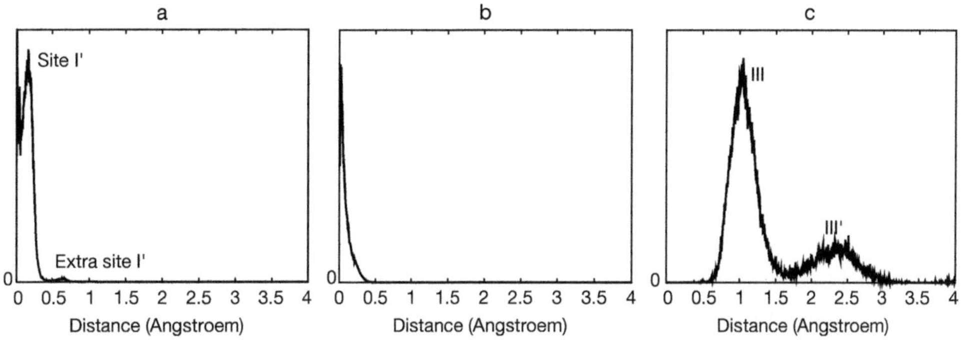

Figure 4.8 Distance histograms between the cations and the crystallographic sites in $Na_{96}X$. a) site I'; b) site II; c) site III. Arbitrary units on the vertical axis.

- sites II remain fully occupied, for both models;
- two different sites I' are found, though the one at ~ 0.7 Å from the crystallographic position described by Vitale [Vitale *et al.* 1997], located on the 111 axis but towards the hexagonal window separating the sodalite cage in which the cation lies from the neigbouring hexagonal prism, is very scarcely populated;
- two different sites III are found. The less populated is a III' site, translated from the site III crystallographic position of Vitale [Vitale *et al.* 1997] towards a 12-membered ring. The relative occupancy of sites III' over sites III is higher for $Na_{86}X$ than for $Na_{96}X$. Thus the occupancy of sites III' effectively depends on the number of cations, i. e. the Si/Al ratio. There is no evidence of multiple sites III'.

The results are summarised in Table 4.4, where they are compared to simulation results obtained by Jaramillo [Jaramillo and Auerbach 1999], and experimental data of Vitale [Vitale *et al.* 1997] and Olson [Olson 1995].

Table 4.4 Distribution of cations in Faujasites $Na_{86}X$ and $Na_{96}X$, compared to previous simulations and experimental data.

Site	$Na_{86}X$			$Na_{96}X$		
	Buttefey 2002 (simulation)	**Jaramillo 1999 (simulation)**	**Olson 1995 (experiment)**	**Buttefey 2002 (simulation)**	**Jaramillo 1999 (simulation)**	**Vitale 1997 (experiment)**
I	0	1	2.9	0	3	0
I'	32	31	29.1	32	29	32
II	32	32	31.0	32	32	32
III	18	7	0	31	0	0
total III	22	22	29.8	32	32	32
III'a		1	10.6		5	0
III'b	4	14	8.6	1	27	32
III'c		0	10.6		0	0
Total	86	86	92.8	96	96	96

To sum up this study, it can be stressed that the simple model proposed, which is based on the principle that cationic distribution in the crystallographic sites is governed essentially by the electrostatic repulsion between cations, is validated by molecular simulation results, using the parallel tempering technique, for sodium Faujasites with a Si/Al ratio between 2 and 3. Moreover, the close agreement between the simulation results and experimental data validates the computational method, and makes it possible to draw reliable predictions for zeolites with Si/Al ratios outside the 2-3 range.

For Na Faujasites with more than 64 cations, the simple model adopted, which does not differentiate silicium and aluminium atoms but uses "mean T-atoms", makes it possible to predict the existence of the III' type site, but not to reproduce its occupancy ratio, nor the existence of several III' sites.

Mellot-Draznieks *et al.* [2001] carried this study further on the specific case of $Na_{86}X$, using an alternative model for the zeolite: Si and Al atoms are explicitely distinguished, and there is a dispersion-repulsion term between these framework atoms and the sodium cations. The partial charges on the atomic species are: O = – 1.2, Na = 1, Si = 2.4 and Al = 1.2, on the basis of previous simulation work on NaX faujasites [Vitale *et al.*, 1999]. Ten randomly chosen Al atoms were substituted with Si atoms, starting from a framework model with a strict alternation of AlO_4 and SiO_4 tetrahedra. The effect of the Si/Al distribution was assessed by trying several different configurations and was found to be negligible. 32 cations were placed in sites I' and 32 in sites II, while random translational moves of the remaining 22 Na cations were adjusted in order to obtain an acceptance rate of 50%.

The simulated pair distribution functions (distance histograms) between the 22 "mobile" sodium cations and the O, Si and Al atoms of the framework reveal that all the cations are located in sites III', in excellent agreement with recently published crystal structures of NaX [Vitale *et al.*, 1999; Zhu and Seff, 1999]. Moreover, two types of sites III' are found, in agreement with the experimental results of Zhu and Seff [1999].

This study highlights two points: a correct and detailed description of the sodium cation's distribution cannot be obtained if Si and Al atoms are not explicitely distinguished in the model, or by considering only the energy of the crystallographic cationic sites: indeed, location of the cations in the zeolite structure must be regarded as a cooperative process, resulting both from short-range interactions with the framework and cation-cation repulsive interactions.

4.2.2 Adsorption Selectivity of Metaxylene vs Orthoxylene

As an application of the study presented above, the adsorption selectivity of NaY Faujasites towards metaxylene and orthoxylene for various Si/Al ratios has been simulated by molecular modeling.

Separation of C_8 aromatic isomers is performed on an industrial scale by adsorption on Faujasite type zeolites. The valuable compound is paraxylene, which is used in a great variety of petrochemical processes. Orthoxylene has also, on a lesser scale, industrial applications, whereas the other two isomers, metaxylene and ethylbenzene, are usually recycled through an isomerisation unit. To increase the efficiency of the global separation process, it may be interesting to include a device which separates orthoxylene from a meta- and orthoxylene flux in which paraxylene is almost entirely absent. Very scarce experimental data on these systems are available in the literature (U.S. patents by Neuzil [1971 and 1982] and by Kulprathipanja [1995 and 1999]). They show an increase of selectivity at saturation towards metaxylene when the Si/Al ratio increases, from NaX zeolites (Si/Al ratio 1.25) to NaY (Si/Al ratios up to 2.7).

Models of NaY zeolites obtained with the above-described parallel tempering method have thus been investigated in order to check if the experimental results can be reproduced, and to explore the behaviour of solids with higher Si/Al ratios. The interaction potential is the TrAZ force field, developed and optimised to reproduce the heat of adsorption of metaxylene in NaY zeolite at low coverage, with no further adjusment or optimisation of the zeolite-adsorbate interaction parameters it includes [Lachet *et al.*, 1998 and 1999].

The TrAZ (**Tr**ansferable for **A**dsorption in **Z**eolites) force field contains a Buckingham dispersion-repulsion term, which means that the repulsion is described by an exponential rather than a (r^{-12}) term. It also takes into account the Coulombic energy, the charges on the zeolite atoms being taken from Uytterhoeven and Mortier [1992] and those on the xylene molecules from the models of Jorgensen and Nguyen [1993], and the polarisation energy, calculated by using the first term of the multipolar development of the perturbation theory (see Eq. 2.29 in Section 2.2.2). The Ewald summation method is used to calculate the electrostatic energy (see Section 2.4.1).

Figure 4.9 shows the adsorption isotherms of pure meta- and orthoxylene in $Na_{48}Y$ Faujasite at 423 K. There are no experimental data on orthoxylene to which the simulation results can be compared. But the TrAZ force field has been shown to be transferable to the paraxylene molecule [Lachet *et al.*, 1998 and 1999], and can thus reasonably be thought to be transferable to the ortho isomer as well: here simulation is predictive.

At low coverage, $Na_{48}Y$ adsorbs more orthoxylene than metaxylene; the opposite happens at saturation. A careful examination of the position of the adsorbed molecules shows that whatever the isomer (para, meta or ortho), they always face a site II cation. The potential energy distribution of the three isomers adsorbed in this position is shown in Figure 4.10. It appears that the adsorption of orthoxylene is slightly favoured as compared to metaxylene, and sensibly more than paraxylene. This can be explained by the dipole moments of the three isomers (respectively 0 D for para, 0.36 D for meta, and 0.62 D for orthoxylene), which justify the behaviour observed at low coverage, as well as the preferential adsorption of metaxylene relatively to paraxylene on NaY zeolites. At saturation, geometrical hindrance of the adsorbed molecules has to be taken into consideration to explain the favourable adsorption of metaxylene.

Co-adsorption of meta and orthoxylene, at 423 K and for equimolar mixtures, has been investigated in NaY zeolites with Si/Al ratios of 2.37 and 3 (57 and 48 Na cations respectively) in order to compare the simulation results with the few experimental data available. The same simulations have then been carried out on NaY Faujasites with higher Si/Al ratios (5.4 and 11.8, corresponding to 30 and 15 cations respectively). Finally, an experiment was carried out at 423 K on a NaY Faujasite with a Si/Al ratio of about 6, using the breakthrough curve technique with isooctane as solvent, associated with gas chromatography. With these experimental conditions, saturation of the adsorbent is ensured.

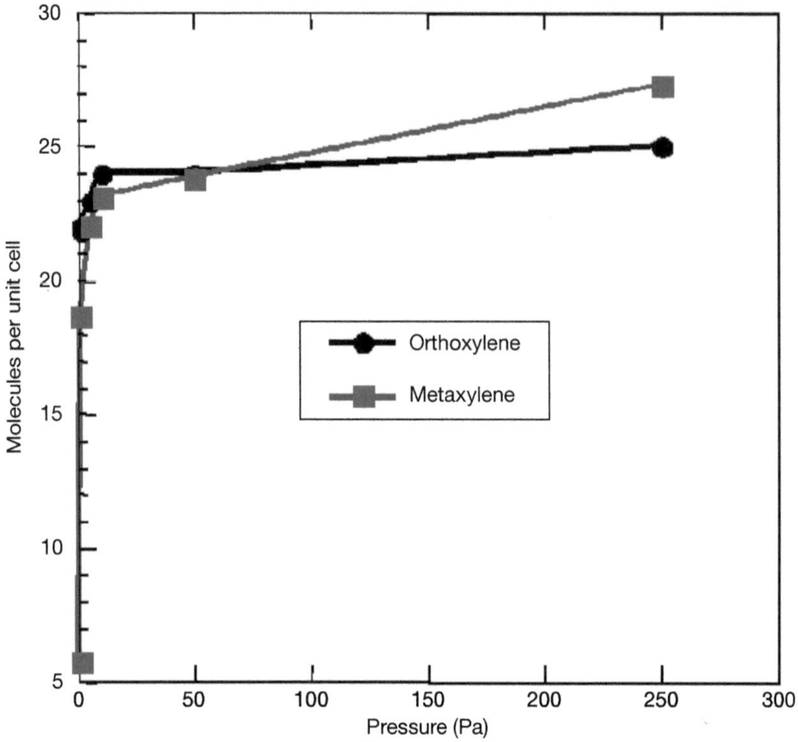

Figure 4.9 Simulated adsorption isotherms of orthoxylene and metaxylene in faujasite $Na_{48}Y$ at 423 K. Lines are linear interpolations between simulation results.

The results are presented in Figure 4.11. The simulation reproduces quite well the experimental selectivity for NaY Faujasites, which exhibit a molar selectivity close to 2 in favour of metaxylene. The model predicts an inversion of selectivity for higher Si/Al ratios, orthoxylene becoming preferentially adsorbed. The experiment carried out with a NaY zeolite of Si/Al ratio around 6 confirms this prediction in a pleasing fashion.

The study has been extended to lower filling ratios. Figures 4.12a and 4.12b show the evolution of the selectivities when approaching saturation, on zeolites $Na_{48}Y$ and $Na_{15}Y$ respectively. This has been performed by forbidding creation or destruction moves, and allowing only "exchanges" of one isomer into another, apart from the usual translation and rotation Monte Carlo moves. Whereas selectivity remains in favour of orthoxylene for $Na_{15}Y$ at all coverages, for $Na_{48}Y$ it is in favour of orthoxylene at lower coverages but shifts towards preferential metaxylene adsorption as saturation is approached.

This is consistent with the adsorption energies of the two isomers – somewhat higher for orthoxylene than for metaxylene. As the loading increases, favourable metaxylene-metaxylene interactions in $Na_{48}Y$ cause the selectivity to change in favour of this isomer. In $Na_{15}Y$, where only 7 cations are present in the 8 supercages of the unit cell, the arrangement of

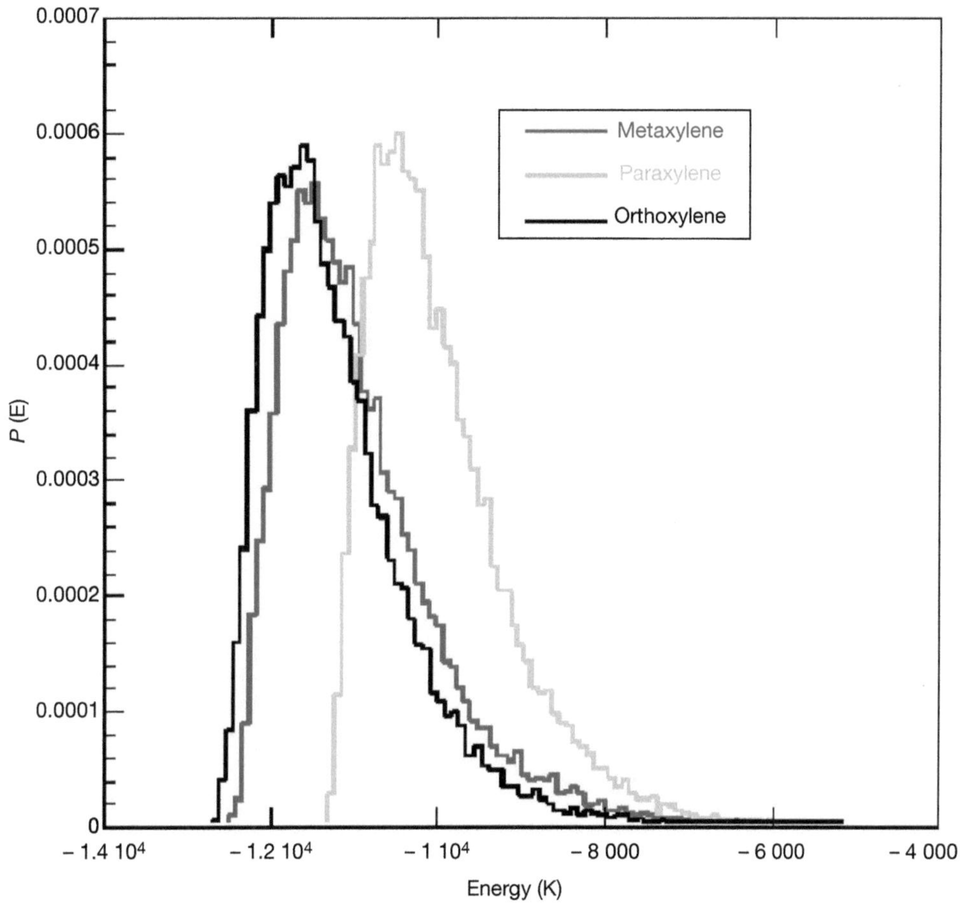

Figure 4.10 Potential energy histograms of one xylene molecule facing a site II cation in $Na_{48}Y$ at 423 K.

adsorbed molecules is entirely different, favouring this time orthoxylene-orthoxylene interactions, and thus the selectivity is in favour of this isomer.

4.2.3 Adsorption of Water in Faujasites

As seen in Section 3.6, molecular simulation of the water molecule is never easy. The main problem lies in the choice of the intermolecular interaction potential. For instance, description of the liquid-vapor equilibrium is better achieved with a simple expression omitting polarisation than with a more "realistic" force field including an expression for polarisation energy. Simulation of water in cationic zeolites is another complex problem.

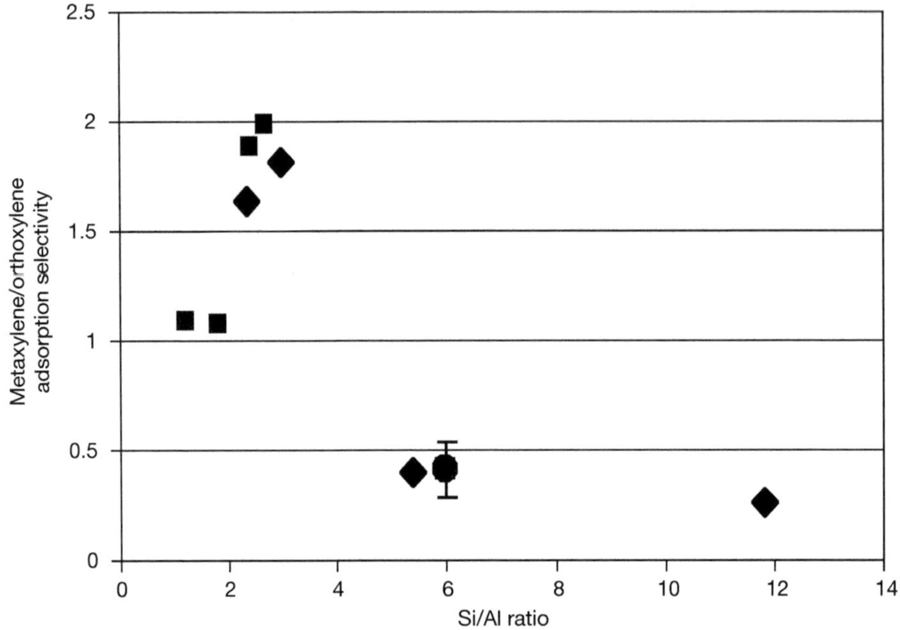

Figure 4.11 Calculated and experimental metaxylene/orthoxylene adsorption selectivities at 423 K in Na-faujasites with various Si/Al ratios. Diamonds: Monte Carlo simulation results. Squares: experimental results. Circle: experimental result obtained by the breakthrough curve technique on Na-faujasite with Si/Al = 6.

In the present work, the TIP4P potential has been used to represent the water molecule. It consists in a Lennard-Jones centre and 4 charges optimised to represent the liquid-vapor equilibrium of water [Jorgensen *et al.*, 1983], and the parameters can be found in Table 3.15. The Faujasite zeolites investigated are described by the TrAZ potential [Lachet *et al.*, 1998], in which the polarisation part has been omitted in order to avoid counting the polarisation energy between water molecules and the zeolitic framework twice (once explicitly with the TrAZ force field, and once implicitly through the charge distribution of the TIP4P potential). The exact Ewald summation method has been employed to calculate electrostatic energy.

To start with, adsorption of water in the $Na_{48}Y$ Faujasite has been investigated. In this zeolite, the parallel tempering molecular simulation method predicts a total occupancy of cationic sites I and II. As this result has been obtained on rigorously dry material, the effect of the cationic distribution has been investigated. As a matter of fact, there is some experimental evidence of cation migration upon adsorption, be it of water molecules or, for instance, aromatic hydrocarbons [Pichon *et al.*, 1999 and 2000]. In our models however, the counter-charge ions are described as sitting in fixed positions. To assess the effect of cation mobility, one would need to use a parallel tempering method in a specific statistical ensemble (grand canonical for the water molecules, the number of which is allowed to fluctuate, and canonical for the cations, to ensure that the system remains electrically neutral).

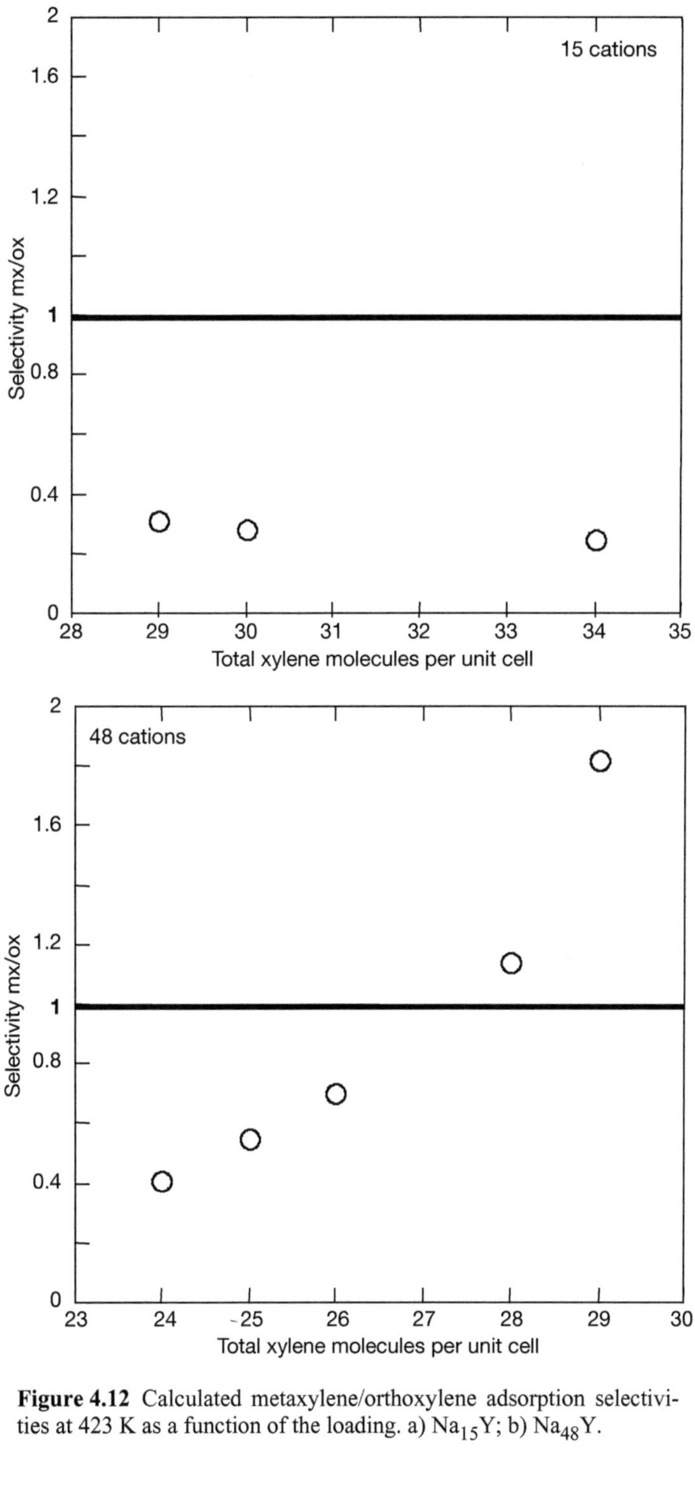

Figure 4.12 Calculated metaxylene/orthoxylene adsorption selectivities at 423 K as a function of the loading. a) $Na_{15}Y$; b) $Na_{48}Y$.

Apart from the cationic distribution predicted for the dry zeolite (16 cations in site I and 32 in site II), a distribution in which all the sites II are occupied, but the 16 others are distributed over the 32 available sites I', has been built. To achieve some heterogeneity of the system, the chosen distribution is such that over the 8 sodalite cages of the unit cell, 4 have 2 cations, 2 have 4 cations and 2 have none. The adsorption isotherms of water at 423 K have been calculated and are reported in Figure 4.13 together with the experimental isotherm [Moïse, 1999]. As can be seen, the influence of the cation distribution on the number of adsorbed water molecules is quite significant.

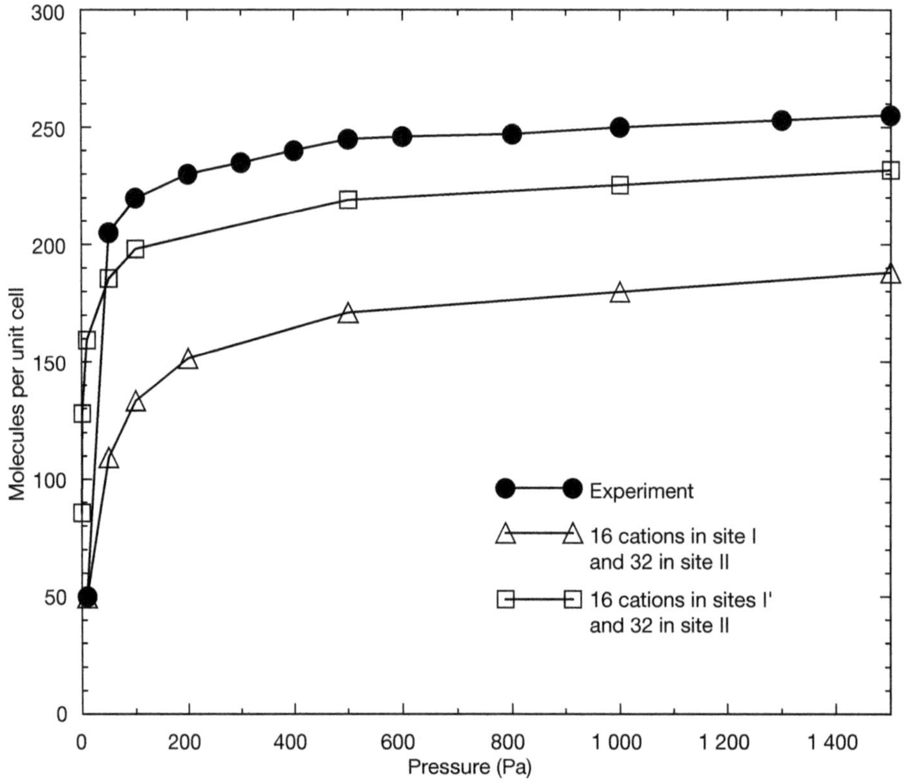

Figure 4.13 Simulated and experimental adsorption isotherms of water in $Na_{48}Y$ at 423 K.

Careful examination of the location of the adsorbed water molecules has been performed on these two models, by drawing pair distribution functions or "distance histograms" throughout a canonical Monte Carlo simulation. 39 water molecules have been placed in an unit cell of $Na_{48}Y$ (second cationic distribution, with occupation of the sites I'). The hydrogen-oxygen pair distribution function between two distinct adsorbed water molecules is shown in Figure 4.14a. Two peaks, at 1.9 and 3.4 Å, are characteristic of hydrogen bonds between adsorbed water molecules.

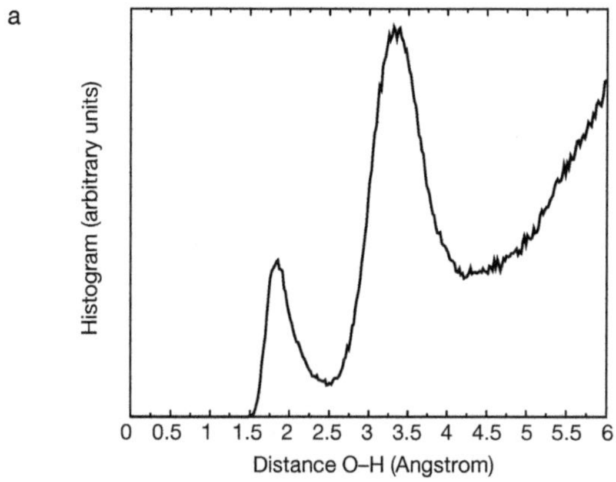

Figure 4.14a Intermolecular Hydrogen-Oxygen distance histogram for water adsorbed in $Na_{48}Y$ at 423 K.

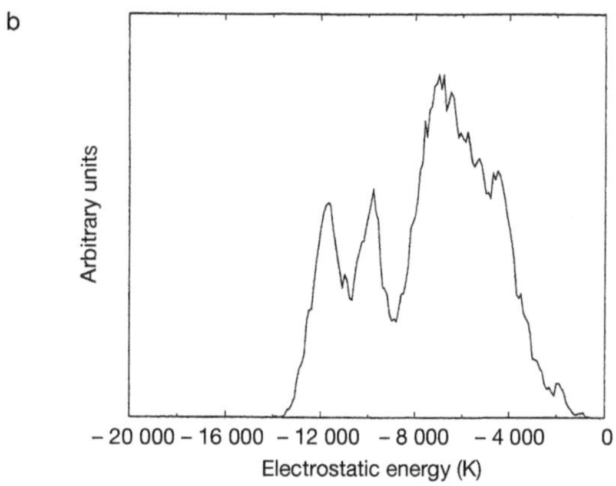

Figure 4.14b Electrostatic energy histogram for 39 water molecules in one elementary cell of $Na_{48}Y$ at 423 K.

To evaluate precisely the geometry of the adsorbed water molecules, a careful examination of their electrostatic energy histograms in their adsorption sites has been undertaken, on a simulation performed at very low temperature (10 K) to avoid site-to-site migration (Figure 4.14b). It shows 4 peaks (the large peak around – 7 000 K can be deconvoluted into two peaks, one at – 7 500 K and one at – 5 500 K), suggesting that 4 different adsorption sites co-exist.

Of these 4 sites, 2 are located in the supercages: the one corresponding to the peak at – 7 500 K on the electrostatic energy histogram belongs to a water molecule sitting at ca. 3.0 Å of a site II cation, and there may be up to three water molecules for each site II cation. The second (electrostatic energy of *ca.* – 5 500 K) belongs to a water molecule located near a cationic site III (unoccupied in the $Na_{48}Y$ Faujasite, though).

The two other adsorption sites for water molecules are located in the sodalite cages, and are more energetically favourable (– 11 000 K and – 12 500 K respectively), and almost identically populated. The oxygen atoms of water molecules are located at 1.6 Å and 2.4 Å from the cationic sites I'.

No other adsorption site for water molecules has been observed, even at higher loadings.

For the system containing 39 water molecules and for the model with sodium cations in sites I', about 55% are located close to cations in site II, 20% in sites III, and – due to the smaller size of the sodalite cages not leaving much place for the water molecules – 25% close to cations in site I'. Examination of the distribution at high loadings indicates that the sodalite cages which have 4 cations do not accommodate any water molecules.

For the model with cations in site I, it has been observed that no water molecule is adsorbed in the sodalite cages, probably because of the absence of favourable interactions with cations therein. This is why the water adsorption isotherm is lower and further away from the experimental one for this model than for the asymmetric one, and suggests that cation migration from sites I to sites I' could occur during water adsorption. This has been shown on BaX zeolite by X-ray diffraction experiments [Pichon *et al.*, 2000].

There are slightly fewer water molecules adsorbed in the supercages when cations are in site I, than when they are in site I'. This means that the distribution of cations outside of the supercages has a direct, and long-range effect on adsorption in the supercages. Figure 4.15 shows the average adsorbate-zeolite interaction energy per adsorbed molecule as a function of the loading, for the two models. It is more negative when the cations are in site I', which explains the higher loading for identical pressures. An explanation can be proposed in terms

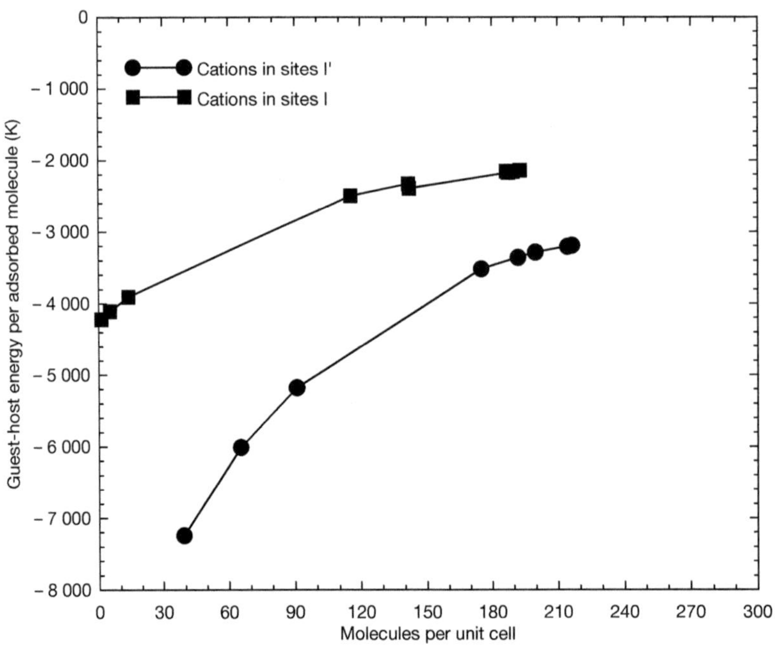

Figure 4.15 Total water-zeolite interaction energy as a function of the loading at 423 K in $Na_{48}Y$.

of the electric field prevailing in the solid: as we use periodic boundary conditions, the high symmetry of the model in which all sites I and II are occupied decreases the intensity of the electric field, and thus the electrostatic interaction energy with polar molecules such as water.

4.2.4 Co-adsorption of Water and Xylenes in NaY Faujasite

It is a well-known fact that adsorption properties in zeolites are strongly correlated to the nature, location and distribution of the non-framework cations. Most models used in molecular simulation describe all the atoms of the solid as fixed, including the charge-compensating cations. But there is some evidence that cation redistribution occurs upon adsorption of polar molecules. For instance, Mellot-Draznieks *et al.* [2003] have recently carried out neutron scattering experiments of $CFCl_3$ in NaY Faujasite and observed cation redistribution and the appearance of a new cationic site.

Here, we consider the case of NaY Faujasite with a Si/Al ratio of 3, which corresponds to 48 cations per unit cell. As seen above (Section 4.2.1), the predicted cation distribution of the dry and empty material corresponds to a full occupancy of sites I and II, sites I' and III remaining free of cations. Beauvais *et al.* [2004b] have used the Parallel Tempering method described above to predict the stable cation distribution at room temperature for $Na_{48}Y$ containing from 0 to 370 water molecules per unit cell. The solid framework, i.e. Aluminium, Silicium and Oxygen atoms, were fixed throughout the simulations, whereas the Sodium cations were allowed translation moves in the canonical ensemble. The TIP4P effective potential [Jorgensen *et al.*, 1983] was used to describe the interactions of water molecules between themselves and with the atoms of the solid.

Results for the computed cation distributions at 300 K are shown in Figure 4.16. For low water contents (under 60 water molecules per unit cell, which roughly corresponds to 2 water molecules per site II cation), the water molecules solvate the cations in sites II. Above 60 water molecules per unit cell, a redistribution of cations in sites I and I' occurs: cations progressively shift from sites I to neighbouring sites I', this shift being accompanied by a progressive occupancy of sodalite cages by the water molecules.

An explanation of the processes involved can be proposed in the following terms: water molecules are preferentially adsorbed in sites II, where they solvate the sodium cations. By the time each site II cation is solvated by 2 water molecules, it becomes energetically favourable for water to solvate other cations. The hexagonal prism connecting two sodalite cages, in which are located the site I cations, being too small to accommodate a water molecule, the site I cations progressively move to sites I', enabling water molecules to adsorb in sodalite cages.

The same Parallel Tempering Monte Carlo simulations were then carried out in the presence of pre-adsorbed xylene molecules. 4 metaxylene molecules, corresponding to the maximum loading at room temperature, were pre-adsorbed in $Na_{48}Y$, and a various amount of water molecules was then introduced. The results are shown in Figure 4.17.

Since sodalite cages are too small to accommodate xylene molecules, there is a steric effect that favours xylene adsorption in front of site II cations, in the supercages. Water molecules

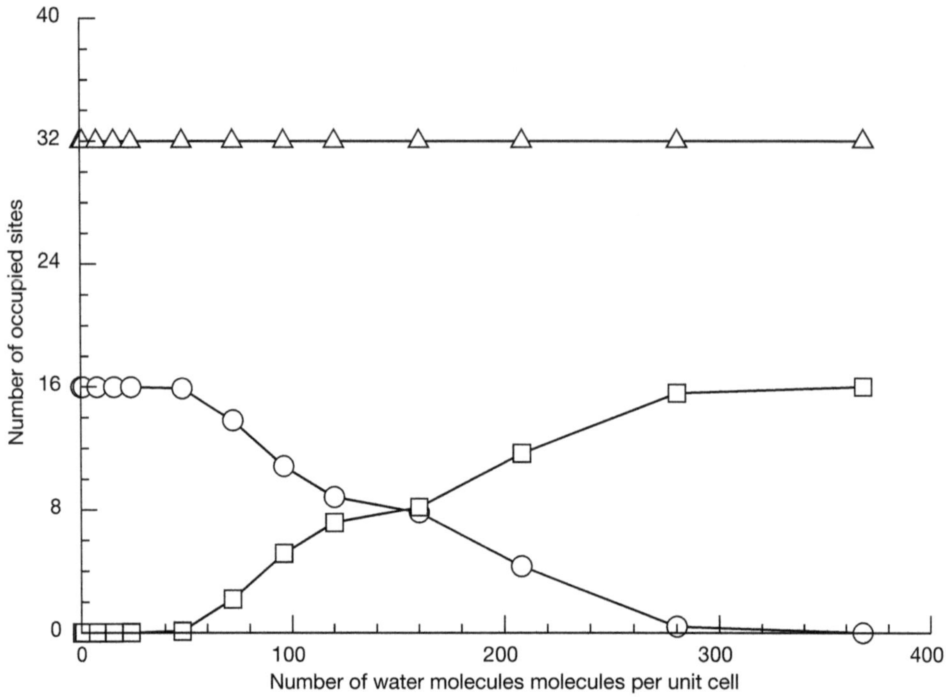

Figure 4.16 Occupation of the sodium crystallographic sites in $Na_{48}Y$ as a function of the water content. Triangles: sites II, circles: sites I, squares: sites I'.

must then adsorb in the sodalite cages, where free space remains, and as a consequence, cation migration from sites I to sites I' occurs at a much lower water loading than in absence of pre-adsorbed metaxylene.

This phenomenon of cation redistribution upon water adsorption could contribute to explain the variation of adsorption selectivity which has been observed experimentally in certain systems. For instance, Pichon *et al.* [1999] observed that BaX zeolite selectivity towards paraxylene versus metaxylene increases when the solid is partially hydrated. In the NaY zeolite studied by Beauvais and co-workers, preliminary simulations show that the adsorption selectivity, in favour of metaxylene for the dry solid, decreases by a factor of 4 when a little amount of water (roughly 2% weight) is added to the system. This effect could be related to the existence of an adsorption site for paraxylene closer to the centre of the supercages, which does not exist for metaxylene, and has been observed by Pichon *et al.* [1999] using X-Ray diffraction.

The study reported here shows that molecular simulation, especially when allied to experimental methods (such as crystallographic techniques), makes it possible to gain valuable insight into the molecular mechanisms of adsorption, currently explaining and hopefully in the future predicting adsorption selectivities in such complex systems as hydrated zeolites.

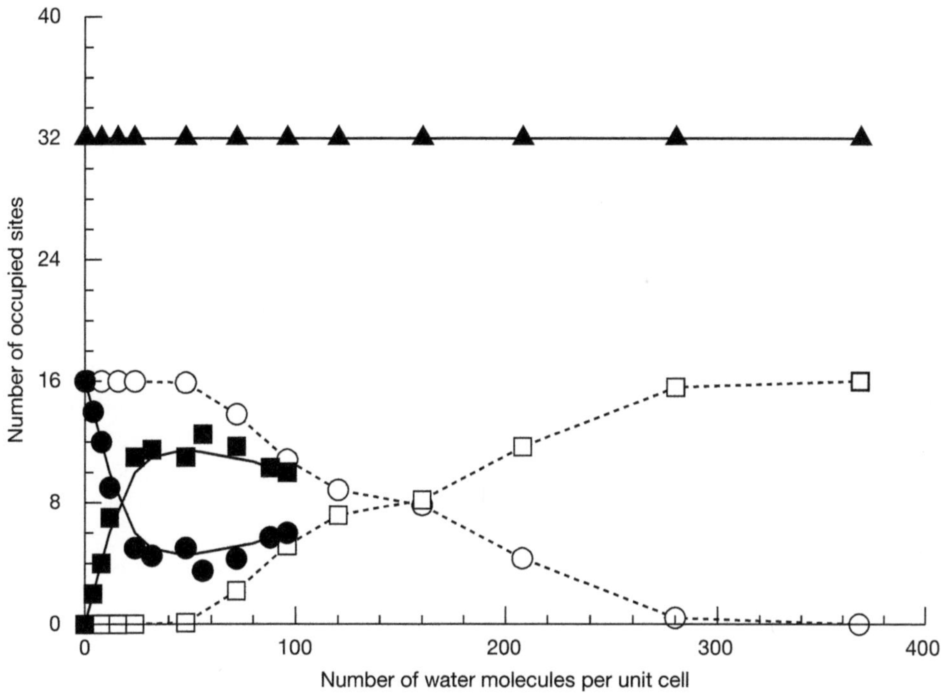

Figure 4.17 Occupation of the sodium crystallographic sites in $Na_{48}Y$ with 4 molecules of metaxylene pre-adsorbed, as a function of the water content. Filled symbols: triangles: sites II, circles: sites I, squares: sites I'. Open symbols correspond to the data in Figure 4.16 (the same zeolite with no metaxylene adsorbed) and are presented for comparison.

4.3 OPTIMISATION OF INTERACTION PARAMETERS SPECIFIC TO ZEOLITES

Modelling of fluid adsorption properties by molecular simulation requires an efficient potential model to describe both the intermolecular interactions between the adsorbed species and the adsorbate-adsorbent interactions. The potential model or force field should have several qualities:

- it should be realistic, i.e. provide an accurate molecular description of the system's thermodynamics;
- its formal expression should be simple enough to ensure that computer needs are not overwhelming;
- it should be accurate enough for the intended degree of precision;
- it should be transferable from one system to another, i.e. once its parameters have been optimised for a set of well-chosen systems, it should not be necessary to modify them for simulating a similar but distinct system.

This last requirement leads to the definition of a force field specific to the solids and independent of the nature of the adsorbed species, besides the force field used to describe the sorbate-sorbate interactions. For these latter interactions, it is convenient and consistent to use the models developed to represent fluid phase properties, since 1° this requires no extra development, these force fields being well described in the literature, and 2° computation of the chemical potentials of the adsorbates in the fluid phase is a pre-requirement for GCMC calculations. One then makes use of mixing rules between the parameters specific to the zeolite and the "adsorbate-adsorbate" force field parameters to compute the zeolite-adsorbate interaction energy.

To calculate the dispersion-repulsion part of the zeolite-adsorbate interactions, one generally adopts a simplified scheme known as the Kiselev approximation [Bezus *et al.*, 1978]. It consists in computing only the interactions between the atoms or pseudo-atoms of the adsorbates and the extra-framework atoms and oxygen atoms of the zeolitic framework, but not with the "T-atoms" (i.e., silicium and aluminium). This simplification is justified by the fact that adsorption takes place in the channels or cages of the solids, and thus adsorbed molecules are much closer to oxygen atoms than to the T-atoms. The parameters obtained in this way, to represent the dispersion-repulsion energy between the adsorbates and the oxygen atoms, are effective parameters, in that they implicitely incorporate the dispersion-repulsion energy between adsorbates and T-atoms.

Several authors have developed efficient parameterisation for interaction models devoted to the representation of fluid properties. All Atoms (AA) force fields have been used successfully for adsorption simulation of alkane molecules [Macedonia and Maginn, 1999; Vlugt *et al.*, 1999], mostly in silicalite. But as the zeolite-adsorbate interactions are not computed from pure zeolite parameters combined with pure adsorbate parameters through a mixing rule, there can be no question of transferability. To tackle other structures, such as Faujasites or type A zeolites for instance, or other adsorbate (sulphur, oxygen or nitrogen atoms, olefins, aromatics or cyclic hydrocarbons etc.), one would first have to optimise *ex nihilo* complete new sets of parameters.

In the work presented here, the AUA-4 force field (see Sections 2.2 and 3.1) is extended to adsorption in zeolites. The goal is to take advantage of the good transferability properties of this potential to predict adsorption properties without any readjustment of the parameters. The AUA-4 force field is based on Lorentz-Berthelot combining rules to obtain cross potential Lennard-Jones parameters from the individual molecular group values.

Since within the Kiselev model, the sorbate-zeolite interactions are dominated by dispersive interaction with oxygen atoms, we just have to optimise a set of effective Lennard-Jones parameters for the oxygen atom (i.e., σ_{O-O} and ε_{O-O}), that can be directly combined to the adsorbate atoms Lennard-Jones parameters to obtain the sorbate-zeolite force field. This procedure makes it possible in principle to predict adsorption properties of any molecule for which AUA-4 force field have already be optimised on the basis of fluid phase properties data, i.e. linear alkanes [Ungerer *et al.*, 2000], branched alkanes [Bourasseau *et al.*, 2002], cyclic alkanes [Bourasseau *et al.*, 2002], alkenes [Bourasseau *et al.*, 2003], and aromatic molecules [Contreras-Camacho, 2002].

The neutral (i.e., non cationic) zeolite silicalite was selected because, within the semi-empirical potential model that we use, only the oxygen parameters have to be adjusted (no charge-compensating cation is present), and also because there is sufficient experimental and simulation data to compare with. We have chosen to use experimental data on the adsorption of butane to adjust the zeolite oxygen parameters and then to directly transfer the force field to other hydrocarbon molecules (such as other linear alkanes, branched alkanes, alkenes…), and thus check the transferability of the optimised force field to other adsorbates.

The transferability of the potential model to other zeolite materials has also been studied: to describe adsorption in Faujasite type zeolites (with a Lennard-Jones type of potential, and without explicitely taking into account the coulombic interactions), one just needs to determine parameters for the counter-charge cations, and use these together with the zeolite oxygen parameters optimised on silicalite with Lorentz-Berthelot combining rules (see Section 4.4.7).

The crystal structure of silicalite-1 zeolite has been derived from X-ray diffraction experiments [Olson *et al.*, 1981]. The space group is orthorhombic with unit cell parameters a = 20.07 Å, b = 19.92 Å, and c = 13.42 Å. The silicalite framework contains interconnected pore channels of two types: straight channels in the direction of the *y* axis and zigzag channels in the direction of the *x* axis (Fig. 4.18), crossing at intersections. There are four straight channels, four zigzag channels, and four intersection sites per unit cell of silicalite and they represent a volume fraction of ~ 33, ~ 45 and ~ 22% respectively.

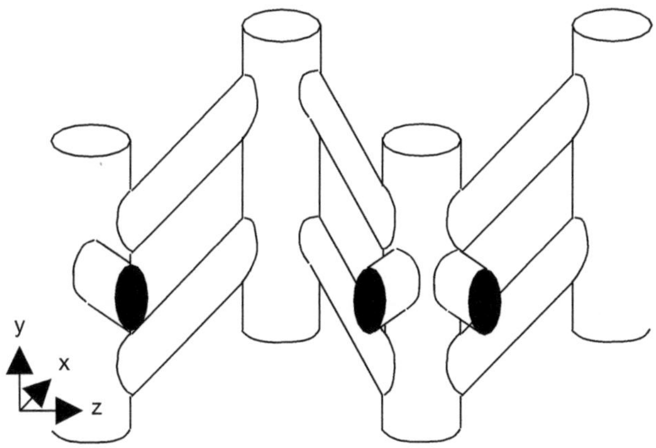

Figure 4.18 Schematic view of the straight and zigzag channels arrangement in silicalite-1 [Pascual *et al.*, 2003], reproduced by permission of the PCCP Owner Societies.

As usual in grand canonical Monte Carlo simulations of adsorption, the zeolite framework is considered as rigid and the zeolite-adsorbate interactions are calculated on a grid of points prior to simulations, so that only linear interpolations are needed in the course of the GCMC calculation (see Section 4.1.2). The grid mesh is about 0.2 Å in the three space directions. The simulation box is composed of two (1 × 1 × 2) or eight (2 × 2 × 2) silicalite unit cells, and periodic boundary conditions are used. Intermolecular interactions are calculated with a cut-off distance fixed at 9.96 Å (= b/2).

To determine transferable parameters for the oxygen atom of the zeolite, the method proposed by Ungerer was used [Ungerer *et al.*, 2000] based on the minimisation of a dimensionless error criterion, as described in Section 2.4.4.

This method makes it possible to optimise several different parameters from one set of reference data. We have decided to fit the two potential parameters using experimental butane adsorption isotherms data only [Pascual *et al.*, 2003]. Four butane experimental data have been included in the optimisation process: one low and one high pressure adsorption data at two different temperatures (277 K and 353 K). The statistical uncertainties were estimated to be twice as large at low pressure than at high pressure. The initial Lennard-Jones parameter set was taken from the simulation data of Cheetham [Boutin *et al.*, 2001] on argon in $AlPO_4$-5 zeolite. One optimisation cycle is composed of three simulations to calculate derivatives and of one minimisation process to determine a new set of parameters. Four cycles were necessary to reach a good precision, for a final result of σ_{O-O} = 3.00 Å and ε_{O-O} = 93.53 K. This final set leads to the best global dimensionless error criterion taking into account the statistical uncertainties. It should be stressed that the parameter values obtained for the zeolitic oxygen atom are close to those optimised for oxygen in the CO_2 or H_2O molecules, which endows a physical coherence to these parameters.

The simulation results obtained with this new set of parameters are compared with experimental data of Sun *et al.* [1998] on Table 4.5. A combination with AUA-4 parameters leads to σ_{O-CHi} parameters that increase with increasing number of hydrogens (σ_{O-CH} = 3.18 Å and σ_{O-CH3} = 3.30 Å), as expected from simple physical considerations. These values are smaller than the single σ (= 3.60 Å) Lennard-Jones parameter used in the previously published force field for adsorption of alkanes in silicalite by Vlugt *et al.* [1999] and June *et al.* [1992]. The cross energetic Lennard-Jones parameters in the present force field are larger by some 20-30 % than those proposed by Smit and co-workers.

Table 4.5 Optimisation process of zeolitic oxygen Lennard-Jones parameters to be used with the AUA-4 force field.

			Molecules per unit cell			
			***T* = 277 K**		***T* = 353 K**	
	σ_{O-O} (Å)	**ϵ_{O-O} (K/*k*)**	***P* = 40 Pa**	***P* = 10⁵ Pa**	***P* = 800 Pa**	***P* = 10⁵ Pa**
Experiment			**7.35**	**9.99**	**3.53**	**8.81**
Initial set	2.65	128.13	8.26	10.72	4.57	9.79
First set	2.85	106.91	7.97	10.16	3.92	9.45
Second Set	2.91	100.11	7.70	10.13	3.45	9.41
Third Set	2.97	95.47	7.62	10.12	3.33	9.30
Final Set	**3.00**	**93.53**	**7.37**	**10.02**	**2.90**	**9.15**

4.4 ADSORPTION ISOTHERMS AND SELECTIVITIES OF HYDROCARBONS ON SILICALITE

The normal alkanes chemical family is studied to check the validity of the Lennard-Jones potential parameters for the zeolitic oxygen atom, which have been optimised on 4 data points relative to *n*-butane (see above). All isotherms represented here have been plotted as number of molecules adsorbed per silicalite unit cell, as a function of pressure drawn on a logarithmic scale.

Adsorption isotherms are computed using biased Grand Canonical ensemble (μ, V, T) simulations. Translation and rotation moves allow respectively a displacement and a rotation of an entire molecule. Maximum displacement and angle are adjusted during the simulation so that the acceptance probabilities reach fifty percent. Flip, reptation and regrowth moves are used to change internal conformation and to displace a part of a molecule inside zeolite pores. Moreover, the regrowth procedure of branched molecules was implemented using a canonical reservoir of branch conformations [Macedonia and Maginn, 1999]. Insertion and deletion moves are attempted in order to allow the number of molecules inside the zeolite to fluctuate. Each move has an occurrence probability (frequency) which depends on the alkane considered.

Simulations were performed during several millions steps in order to attempt at least 2 10^5 moves per molecule during the stabilisation period and more than 3 10^5 moves per molecule for statistical averages.

The chemical potential is calculated from the fugacity y_iP of the vapour phase by using the ideal gas relation (4.1). It has been checked whether the ideal gas approximation was satisfactory for alkanes for the thermodynamic conditions used here. If not, the chemical potentials of the species have been calculated by the Widom test insertion method in the *NPT* statistical ensemble, as described in Section 2.3.5. Heats of adsorption have been computed according to the expression presented above (see Section 4.1.4).

4.4.1 Linear Alkanes

Pure component adsorption isotherms of methane, ethane, propane, butane, pentane, hexane, heptane, octane, and nonane were simulated in the temperature range from 277 to 374 K.

For methane, ethane, propane, and butane, simulations were carried out at 277 K, 308 K, 325 K, and 353 K and compared with experimental results of Sun *et al.* [1998] and Abdul-Rehman *et al.* [1990] (Fig. 4.19 to 4.22). Sun and co-workers' experimental data plotted here have been converted by assuming a perfectly crystallised zeolite sample whereas Abdul-Rehman and co-workers' have been corrected by considering a 20% weight amount of clay binder. It should be stressed that Abdul-Rehman's temperatures differ slightly from those of Sun and the simulated ones: 275 K instead of 277 K, 300 K instead of 308 K, and 350 K instead of 353 K. The somewhat lower temperatures can partly explain the slightly higher adsorbed quantities observed for Abdul-Rehman's experiments, as compared to those of Sun and the simulations.

- For methane, simulations are in perfect agreement with experimental data (Fig. 4.19). Simulations have been carried out up to a pressure of 100 MPa, difficult to obtain experimentally. Saturation is not achieved even at 20 molecules per unit cell. These perfectly predictive calculations are the mark of the transferability of the optimised model, since adsorption data concerning the CH_4 group were not used in the optimisation procedure.

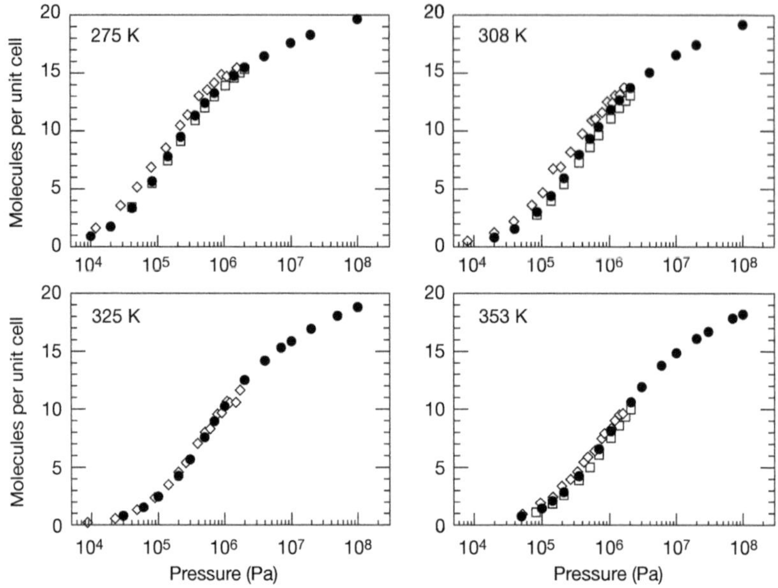

Figure 4.19 Adsorption isotherms of methane in silicalite-1. Open symbols: experimental data [squares: Sun *et al.* 1998; diamonds: Abdul-Rehman *et al.* 1990]. Filled symbols: Monte Carlo simulation. After [Pascual *et al.*, 2003], reproduced by permission of the PCCP Owner Societies.

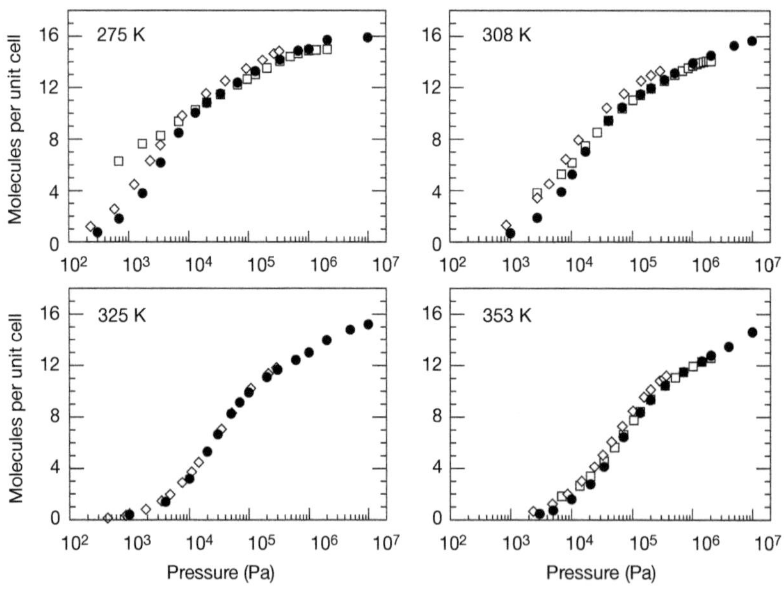

Figure 4.20 Adsorption isotherms of ethane in silicalite-1. Open symbols: experimental data [squares: Sun *et al.* 1998; diamonds: Abdul-Rehman *et al.* 1990]. Filled symbols: Monte Carlo simulation. After [Pascual *et al.*, 2003], reproduced by permission of the PCCP Owner Societies.

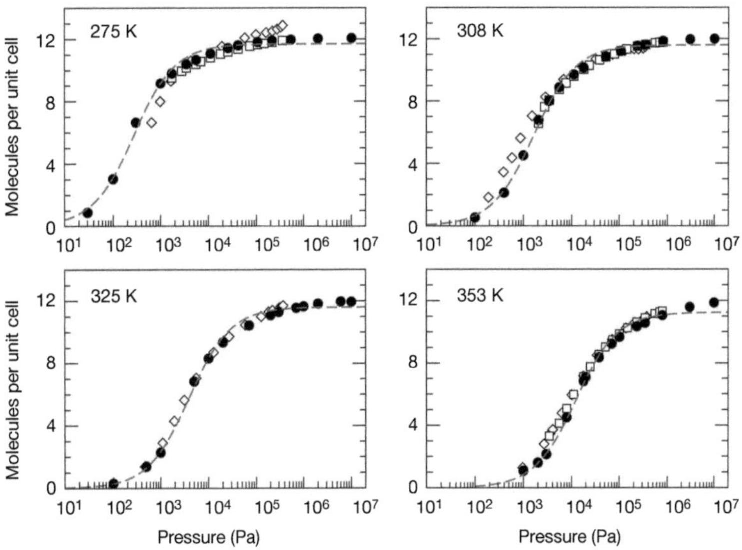

Figure 4.21 Adsorption isotherms of propane in silicalite-1. Open symbols: experimental data [squares: Sun *et al.* 1998; diamonds: Abdul-Rehman *et al.* 1990]. Filled symbols: Monte Carlo simulation. Langmuir fitting curves are plotted (dotted lines). After [Pascual *et al.*, 2003], reproduced by permission of the PCCP Owner Societies.

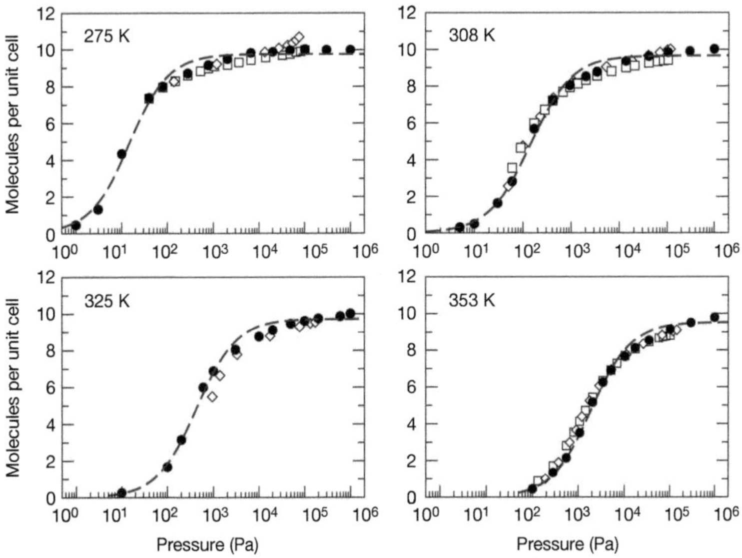

Figure 4.22 Adsorption isotherms of butane in silicalite-1. Open symbols: experimental data [squares: Sun *et al.* 1998; diamonds: Abdul-Rehman *et al.* 1990]. Filled symbols: Monte Carlo simulation. Langmuir fitting curves are plotted (dotted lines). After [Pascual *et al.*, 2003], reproduced by permission of the PCCP Owner Societies.

- For ethane, simulations underestimate the experimental data of Sun *et al.* [1998] at low pressure (Fig. 4.20). It is worth noticing a slight disagreement between experimental data: Sun and co-workers' experiments made at 277 K report higher adsorbed quantities than Abdul-Rehman at 275 K [Abdul-Rehman *et al.*, 1990], whereas the opposite would be expected. Saturation is achieved at about 16 molecules per unit cell.
- For propane and butane (Fig. 4.21 and 4.22 respectively), simulation data are also in perfect agreement with experimental data. Saturation corresponds to 12 propane molecules per unit cell whatever the temperature, one in each adsorption site (4 in the straight channels, 4 in the zigzag channels, and 4 in the intersections). For butane, saturation corresponds to only 10 molecules per unit cell: to understand the filling mechanisms of the silicalite pores, a systematic investigation of the sitting of the adsorbed molecules has been undertaken.

The occupancies at high and low loadings are listed in Table 4.6 for the three types of adsorption sites.

Table 4.6 Occupancy probability (%) of mass centres in the straight channels (SC), zigzag channels (ZC) or intersections (I) of silicalite at 308 K. <*N*> is the mean number of adsorbed molecules in the corresponding site, per unit cell.

	Very low coverage				Saturation			
	<*N*>	**SC**	**ZC**	**I**	**<*N*>**	**SC**	**ZC**	**I**
Volume fraction of the site (%)		33	45	22		33	45	22
Methane	0.8	32	47	21	19.1	25	48	27
Ethane	1.1	35	48	17	15.9	25	50	25
Propane	0.5	27	40	33	12.0	33	36	31
Butane	0.5	23	40	37	10.0	39	40	21

Silicalite provides a homogeneous energetic environment for small molecules so the distribution of mass centres of methane tends to be uniform; the occupancy ratios agree with the volume fractions of each type of adsorption sites (Table 4.6). A slight preference for the zigzag channels is observed, these being relatively more populated at low pressure. At a loading of 16 molecules per unit cell, adsorbate-adsorbate interactions prevail and the location of methane molecules is perfectly defined: one molecule in each straight channel and intersection, whereas zigzag channel sites can accommodate two molecules. At higher loadings, straight channel and intersection sites also begin to accommodate two molecules.

For ethane, a specific ordering appears: molecules are stacked in the middle of zigzag channels. This is in agreement with an observation made by Du *et al.* [1998]: the size of the ethane molecules is such that they can form a structure commensurate with the zeolite structure, which means that the sum of the size of two ethane molecules and their repulsion distances is equal to the length of a zigzag channel.

For propane, this ordering still exists, but the molecular length of propane implies that only one molecule can be located in a site, even the zigzag channel. Thus, the maximum adsorbed quantity is 12 molecules per unit cell. AUA simulations effectively predict a saturation plateau for 12 molecules per unit cell and an occupancy of 33, 36, and 31% respectively for straight channels, zigzag channels, and intersections.

The molecular length of butane is approximately equal to the length of a straight channel. At low coverage (0.5 molecules per unit cell), butane molecules are preferentially localised in the zigzag channels. At high loadings (10 molecules per unit cell), only half of the intersections are filled, whereas zigzag and straight channels are all accupied by one butane molecule. A butane molecule in an intersection strongly interacts with molecules in the straight channels and this results in a displacement of molecules in the straight channels toward the next intersections. This move blocks an adsorption site and leads to a theoretical maximum of ten adsorption sites per unit cell.

For the small alkanes (C_1 to C_4), a slight preference for the zigzag channels is observed at low loadings. This may simply be due to entropic effects since the volume of the zigzag channel is larger than that of the two other adsorption sites, straight channels and intersections.

For medium-length linear alkanes from pentane to nonane, adsorption isotherms at 300 K are plotted in Figures 4.23 to 4.27. The adsorption isotherm of pentane at 300 K is compared with experimental results of Sun *et al.* [1996] and with UA simulation results of Vlugt *et al.* [1999]. The computed pentane adsorption isotherm overestimates the experimental adsorbed quantities at low pressure, just like the UA simulations.

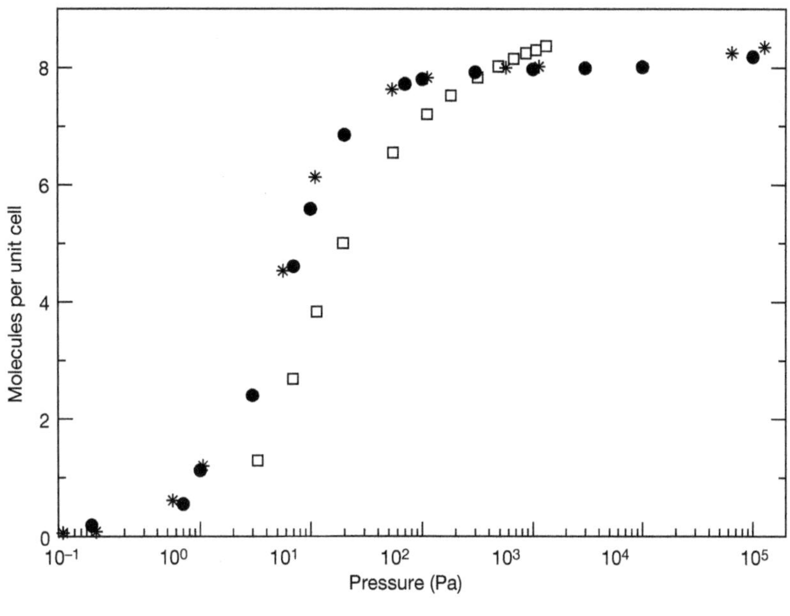

Figure 4.23 Adsorption isotherm of pentane in silicalite-1 at 300 K. Filled circles: Monte Carlo simulation. Open squares: experimental data from Sun *et al.* [1996]. Stars: previous simulations of Vlugt *et al.* [1999]. After [Pascual *et al.*, 2003], reproduced by permission of the PCCP Owner Societies.

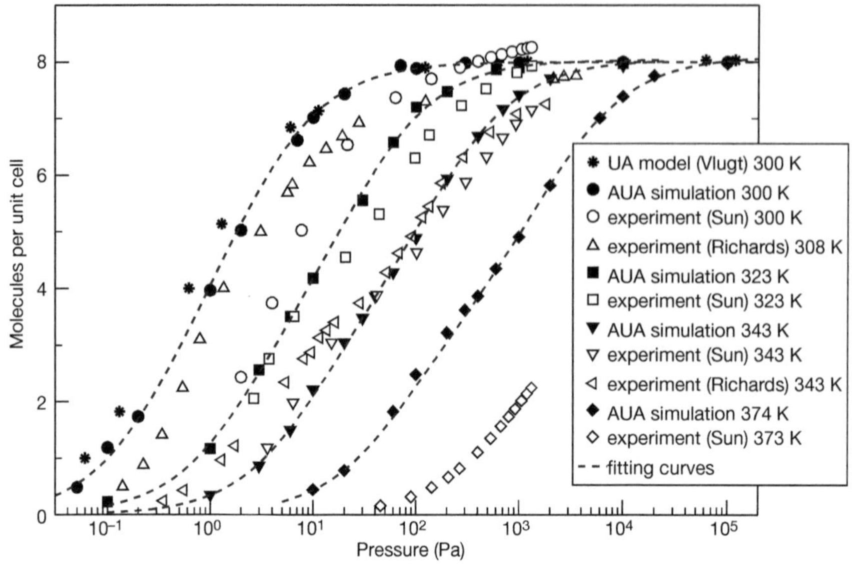

Figure 4.24 Adsorption isotherms of hexane in silicalite-1 at 300, 323 K, 343 K, and 373 K. Stars: UA simulations of Vlugt *et al.* [1999] at 300 K. Dotted lines: Langmuir fitting curves. After [Pascual *et al.*, 2003], reproduced by permission of the PCCP Owner Societies.

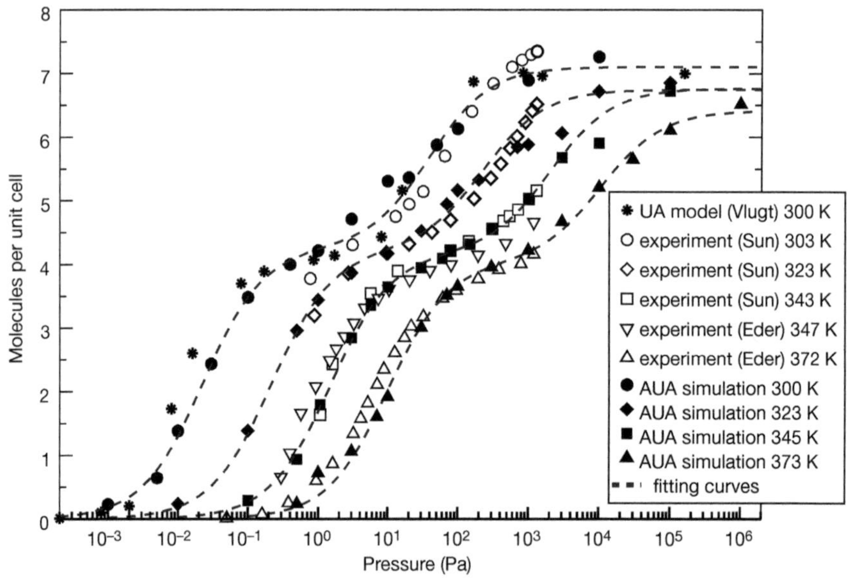

Figure 4.25 Adsorption isotherms of heptane in silicalite-1 at 300, 323 K, 345 K, and 373 K. Stars: UA simulations of Vlugt *et al.* [1999] at 300 K. Dotted lines: Langmuir fitting curves. After [Pascual *et al.*, 2003], reproduced by permission of the PCCP Owner Societies.

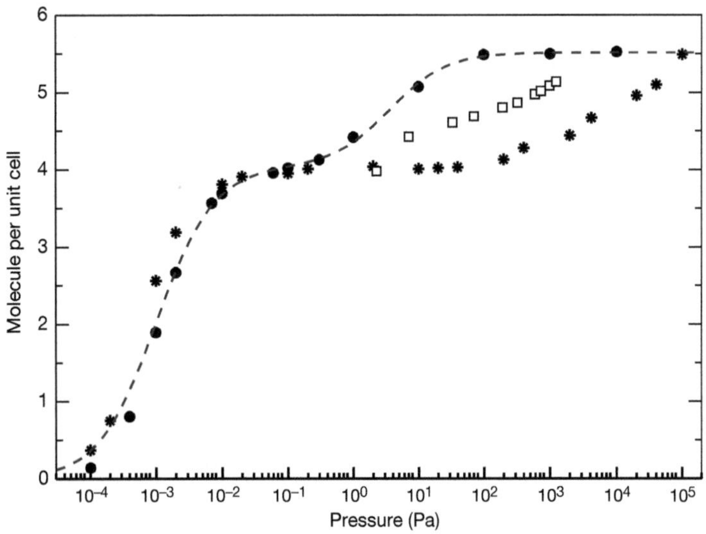

Figure 4.26 Adsorption isotherm of octane in silicalite-1 at 300 K. Filled symbols: Monte Carlo simulations. Open squares: experimental data from Sun *et al.* [1998]. Stars: UA simulations of Vlugt *et al.* [1999]. Dotted line: Langmuir fitting curve. After [Pascual *et al.*, 2003], reproduced by permission of the PCCP Owner Societies.

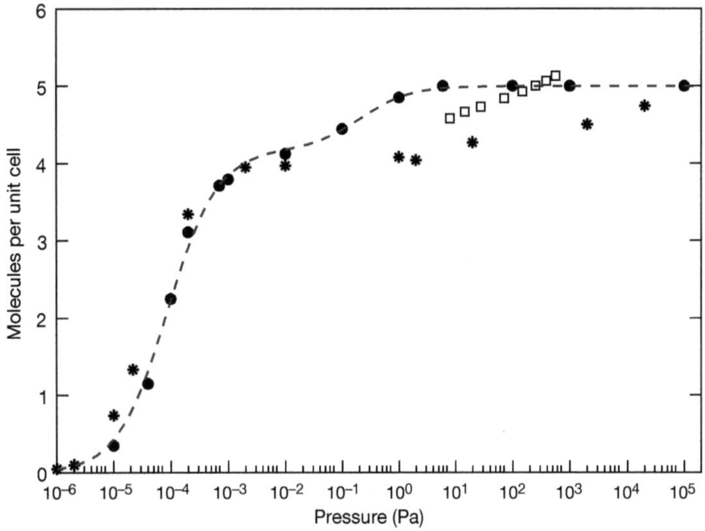

Figure 4.27 Adsorption isotherm of nonane in silicalite-1 at 300 K. Filled symbols: Monte Carlo simulations. Open squares: experimental data from Sun *et al.* [1998]. Stars: UA simulations of Vlugt *et al.* [1999]. Dotted line: Langmuir fitting curve. After [Pascual *et al.*, 2003], reproduced by permission of the PCCP Owner Societies.

For hexane (Fig. 4.24), simulated adsorptions carried out at 300 K, 323 K, 343 K and 373 K are compared with Sun *et al.* [1996] and Richards and Rees [Richards and Rees, 1987] experiments as well as the UA simulation of Vlugt *et al.* [1999]. At 300 K, the AUA simulation provides an isotherm slightly closer to experimental data than the UA simulation. At 374 K, a surprising strong discrepancy is observed between simulation and experiments. Furthermore, the simulation results fully agree with the calculations of Vlugt and co-workers. In Section 4.4.3., the internal consistency of the simulations performed at different temperatures is demonstrated, by the use of a Langmuir type fit.

For heptane (Fig. 4.25), simulated adsorptions carried out at 300 K, 323 K, 345 K and 373 K are compared with Sun *et al.* [1996] and Eder and Lercher [1997a, 1997b] experiments as well as the UA simulation of Vlugt *et al.* [1999] at 300 K. Similarly to the UA simulations, the AUA model reproduces the step in the adsorption isotherms and agrees with experiments. For alkane carbon numbers over 7, insertion in silicalite pores at high coverage becomes more difficult and the insertion move acceptance probability is very weak (~ 0.01%). The number of Monte Carlo steps was thus increased up to 10^7. The step in the isotherms is observed at four molecules per unit cell.

For octane and nonane (Fig. 4.26 and 4.27 respectively), simulations were performed at 300 K and compared with Sun and co-workers' experiments [Sun *et al.*, 1996] and UA simulations [Vlugt *et al.*, 1999]. Stepped isotherms are observed both in simulations and experiments. The UA simulation underestimates the experimental data at high pressure whereas the AUA simulation overestimates it. This might be due to simulation convergence difficulties: over four molecules of octane or nonane per unit cell, the insertion move is rarely accepted because insertion of an extra molecule resuires a strong rearrangement of the other adsorbed molecules, and simulations have to be performed over several millions of steps.

The occupation of pore sites by these medium linear alkanes has also been studied. From hexane to nonane, Jacobs *et al.* [1981] have shown that straight channels are preferentially occupied. A different point of view is advanced by Zhu *et al.* [2001] and Vlugt *et al.* [1999], according to whom, hexane and heptane molecules uniformly fill the silicalite channels below four molecules per unit cell whereas zigzag channels are preferentially occupied over a global loading of four molecules per unit cell. From pentane to nonane, both centres of mass and heads (terminal methyl groups) distributions have been calculated in this work. In the case of pentane and hexane, molecular length (5.1 and 6.4 Å respectively) is greater than the straight channel length (4.9 Å) but lower than the zigzag channel's (7.7 Å). Thus, whenever straight channels are occupied, intersections are blocked and as a result, only eight sites are available per unit cell of silicalite.

At low coverage, the distribution of pentane and hexane molecules is close to being homogeneous even though the centres of mass are preferentially located in intersections. At high coverage, pentane and hexane molecules are stacked in zigzag channels. In the case of heptane, this localisation is less precise because the molecular length (7.8 Å) is larger than the zigzag channel length. Thus, a heptane molecule localised in a zigzag channel partly blocks the neighbouring intersection. When all zigzag channels are filled, all remaining sites are blocked. This explains the existence of an intermediate plateau in the adsorption isotherm. In order to insert more molecules, a significant rearrangement of the adsorbed molecules is necessary, and this becomes favourable only at higher chemical potentials. This qualitative explanation of the observed stepped isotherm is similar to the one proposed previously by Smit and co-workers [Vlugt *et al.,* 1999].

For octane and nonane, the favoured conformation is an intermediate position occupying a zigzag or a straight channel, the neighbouring intersection, and again a zigzag or a straight channel ("ZIZ", "SIS" or "ZIS" conformations). This observation agrees with low resolution simulations performed by Maginn and co-workers [Maginn *et al.* 1995].

4.4.2 Branched Alkanes

For branched alkanes, simulated adsorption isotherms were carried out at 300 K. Stepped isotherms were found with a maximum loading which strongly decreases with the carbon number. A detailed study of isobutane (i.e. 2-methylpropane) was performed in the temperature range from 300 K to 353 K (Fig. 4.28). Simulated isotherms are compared with Sun and co-workers' experimental data [Sun *et al.,* 1998]. The agreement between simulation and experiment is satisfactory and once again demonstrates the transferability of the AUA potential, since the CH force centre was not included in the optimisation process. The 2-methylbutane, 2-methylpentane, and 2-methylhexane, adsorption isotherms performed at 300 K are consistent with UA simulation data [Macedonia and Maginn, 1999] (Fig. 4.29).

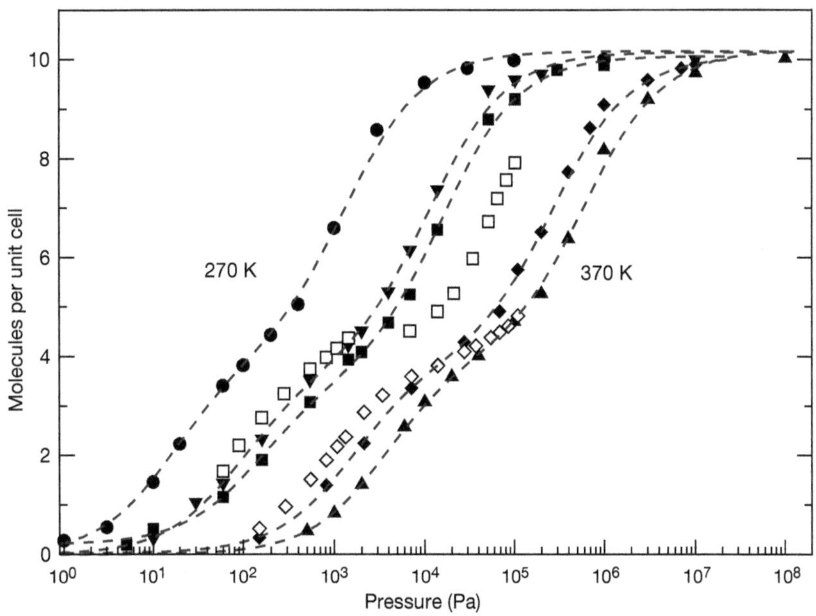

Figure 4.28 Adsorption isotherms of isobutane in silicalite-1 at 270, 300, 308, 353, and 370 K. Filled symbols: Monte Carlo simulations. Experimental data from Sun *et al.* [1998] at 308 K (open squares) and 353 K (open diamonds). Dotted line: Langmuir fit. After [Pascual *et al.*, 2003], reproduced by permission of the PCCP Owner Societies.

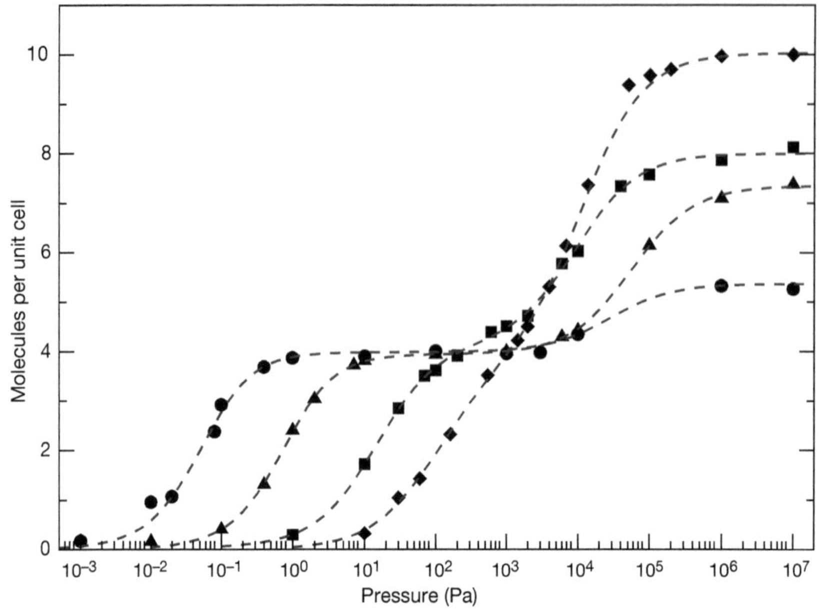

Figure 4.29 Adsorption isotherms of 2-methylpropane (diamonds), 2-methylbutane (squares), 2-methylpentane (triangles), and 2-methylhexane (circles) in silicalite-1 at 300 K. Dotted lines: Langmuir fits. After [Pascual *et al.*, 2003], reproduced by permission of the PCCP Owner Societies.

A study of the location of branched molecules in silicalite was carried out at 300 K. In the case of isobutane and 2-methylbutane molecules, only intersections are occupied at low coverage. Over four molecules per unit cell, the zigzag channels are preferentially occupied and only at high coverage are the straight channels filled. This successive filling of silicalite sites is shown in Figure 4.30. For 2-methylpentane and 2-methylhexane molecules, intersection sites are preferentially occupied. This is shown by the centre of mass distribution (97% of the molecules have their centre of mass in intersections at a loading of 0.41 molecules per unit cell). Nevertheless, the main chain is too long to sit in an intersection and occupies also the neighbouring zigzag or straight channels (for 0.41 molecules per unit cell, 55% of the methyl heads are found in intersections, 20% in straight channels, and 25% in zigzag channels).

4.4.3 Isotherm Fit Using the Langmuir Formalism

Simulated adsorption isotherms were fitted with a Langmuir type model. Even though the very severe hypotheses of the Langmuir model might appear not to be adapted to adsorption in micropores, this very simple formalism enables to fit rather well the simulated isotherms and provides some information on the "effective site" enthalpy and entropy.

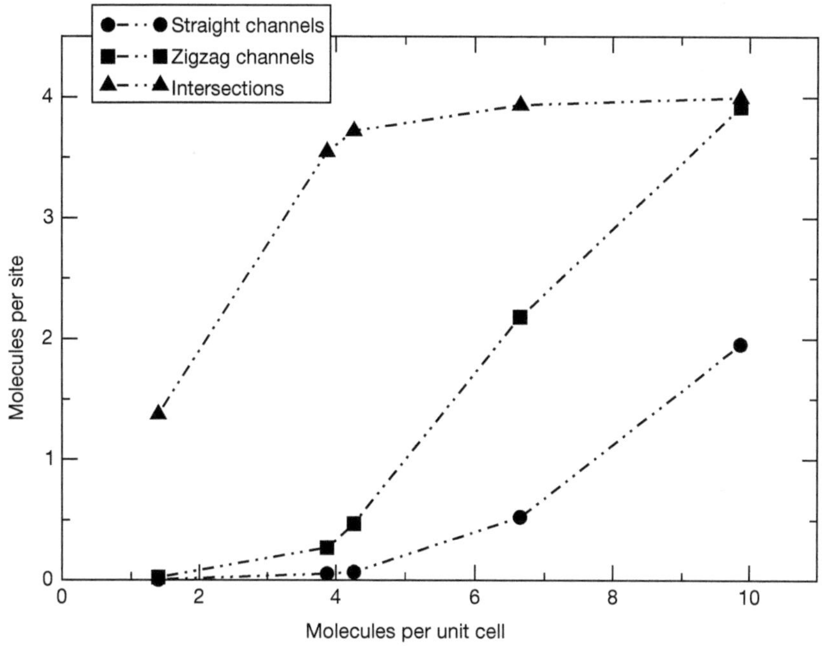

Figure 4.30 Site occupancy (mass centers) of isobutane in silicalite-1 at 353 K, versus loading. After [Pascual *et al.*, 2003], reproduced by permission of the PCCP Owner Societies.

The single site Langmuir (SSL) model has thus been used to fit simulation results. It can be written as:

$$q = \frac{q_s kP}{1+kP} \tag{4.8}$$

where q is the number of adsorbed molecules, q_s the number of adsorbed molecules at saturation, k the adsorption equilibrium constant, and P the pressure. The equilibrium constant can be expressed with the van't Hoff equation in order to link fitted parameters and thermodynamic quantities of adsorption process:

$$k = \exp\left(-\frac{\Delta G^o_{\text{ads}}}{RT}\right) = \exp\left(\frac{\Delta S^o_{\text{ads}}}{R}\right)\exp\left(-\frac{\Delta H^o_{\text{ads}}}{RT}\right) \tag{4.9}$$

If the sites are energetically identical and interactions between adsorbed molecules are neglected, then the heat of adsorption is independent of coverage and the isosteric heat of adsorption Q_{st} is the same as the limiting heat of adsorption $-\Delta H^o_{\text{ads}}$. The plot of ln(k) as a function of $1/T$ is thus a way to determine the isosteric heat of adsorption:

$$\ln(k) = \frac{\Delta S^o_{\text{ads}}}{R} - \frac{\Delta H^o_{\text{ads}}}{RT} \tag{4.10}$$

Figure 4.31 shows plots of ln(k) as a function of $1/T$ for propane, butane, hexane and heptane. The slopes of the lines and extrapolation to $1/T = 0$ yield isosteric heats of adsorption of 38.9 and 49.1 $kJ \cdot mol^{-1}$ for propane and butane respectively, and adsorption entropies of – 189 and – 202 $J \cdot K^{-1} \cdot mol^{-1}$. The decrease of adsorption entropy with increasing carbon number is well represented, but on the other hand, simulation greatly overestimates experimental data, which Zhu *et al.* [2001] have determined experimentally for adsorption of propane and butane in silicalite at – 71 and – 87 $J \cdot K^{-1} \cdot mol^{-1}$ respectively.

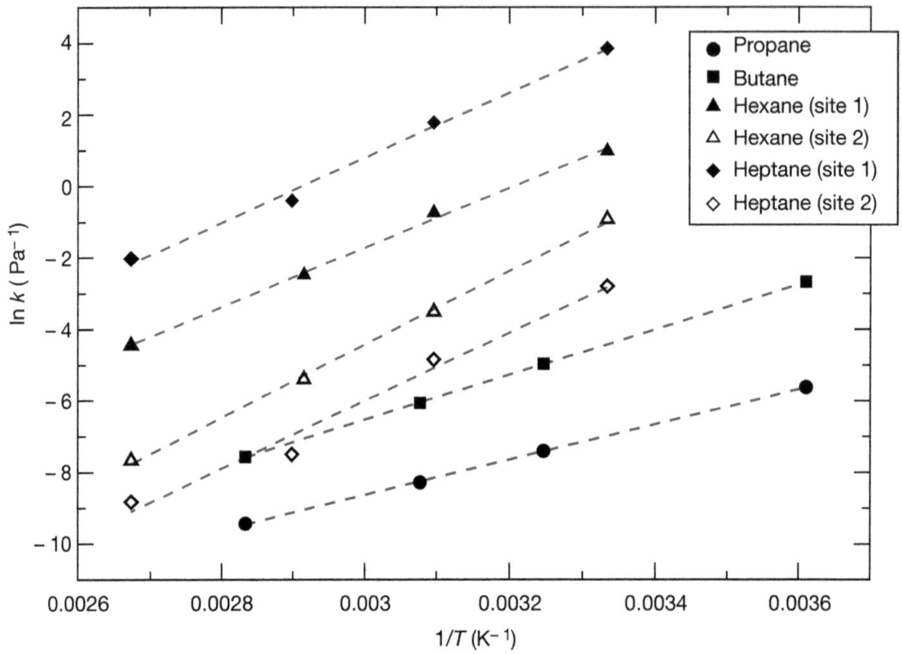

Figure 4.31 ln(k) as a function of $1/T$ for propane, butane, hexane and heptane adsorbed in silicalite-1. After [Pascual *et al.*, 2003], reproduced by permission of the PCCP Owner Societies.

For methane and ethane, Single Site Langmuir (SSL) fits are not suitable because adsorbates progressively reorganise upon pore filling and the adsorption sites are not well defined: the Langmuir hypotheses do not hold. However, if we apply the SSL formalism on the first part of the isotherm for which the number of molecules per unit cell is lower than 12, we can extrapolate adsorption enthalpies and entropies for the first equivalent site. For ethane, this yields an adsorption enthalpy of – 30.8 $kJ \cdot mol^{-1}$ and an adsorption entropy of – 180 $J \cdot K^{-1} \cdot mol^{-1}$.

For linear alkanes from hexane to nonane, adsorption isotherms exhibit an inflexion shape. In these cases, a dual site Langmuir (DSL) formalism must be used to fit the adsorption isotherms [Smit, 1995]. This model describes pairs of non equivalent independent sites, with no interaction between adsorbed molecules:

$$q = \frac{q_{s1}k_1P}{1+k_1P} + \frac{q_{s2}k_2P}{1+k_2P} = \frac{(q_{s1}k_1 + q_{s2}k_2)P + (q_{s1} + q_{s2})k_1k_2P^2}{1+(k_1 + k_2)P + k_1k_2P^2} \tag{4.11}$$

where q_{s1} and q_{s2} represent respectively the maximum number of molecules adsorbed in site 1 and site 2, and k_1 and k_2 the equilibrium constants for site 1 and site 2. By convention, the site 1 is assumed more attractive than site 2, that is to say k_1 is greater than k_2.

For *n*-hexane and *n*-heptane, $\ln(k_1)$ and $\ln(k_2)$ as a function of $1/T$ are shown in Figure 4.31. The linearity of these plots indicates that meaningful parameters can be obtained from the DSL model for C_6 and C_7. The derived values of the adsorption enthalpies and entropies are listed in Table 4.7. Adsorption enthalpies of both sites are quite similar whereas adsorption entropies are quite different. This is consistent with the presence of two sites (zigzag and straight channels), similar from an energetic point of view. These sites are considered as equivalent for shorter linear alkanes. However, the entropic effects are very different in these two sites. In zigzag channels, the heptane molecule has a higher packing efficiency than in straight channels.

Table 4.7 Adsorption enthalpies and entropies obtained from Langmuir fits of the isotherms.

	Site 1 – ΔH°_{ads}	Site 1 – ΔS°_{ads}	Site 2 – ΔH°_{ads}	Site 2 – ΔS°_{ads}	Q°_{st}
	$kJ \cdot mol^{-1}$	$J \cdot K^{-1}\ mol^{-1}$	$kJ \cdot mol^{-1}$	$J \cdot K^{-1}\ mol^{-1}$	$kJ \cdot mol^{-1}$
Propane	38.9	189	–	–	39
Butane	49.1	202	–	–	49
Isobutane	45.6	190	51.3	249	45
Hexane	69.2	222	85.2	292	67
Heptane	75.6	220	78.9	287	75

The DSL model is used successfully from hexane to nonane, but fails for pentane. Since the size of the pentane molecule closely corresponds to the length of the straight channel, two adjacent sites can be simultaneously occupied and the interaction between adsorbed pentane molecules is important, leading to some rearrangements. The hypothesis of independence of the adsorption sites of the Langmuir model is not respected.

In the same way, branched alkanes isotherms were fitted using the DSL model. Smit and co-workers have shown that the adsorption of branched alkanes preferentially takes place in intersections because of the geometry of these molecules [Krishna *et al.*, 2002; Smit, 1995; Paschek and Krishna, 2001; Krishna *et al.*, 1999]. At high loading (greater than 4 molecules per unit cell), isobutane molecules fill both straight and zigzag channels. The AUA simulation reveals a slightly different picture: three different sites are successively occupied by the adsorbed molecules (Fig. 4.30). At low coverage, only intersections are occupied, under medium pressure some molecules are present in the zigzag channels, and closer to saturation straight channels are also filled. From this observation, the DSL model is applied by supposing that straight and zigzag channels are equivalent for branched molecules. This approach yields good results (see the fits – the dotted lines in Figures 4.28 and 4.29). Indeed, a "triple site" Langmuir fit leads to the conclusion that two sites are in effective similar.

4.4.4 Heats of Adsorption

Heats of adsorption of alkanes, quantities that have not been used in the optimisation of the AUA force field, have been determined from Monte Carlo simulations with the fluctuations method detailed in Section 4.1.4. In Figure 4.32, simulated isosteric heats of adsorption are compared with various experimental data determined by gravimetry [Sun *et al.*, 1998; Richards and Rees, 1987; Sun *et al.*, 1996], calorimetry [Abdul-Rehman *et al.*, 1990; Thamm *et al.*, 1983], chromatography [Hufton and Danner, 1993] or microbalance measurements [Zhu *et al.*, 1998 and 2001]. The heats of adsorption of linear alkanes obtained by AUA simulations are in good agreement with available experimental data. The Q^0_{st} calculated by the fluctuations method are directly comparable with values determined from Langmuir fits (see above, Section 4.4.3). The good agreement between the two sets of values illustrates the consistency of our simulations. A linear correlation between the isosteric heat of adsorption and the carbon number of the adsorbate has been found:

$$Q^0_{st} = 9.5N_C + 10.8 \tag{4.12}$$

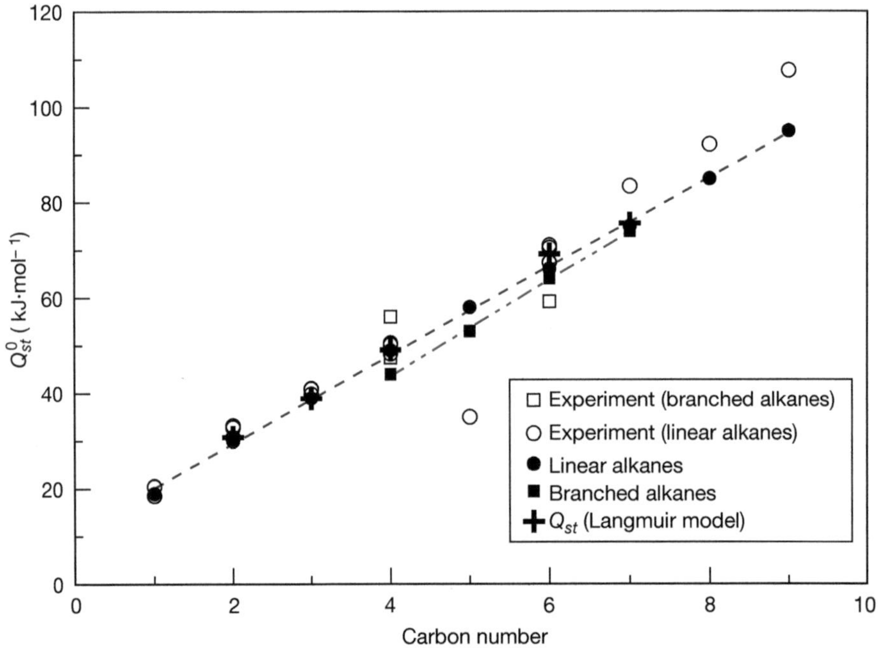

Figure 4.32 Adsorption heats of linear and branched alkanes in silicalite-1 at 300-308 K. simulation results obtained with the fluctuations method (filled symbols), derived from the Langmuir theory (crosses), and experimental data (open symbols). After [Pascual *et al.*, 2003], reproduced by permission of the PCCP Owner Societies.

The slope of 9.5 $kJ \cdot mol^{-1}$ per carbon atom agrees with experimental results of Hufton and Danner [1993], Richards and Rees [1987], Zhu *et al.* [1998], and Abdul-Rehman *et al.* [1990] where the increase is approximately 10 $kJ \cdot mol^{-1}$ per additional CH_2 group. Q_{st} as a function of molecules per unit cell are plotted in Figure 4.33. An increase of heat of adsorption is observed for all linear alkanes due to sorbate/sorbate interactions. Moreover, from hexane to nonane, a change in the slope can be observed for four molecules per unit cell, which corresponds to the existence of a pair of non-equivalent sites. Adsorption enthalpies being different for each site, the zeolite/sorbate contribution is modified when sorbate molecules fill the second site, which causes the slope to increase. For 2-methylalkanes, simulated results are also in a good agreement with the few available experimental data. A linear correlation of Q_{st}^0 with the number of atom units in the sorbate molecule has been made for branched alkanes. An increase of 10.1 $kJ \cdot mol^{-1}$ per carbon atom has been obtained, which is consistent with the result for linear alkanes.

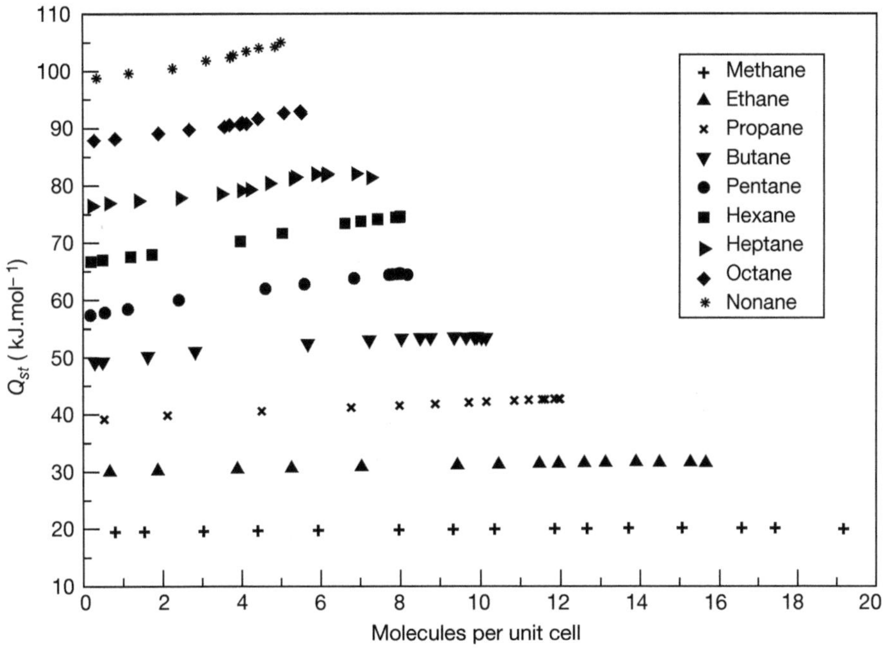

Figure 4.33 Adsorption heats of linear alkanes in silicalite-1 at 300-308 K, as a function of loading. After [Pascual *et al.*, 2003], reproduced by permission of the PCCP Owner Societies.

Once again, these results are of a fundamental interest because they shed light on the transferability of the optimised AUA potential. Indeed, the CH force centre present in branched alkanes was not included in the optimisation process: in this case, simulation is purely predictive. The use of Lorentz-Berthelot combining rules to calculate Lennard-Jones parameters makes it possible without correction to obtain very good results.

4.4.5 Adsorption of Alkenes in Silicalite

As an extension of the work presented above, adsorption of light alkenes in silicalite has been studied by grand canonical Monte Carlo simulation. The double C=C bond of alkenes is described in terms of intermolecular interactions in the AUA-4 force field: parameters for olefinic carbons have been optimised by Bourasseau *et al.* [2003] as seen in Section 3.1.5. In the same manner as above, application of the Lorentz-Berthelot combining rules on the Lennard-Jones parameters of alkenes one one hand and zeolitic oxygen on the other hand, lead to cross parameters which make it possible to compute adsorbate-zeolite interactions. Thus, the present study is entirely predictive: as a matter of fact, absolutely no further optimisation of the parameters has been done.

To start with, we have concentrated on light alpha-olefins and isomers of butene.

The methodology developed for alkanes is reproduced here [Pascual *et al.*, 2004b], with a few minor adjustments such as the elimination of the Monte Carlo reptation move, which would not preserve the molecular structure of the alpha-olefin.

Calculated adsorption isotherms of ethene on silicalite at 250, 306 and 353 K are plotted in Figure 4.34 and compared with experimental data of Choudary and Mayadevi [1993, 1996] at 306 and 353 K and those of Stach *et al.* [1986] obtained at 250, 300 and 350 K. For all the temperatures studied, agreement between simulation and experience is excellent, the more so since experimental points are concentrated at lower loadings, where agreement is most difficult to achieve, as observed in most of the previous simulation studies. Saturation is obtained for a loading of 16 molecules per unit cell, which is equal to the value for ethane.

Ethene and ethane locations in the three types of sites (straight channels, zigzag channels and intersections) are very similar: in both cases, the adsorption sites are identical and two molecules can simultaneously fill the same zigzag channel at high pressure. At very low coverage (0.8 molecule per unit cell), ethene molecules equally fill all the available pore space. The partial occupancies of the three types of sites (31, 47 and 22% respectively for the straight channels, zigzag channels and intersections) correspond to the volume fractions. At saturation, the specific ordering observed for ethane is less pronounced for ethene. This is not surprising since the molecular length of ethene is shorter than that of ethane, and is not commensurate with the zeolite structure.

The first part of the adsorption isotherms, where all sites can be considered as energetically equivalent and interactions between sorbed molecules can be overlooked, makes possible the determination of enthalpy and entropy of adsorption by fitting this region with a single site Langmuir (SSL) model. The results are listed in Table 4.8.

Adsorption isotherms of alpha-olefins from ethene to hept-1-ene have been calculated at 300 K; they are shown in Figures 4.35a and 4.35b and are found to be in good agreement with the available experimental data [Stach *et al.*, 1986]. A comparative study with alkanes adsorption on silicalite (preceding section) reveals strong similarities: the adsorbed quantity at saturation decreases with carbon number and is identical for alkane and alkene with the same carbon number.

For propene (i.e. propylene), a detailed study has been done to determine the enthalpy and entropy of adsorption. The isotherms at 300, 330 and 360 K have been used to evaluate Lang-

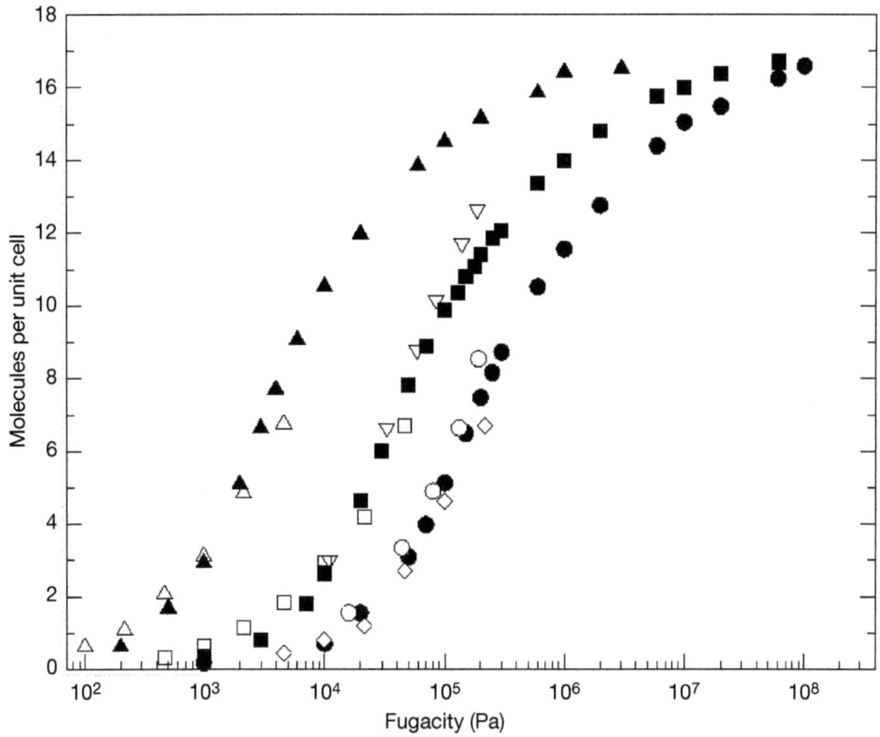

Figure 4.34 Adsorption isotherms of ethene in silicalite-1. Simulation results at 250 K (filled triangles), 306 K (filled squares) and 350 K (filled circles); experimental data of Choudary *et al.* [1993, 1996] at 306 K (open point-down triangles) and 353 K (open circles), and of Stach *et al.* [1986] at 250 K (open point-up triangles), 300 K (open squares), and 350 K (open diamonds). Reprinted with permission from [Pascual *et al.*, 2004b], © 2004 American Chemical Society.

Table 4.8 Enthalpies and entropies of adsorption of some alkanes and alkenes in silicalite, as obtained from the Langmuir model.

	Site 1 – ΔH°_{ads}	Site 1 – ΔS°_{ads}	Site 2 – ΔH°_{ads}	Site 2 -ΔS°_{ads}	Q°_{st}
	kJ·mol^{-1}	J·K^{-1} mol^{-1}	kJ·mol^{-1}	J·K^{-1}·mol^{-1}	kJ·mol^{-1}
Ethane	30.8	181	–	–	30
Ethene	26.6	175	–	–	27
Propane	38.9	189	–	–	39
Propene	39.2	193	–	–	38
Heptane	75.6	220	78.9	287	75
Heptene	74.3	221	72.0	268	73

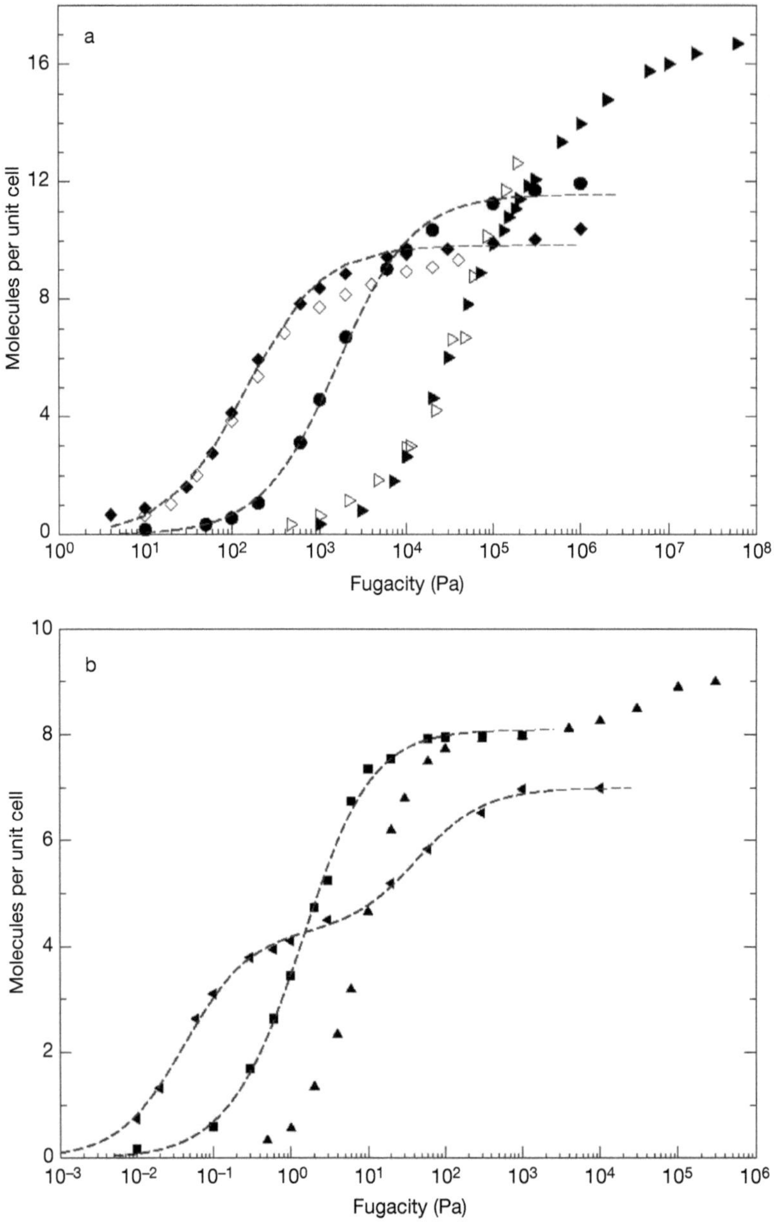

Figure 4.35 Adsorption isotherms of alpha-olefins in silicalite-1 at 300 K. Filled symbols are simulation results, and open symbols are experimental data. Dotted lines are Langmuir fits. Fig. 35-a: ethene (triangles), propene (circles), and but-1-ene (diamonds). Fig. 35-b: pent-1-ene (point-up triangles), hex-1-ene (squares), and hep-1-ene (leftward triangles). Reprinted with permission from [Pascual *et al.*, 2004b], © 2004 American Chemical Society.

muir parameters [Pascual *et al.*, 2004b] (see Table 4.8). Each silicalite channel can accept one propene molecule only, which leads to 12 supposedly equivalent sites that are totally filled at saturation. At low coverage, a weak preference for zigzag channels is observed (43% of the adsorbed molecules). Unfortunately, experimental data on adsorption of alkenes are very scarce, so this model cannot be definitively validated.

For but-1-ene, the saturation plateau is reached at 10 molecules per unit cell: all straight and zigzag channels are occupied, but half of the intersections are unoccupied. The same mechanism is at work here as is explained above (see Section 4.4.1) for butane adsorption: half the intersections are blocked by molecules with their mass centres located in the straight channels. At low coverage, all sites are occupied, with a weak preference for the zigzag channels (39% of the adsorbed molecules).

Pent-1-ene is the shortest molecule for which both Langmuir models (single site and dual site Langmuir models) fail. This is due to the fact that the molecular length of pentene is larger than that of the straight channel, but shorter than that of the zigzag channel. Table 4.9 clearly shows that all the straight channels are filled (43% of centres of mass are located in straight channels) whereas only 5% of terminal methyl groups sit in the straight channels. However, at high pressure pentene molecules leave some free space in intersections which allows for insertion of one new molecule. These moves in cavity pores modify the description of the adsorption sites and the SSL Langmuir model is no longer suitable. This location mech-

Table 4.9 Occupancy probability (%) of mass centres (mc) or terminal heads (hd) in the straight channels (SC), zigzag channels (ZC) or intersections (I) of silicalite at 308 K. <*N*> is the mean number of adsorbed molecules in the corresponding site, per unit cell.

		Low coverage				**Saturation**			
		<*N*>	**SC**	**ZC**	**I**	<*N*>	**SC**	**ZC**	**I**
volume fraction of the site (%)			33	45	22		33	45	22
Ehene	mc	0.8	31	47	22	16.0	25	49	26
Propene	mc	0.6	28	43	29	11.9	33	36	31
but-1-ene	mc	0.9	23	39	37	9.9	25	49	26
	hd		24	40	36		20	45	36
Z-but-2-ene	mc	1.1	17	28	55	9.9	30	40	30
	hd		16	35	49		28	42	30
E-but-2-ene	mc	0.6	17	26	57	9.1	32	41	27
	hd		17	33	50		27	43	30
Isobutene	mc	1.0	6	16	79	9.9	22	39	39
	hd		12	26	62		27	41	32
pent-1-ene	mc	0.6	19	38	43	9.0	43	44	13
	hd		20	43	37		5	47	48
hex-1-ene	mc	0.6	18	32	50	8.0	42	50	8
	hd		26	34	40		14	50	36
hept-1-ene	mc	0.7	19	28	53	7.0	8	56	36
	hd		34	29	37		34	49	17

anism is illustrated by the step observed at high pressure on the adsorption isotherm. At saturation (9.0 molecules per unit cell), all straight channels and zigzag channels are filled (43 and 44% of the adsorbed molecules respectively), as well as one intersection (13%).

An inflexion point appears for hex-1-ene and a two-step isotherm is observed for hept-1-ene. The geometrical considerations put forward for adsorption of alkanes in silicalite [Zhu *et al.*, 1998; Maginn *et al.*, 1995; Pascual *et al.*, 2003] (see Section 4.1.1) turn out to be equally true for alkene molecules. The hexene molecule can form a structure commensurate with the zigzag channel. The heptene molecule is longer than the zigzag channel, so that all the sites are blocked when four heptene molecules are adsorbed in one unit cell of silicalite. The insertion of a fifth molecule requires a rearrangement in the pores which explains the first plateau on the isotherm.

Adsorption isotherms of heptene at 300, 333 and 373 K are plotted in Figure 4.36. From the Langmuir fit parameters, adsorption enthalpies and entropies for both sites are derived and compared to those of heptane (Table 4.8). The first type of sites are similar for heptane and heptene, whereas the second is quite different. At low loadings, the rigid double bond of heptene weakly influences the location of the long alkyl chain, and the same behaviour is observed for heptane and heptene molecules. At high coverage, when insertion of a new molecule requires a strong reorganisation in the cavity pores, the double bond seems to play an important role.

A detailed study of butene isomers has been undertaken to shed light on the capability of the AUA potential to differentiate between molecular isomers behaviour. Adsorption iso-

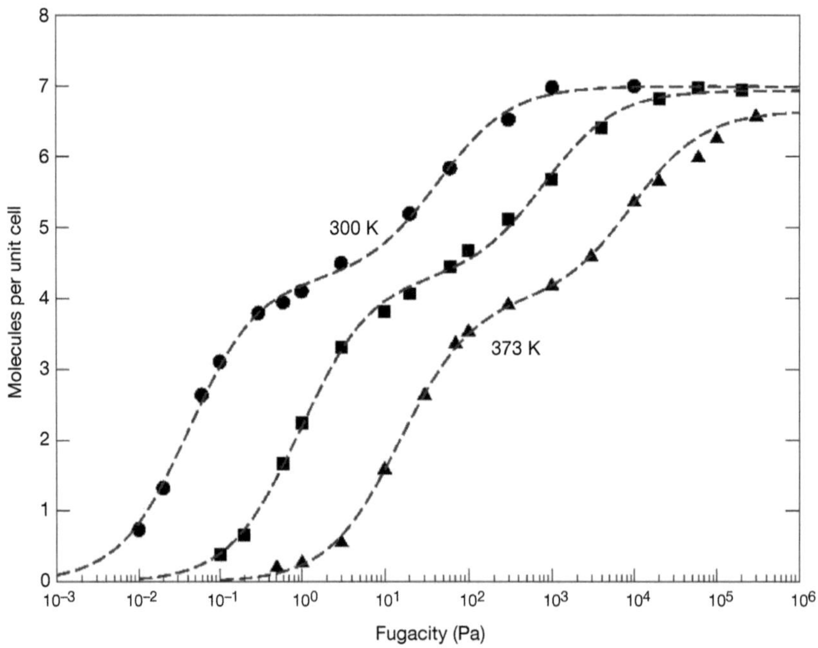

Figure 4.36 Calculated adsorption isotherms of hept-1-ene in silicalite-1 at 300 K (circles), 333 K (squares), and 373 K (triangles). Dotted lines: Langmuir fits. Reprinted with permission from [Pascual *et al.*, 2004b], © 2004 American Chemical Society.

therms of but-1-ene, Z-but-2-ene (i.e., *cis*-but-2-ene), E-but-2-ene (i.e., *trans*-but-2-ene), and isobutene in silicalite at 300 K are shown in Figure 4.37. The maximum loading is similar for the four isomers, but the location mechanisms are quite different. The preferential adsorption site is closely linked to the molecular geometry of the considered isomer. But-1-ene geometry is comparable to that of butane, and the two compounds exhibit a similar behaviour. For but-2-ene isomers, the rigid double bond is situated in the middle of the molecule and decreases its flexibility, so that intersection sites are preferentially filled at low coverages. At saturation, we find three molecules per silicalite unit cell in the straight channels, four in the zigzag channels, and 3 at intersections.

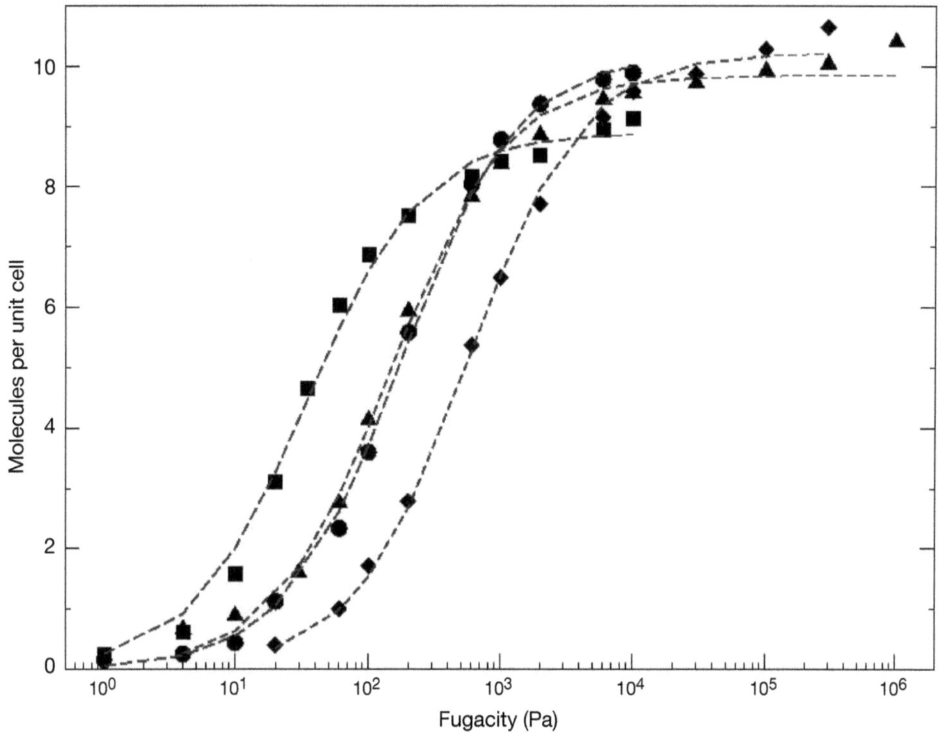

Figure 4.37 Calculated adsorption isotherms of butene isomers in silicalite-1 at 300 K. (Z)-but-2-ene (circles), (E)-but-2-ene (squares), but-1-ene (triangles), and isobutene (diamonds). Dotted lines: Langmuir fits. Reprinted with permission from [Pascual *et al.*, 2004b], © 2004 American Chemical Society.

For isobutene at low coverage (1 molecule per unit cell), 79% of the adsorbed molecules are located at intersections. At saturation (9.9 molecules per unit cell), all intersections and zigzag channels are occupied, but only two straight channels. This distribution is exactly the same as it is for isobutane at saturation (see Section 4.4.2). For this type of molecular geometry, planar and branched, sitting in intersections is more favourable than occupying straight or zigzag channels.

The heats of adsorption obtained by the fluctuations method (see Section 4.1.4.) are plotted in Figure 4.38 as a function of carbon number. Simulated values are in good agreement with the available experimental data [Choudary and Mayadevi, 1993, 1996; Stach *et al.*, 1986], indicating an increase of 9.1 $kJ \cdot mol^{-1}$ per additional CH_2 group on the main chain. Previous works on adsorption of alkanes in silicalite have shown an increase in the heat of adsorption amounting to 9.3 $kJ \cdot mol^{-1}$ per additional CH_2 group. Moreover, the adsorption enthalpies derived through the Langmuir parameters and the van't Hoff equation are consistent with the values obtained by the fluctuations method. Compared to the adsorption enthalpies of linear alkanes, the difference lies within 2-3 $kJ \cdot mol^{-1}$ whatever the carbon number, in favour of the alkanes.

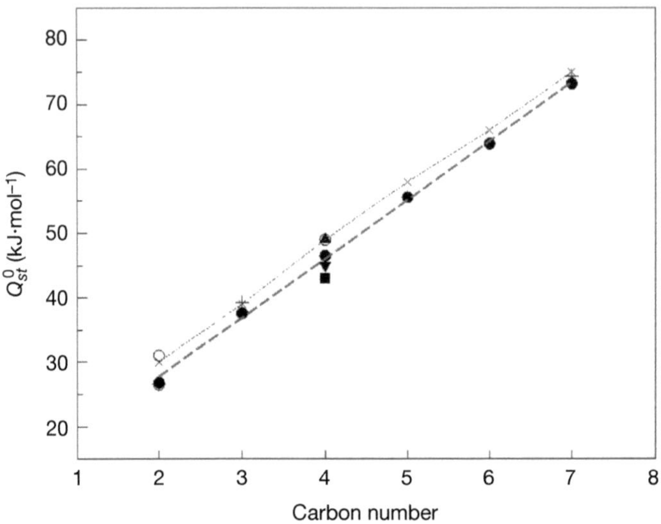

Figure 4.38 Heats of adsorption of alkenes is silicalite-1. Filled symbols are simulation results obtained with the fluctuations method, open symbols are experimental data of Choudary *et al.* [1993, 1996] and Stach *et al.* [1986]. (+): results obtained with the Langmuir theory. (x): heats of adsorption of corresponding linear alkanes, for comparison. Circles: alpha-olefins; square: isobutene; upward triangle: (E)-but-2-ene, downward triangle: (Z)-but-2-ene. Reprinted with permission from [Pascual *et al.*, 2004b], © 2004 American Chemical Society.

4.4.6 Binary Mixture Coadsorption Isotherms

Adsorption selectivity studies have been carried out at various temperatures on alkane-alkene mixtures of same carbon number (Fig. 4.39a, 4.39b, 4.40a, 4.40b, and 4.41) and on a mixture of diastereoisomers (Fig. 4.42). All coadsorption isotherms are plotted as a function of the total fugacity of the mixture. In all cases, the partial fugacities of both compounds in the binary mixture were fixed to the same value. It was checked that this condition leads to equimolar mixtures to an accuracy of within 3% in the temperature and pressure range used for computing the isotherms.

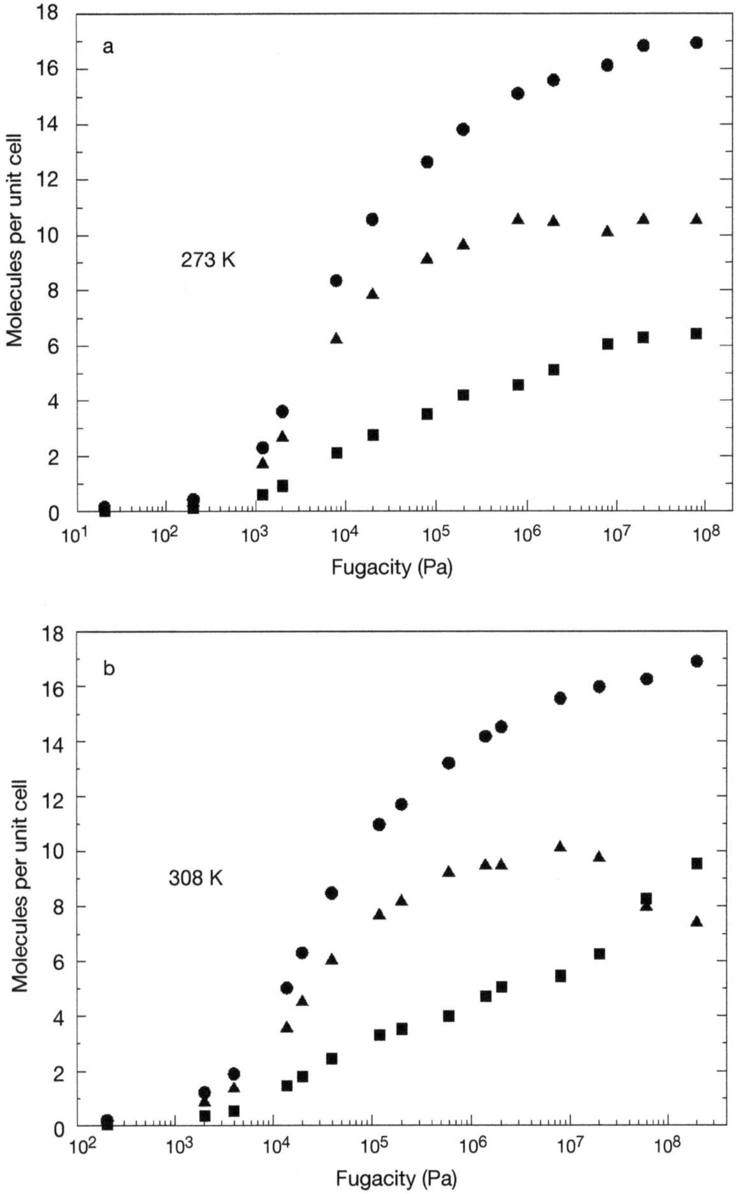

Figure 4.39 Calculated adsorption isotherm of a binary mixture of ethane and ethene at 273 K (a) and 308 K (b), as a function of the total fugacity of the mixture. Squares: ethene; triangles: ethane, circles: total quantity. Reprinted with permission from [Pascual *et al.*, 2004b], © 2004 American Chemical Society.

It can be observed that, at low pressure, silicalite preferentially adsorbs the alkane component in alkane-alkene mixtures of equal carbon number, and the E isomer in the but-2-ene diastereoisomer mixture. The selectivity factor α is simply calculated as the ratio of the number of alkane to alkene molecules in the crystal, since we are dealing with roughly equimolar fluid mixtures. In all cases, selectivity decreases with pressure.

For the ethane-ethene mixture at 273 K (Fig. 4.39a), the selectivity factor α decreases smoothly from a value of 2.9 at 2 000 Pa to 1.6 at 80 MPa. The preferential adsorption of the alkane over the alkene is consistent with the computed heats of adsorption of the two pure components. However, the adsorption entropy of ethane is larger than that of ethene. One can thus expect that, at sufficiently high temperature, the entropy effect compensates the differences in enthalpy. Indeed, at 308 K (Fig. 4.39b), a reversal of selectivity appears at high loadings and at saturation, silicalite preferentially adsorbs the olefinic molecule: $\alpha = 2.3$ at 0.6 MPa and 0.8 at saturation).

For the propane-propene mixture, a similar trend is observed (Fig. 4.40a and 4.40b). However, the difference between the heats of adsorption of propane and propene is smaller than in the case of ethane and ethene. The selectivity factor is also smaller ($\alpha = 1.4$, 1.6 and 1.3 at pressures of 200 Pa, 6 000 Pa and 200 MPa respectively). The entropies of adsorption are very close (– 189 and –193 $J \cdot mol^{-1} \cdot K^{-1}$ for propane and propene respectively) and the selectivity is not significantly modified when the temperature increases. From a qualitative point of view, the effect of a single double bond on the adsorptive properties of hydrocarbons is smoothed out as the number of carbon atoms increases in the molecule.

For the heptane-heptene mixture (Fig. 4.41), the behaviour is more complex due to the existence of two distinct adsorption sites. The first adsorption site is quite similar for the two molecules (75.6 and 74.3 $kJ \cdot mol^{-1}$ for heptane and heptene respectively) and this leads to a weak selectivity for heptane at low coverage ($\alpha = 1.4$ at 2 Pa). However, the second adsorption site is quite different. Although enthalpic effects are strongly in favour of heptane (78.9 and 72.0 $kJ \cdot mol^{-1}$ for heptane and heptene respectively), selectivity tends to decrease because of entropic effects: the important difference between entropies of adsorption of heptane and heptene in the second site (– 287 and –268 $J \cdot mol^{-1} \cdot K^{-1}$ respectively) counterbalances the enthalpic effects. No selectivity is observed at saturation ($\alpha = 1$).

For the but-2-ene diastereoisomer mixture, a selectivity in favour of the E isomer is observed, which decreases with loading (Fig. 4.42). The selectivity factor is 3.9, 3.8, and 1.5 at 12 Pa, 400 Pa and 0.12 MPa respectively. The selectivity at low loadings is driven by energetic effects. The computed heats of adsorption are 50 and 45 $kJ \cdot mol^{-1}$ for the E and Z isomers respectively. As for alkane-alkene coadsorption, the decrease of the selectivity factor at saturation can be interpreted in terms of entropic effects. It can be observed (Table 4.9) that the Z-but-2-ene isomer fits better in the intersection sites than does the E-but-2-ene isomer.

In conclusion to this study of alkane and alkene adsorption in silicalite by Monte Carlo molecular simulation in the Grand Canonical ensemble using the AUA-4 force field and optimised parameters for oxygen to describe the sorbate-zeolite interactions, it has been demonstrated that the difference between adsorption heats is responsible for the observed selectivities at low coverage, whereas towards saturation, selectivity is driven by entropic effects. This is consistent with the "configurational entropy effects" described by Krishna for coadsorption of mixtures of branched and linear alkanes in silicalite [Krishna *et al.*, 2002]. In alkane-alk-

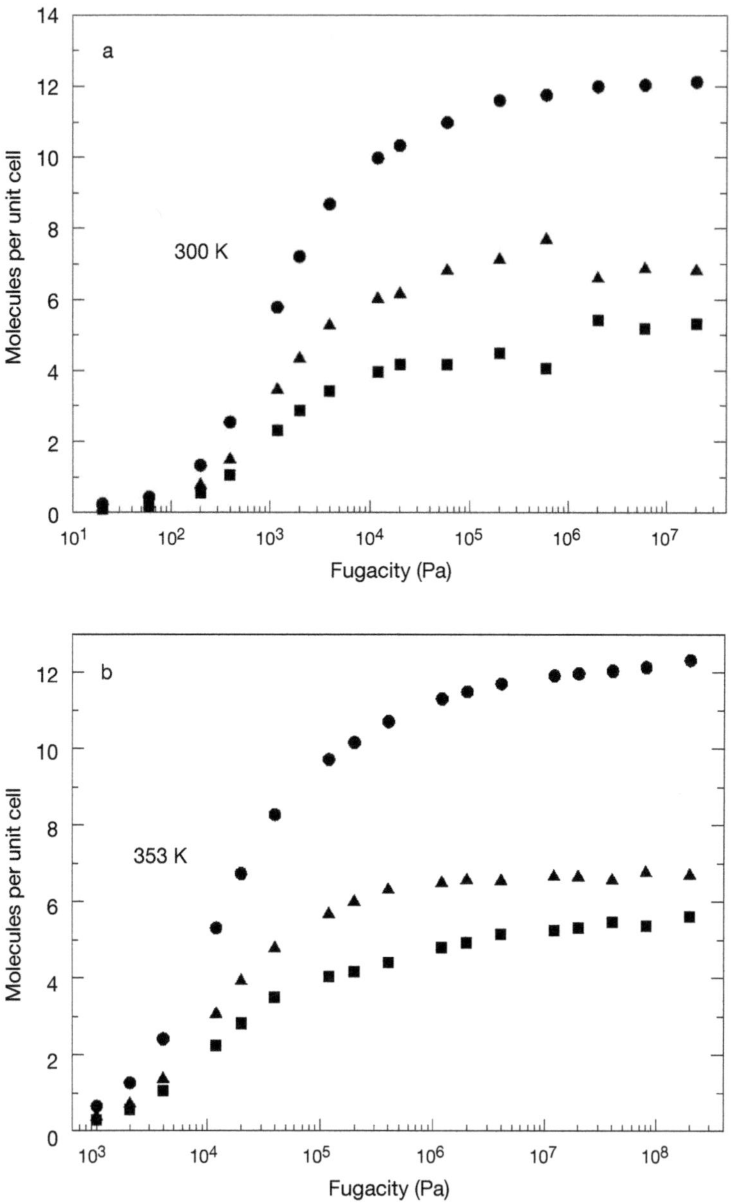

Figure 4.40 Calculated adsorption isotherm of a binary mixture of propane and propene at 300 K (a) and 353 K (b), as a function of the total fugacity of the mixture. Squares: propene; triangles: propane, circles: total quantity. Reprinted with permission from [Pascual *et al.*, 2004b], © 2004 American Chemical Society.

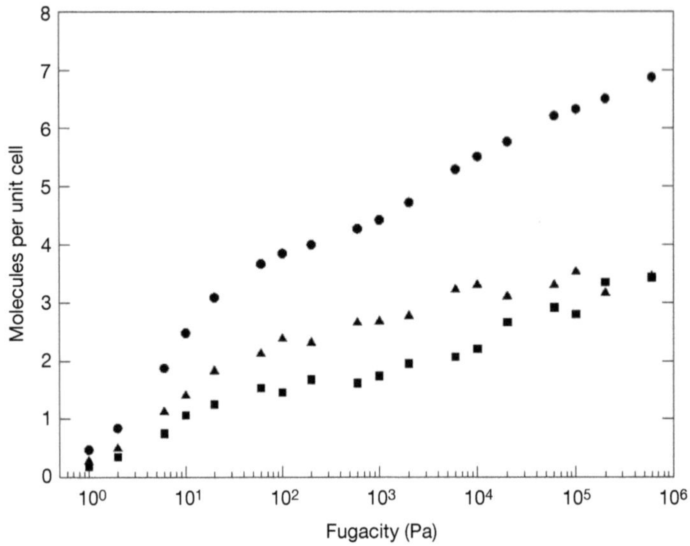

Figure 4.41 Calculated adsorption isotherm of a binary mixture of heptane and heptene at 373 K, as a function of the total fugacity of the mixture. Squares: heptene; triangles: heptane; circles: total quantity. Reprinted with permission from [Pascual *et al.*, 2004b], © 2004 American Chemical Society.

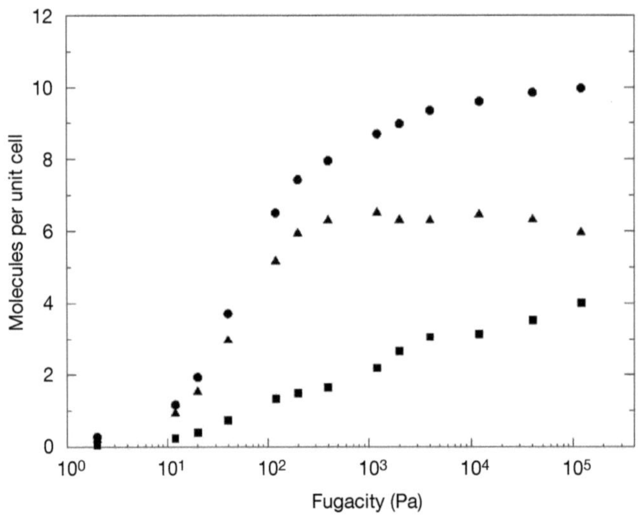

Figure 4.42 Calculated adsorption isotherm of a binary mixture of (Z)- and (E)-but-2-ene at 300 K, as a function of the total fugacity of the mixture. Squares: (Z)-but-2-ene; triangles: (E)-but-2-ene; circles: total quantity. Reprinted with permission from [Pascual *et al.*, 2004b], © 2004 American Chemical Society.

ene mixtures, the configurational entropy effects favour the adsorption of alkanes because these molecules "pack" more efficiently within silicalite pores than their somewhat more rigid alkene homologues. It should be stressed however that these results are predictions which have not yet received any experimental confirmation. Pure component adsorption data at different temperatures enabled us to evaluate the adsorption heat and entropy of the individual components. The computed values are perfectly consistent with the observed selectivities in binary mixtures coadsorption simulations.

To summarize, methods have been developed to build a transferable force field for adsorption in zeolites. Simple combining rules are used to determine cross potential parameters from the optimised parameters for zeolitic oxygen and the AUA-4 force field parameters for hydrocarbon interactions. This allowed us to predict adsorption of linear and branched alkanes and alkenes using only four experimental points for butane in silicalite to adjust the model. The results obtained perfectly match the available experimental data.

In the next section, the force field is applied to aluminosilicate zeolites: Faujasites. Although these are solids in which coulombic interactions are expected to be important, because of the presence of aluminium and counter-charge cations, no electrostatic terms were at present incorporated in the interaction potential description.

4.4.7 Separation of Branched Alkanes on Faujasite Type Zeolites

In this section, the predictive character of molecular modeling is highlighted: simulation has been used as a tool to obtain insight into the behaviour of Faujasite type zeolites exchanged with various charge-compensating cations to assess the effect of the nature of the cation and thus help identify suitable solid candidates for a given adsorptive separation process.

Separation of branched alkane isomers relatively to their degree of branching, for instance separating a methyl-alkane from one of its dimethyl-alkane isomers, can be quite difficult since the boiling points of these compounds may be very close, thus forbidding a separation by distillation processes.

In the present work, adsorption of mixtures of branched alkanes with 6 carbon atoms has been simulated on Faujasite zeolites exchanged with various charge-compensating cations. Adsorbed quantities and selectivities of adsorption have been computed by grand canonical Monte Carlo simulation at 423 K and 2 MPa total pressure for an equimolar binary mixture of 2-methylpentane (2MP) with 2,3-dimethylbutane (23DMB) on $M_{48}Y$ zeolites, where M is a Na^+, K^+, Rb^+ or Cs^+ cation, and $M_{48}X$ zeolites, where M is a Ca^{2+} or Ba^{2+} cation.

Models of X and Y zeolites have been taken from Uytterhoeven *et al.* [1992]. For $M_{48}Y$ solids (M = Na, K, Rb and Cs, note that these zeolites have a Si/Al ratio equal to 3), the crystals are described in the F23 symmetry group, and have a cubic unit cell of parameter a (= b = c) = 25.0184 Å. The distribution of the 48 charge-compensating cations is 16 in sites I, and 32 in sites II, which is in accord with the simulation results presented above (Section 4.2.1). The I and II cationic sites are thus fully occupied, which leaves just one possible configuration for the cations. Let us recall that site I is located at the centre of the hexagonal prism linking two sodalite units, whereas site II sits on the 111 symmetry axis of the solid, into the supercage and facing an hexagonal window of a sodalite cage (see Fig. 4.3). When the mass of the charge-compensating cation increases, as one goes from Na^+ to K^+, Rb^+ and Cs^+, the position of the corresponding site II shifts towards the centre of the supercage (Fig. 4.43).

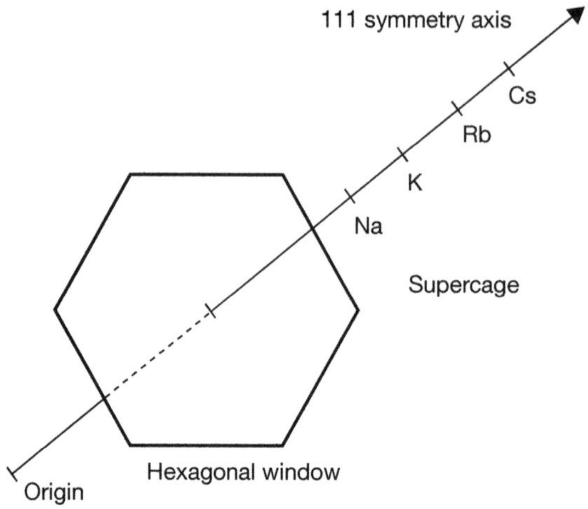

Figure 4.43 Positions of the site II charge compensating cations on the 111 symmetry axis of the faujasite.

$M_{48}X$ Faujasites, with a Si/Al ratio equal to unity and M being a divalent cation such as Ca^{2+} or Ba^{2+}, are described in the Fd3 symmetry group, and have a cubic unit cell of parameter a (= b = c) = 25.5196 Å. The distribution of the 48 charge-compensating cations is 16 in sites I, and 32 in sites II, exactly the same as the $M_{48}Y$ zeolites described in the previous paragraph, and the position of the site II on the 111 symmetry axis is correlated to the size of the counter-charge ion in the same manner as well.

A. Derivation of Force Field Parameters for the Cations

Interactions between adsorbed species are described by the AUA-4 force field [Ungerer *et al.*, 2000; Bourasseau *et al.*, 2002]. For the zeolite-adsorbate interactions, silicium and aluminium atoms of the solid backbone have not been explicitely taken into account. Interactions with the zeolitic oxygen atoms are described by a Lennard-Jones potential, without electrostatic energy terms, and the parameters are those optimised by Pascual (see Section 4.3 above) [Pascual *et al.,* 2003]. The values are $\sigma = 3.00$ Å and $\varepsilon/k = 93.53$ K. Lorentz-Berthelot combining rules on these parameters and the AUA-4 Lennard-Jones parameters have been used to describe adsorbate-oxygen interactions.

Interactions between adsorbates and the cations are also described by a Lennard-Jones interaction potential, with no electrostatic terms. Parameters for the sodium cation have been taken from Dang [1995]: $\sigma = 2.584$ Å and $\varepsilon/k = 50.34$ K. To derive Lennard-Jones parameters for the other cationic species, we have proceeded as follows: first, the geometric parameter σ was estimated through a linear correlation on the ionic diameters D_M of species M, the sodium cation being taken as a reference:

$$\frac{\sigma_{Na} \cdot D_M}{D_{Na}} = \sigma_M \tag{4.13}$$

where M is any of the K^+, Rb^+, Cs^+, Ca^{2+}, Sr^{2+} or Ba^{2+} cations. Values for the ionic radii have been taken from the University of Sheffield internet site http://www.webelements.com, the cations being in their highest coordination state; they are listed in Table 4.10.

To derive the energetic parameter ε, use has been made of the Eq. (2.34) approximative formula, derived from the theory of London, for the dispersion potential between two particules indexed A and B:

$$\langle V(r_{AB}) \rangle_{\text{dispersive}} = -\frac{3}{2}\left(\frac{E_A E_B}{E_A + E_B}\right)\frac{\alpha_A \alpha_B}{(4\pi\varepsilon_0)^2}\frac{1}{r_{AB}^6} \tag{4.14}$$

r_{AB} being the distance between A and B, α_A and α_B the polarisabilities of these particles, and E_A and E_B being approximately equal to the energies of their first electronic transitions [Alberty and Silbey, 1996]. Polarisabilities have been taken from the internet site http://www.webelements.com. The expression (4.14) is equalised to the dispersion part of the Lennard-Jones potential:

$$U(r_{AB})^{\text{LJ}}_{\text{dispersive}} = -4\varepsilon_{ij}\frac{\sigma_{AB}^6}{r_{AB}^6} \tag{4.15}$$

If A and B are the same species (i.e. $E_A = E_B$), we can express the polarisability α of a particle as a function of its Lennard-Jones parameters:

$$\alpha^2 \propto \varepsilon \cdot \sigma^6 \tag{4.16}$$

For any cation M with polarisability α_M, the Lennard-Jones σ parameter is derived by means of Eq. (4.13), and ε by means of (sodium being the reference):

$$\varepsilon_M = \frac{\varepsilon_{Na}\sigma_{Na}^6}{\alpha_{Na}^2} \cdot \frac{\alpha_M^2}{\sigma_M^6} \tag{4.17}$$

Table 4.10 Ionic radii, diatomic polarisabilities, and derived Lennard-Jones parameters for the ionic species considered in this work.

Cation	Ionic radii (Å)	σ (Å)	α (10^{-24}cm^3)	ϵ/k (K)
Na^+	1.32	2.584	23.6	50.340
K^+	1.65	3.230	43.4	44.628
Rb^+	1.75	3.426	47.3	37.241
Cs^+	1.88	3.680	59.6	38.466
Ca^{2+}	1.26	2.467	22.8	62.113
Ba^{2+}	1.56	3.054	39.7	52.284

Interactions between adsorbed species and cations are computed from these parameters and the AUA-4 force field Lennard-Jones parameters through the Lorentz-Berthelot combining rules. Let us stress again that electrostatic energy, permanent or induced, has not been taken into account in the force field. Even if we are dealing with non polar molecules such as alkanes, the zeolite itself, and in particular the cations, are charged species and it is quite possible that polarisation energy plays a non negligible role in the mechanisms of adsorption. This role can not be assessed with the method used here, though.

The chemical potentials of the two adsorbates of the binary equimolar mixture have been computed at 423 K and 2 MPa with the AUA-4 force field in the *NPT* statistical ensemble; it was checked that the mixtures are in the liquid state at these thermodynamic conditions. Grand canonical Monte Carlo simulations in the presence of the solid have then been performed. Monte Carlo steps consisted of 15% classical translation moves, as many classical rotation moves, 20% internal regrowth of molecules, and 50% insertion or deletion moves, these latter chosen with equal probability. At least 2 million Monte Carlo steps were done to achieve stabilisation of the system, and equilibrium was checked by insuring that both number of adsorbed molecules (for each species) and the system's total energy fluctuated around well-defined mean values. Statistical averages were then taken over a minimum of 10 million Monte Carlo steps.

B. Influence of the Nature of the Charge-compensating Cation on the Adsorption Properties

Results of the simulations for adsorption of the 2-methylpentane/2,3-dimethylbutane equimolar mixture (C_6 isomers "2MP" and "23DMB") on X and Y zeolites with 48 charge-compensating cations are presented in Figures 4.44 and 4.45. Figure 4.44 shows the evolution of the 2MP/23DMB adsorption selectivity (as defined by Eq. 4.1) with the ionic radii of the species, and Figure 4.45 shows the adsorbed quantities of both species as well as the total adsorbed quantity.

It appears that the adsorption binary selectivity is correlated to the size of the charge-compensating cation: for the smallest cation (Ca^{2+}), it is in favour of the alpha-methyl isomer 2MP, and as the cationic radius increases, selectivity shifts towards lower values, becoming in favour of the di-methyl isomer 23DMB for the two largest cations: Rb and Cs.

For the Y zeolites, the position of site II counter-charge cations on the 111 symmetry axis of the solid is correlated to the size of the cation: the bigger the size, the further the position shifts from the centre of the hexagonal window of a sodalite cage towards the centre of the supercage. If adsorbed quantities are drawn as a function of the distance of site II from the origin of the crystal (Fig. 4.46), the shape of the curves are similar to those in Figure 4.45.

A striking particularity appearing in Figures 4.45 and 4.46 is that the nature of the charge-compensating cation seems to have an effect only on the 2MP isomer, and none on 23DMB. A tentative explanation can be proposed in terms of entropic effects.

When the size of the cation decreases, and its position shifts from the centre of the supercage towards the hexagonal window of a sodalite unit, the available space for adsorption increases. At constant temperature, the second law of thermodynamics tends to favour the

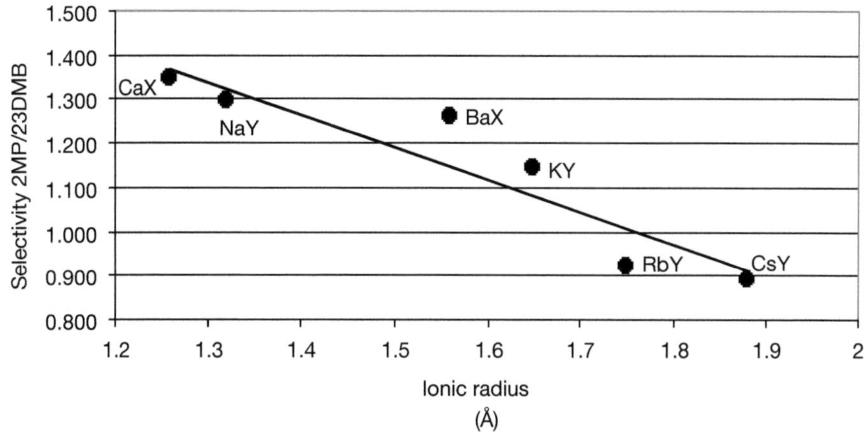

Figure 4.44 Binary adsorption selectivity of 2-methylpentane over 2,3-dimethylbutane for equimolar mixtures at 423 K and 20 bar on various faujasites with 48 charge compensating cations, versus ionic radii of the cations.

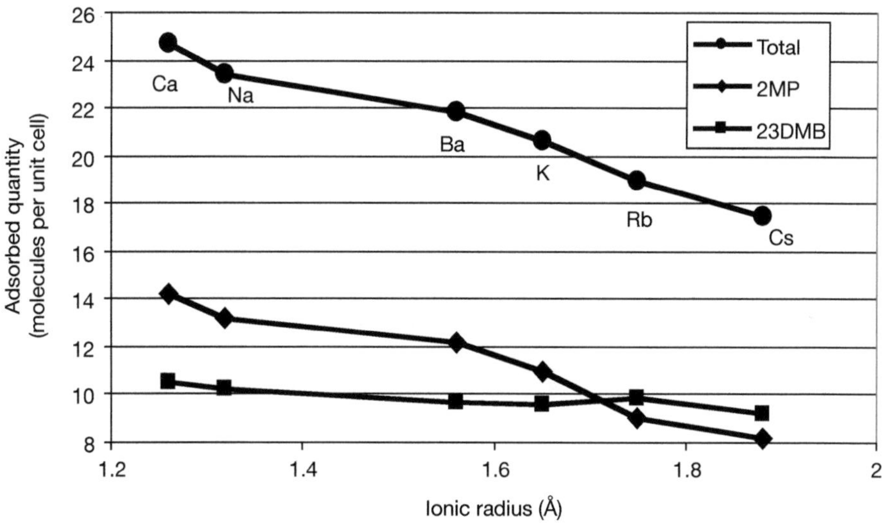

Figure 4.45 Adsorbed quantities of 2-methylpentane (2MP), 2,3-dimethylbutane (23DMB), and total, for equimolar mixtures at 423 K and 20 bar on various faujasites with 48 charge compensating cations, versus ionic radii of the cations.

inclusion of extra molecules more strongly for the more flexible species, relatively to the more rigid. Thus, entropic effects are responsible for an enhanced adsorption of 2MP relatively to 23DMB when the available space increases.

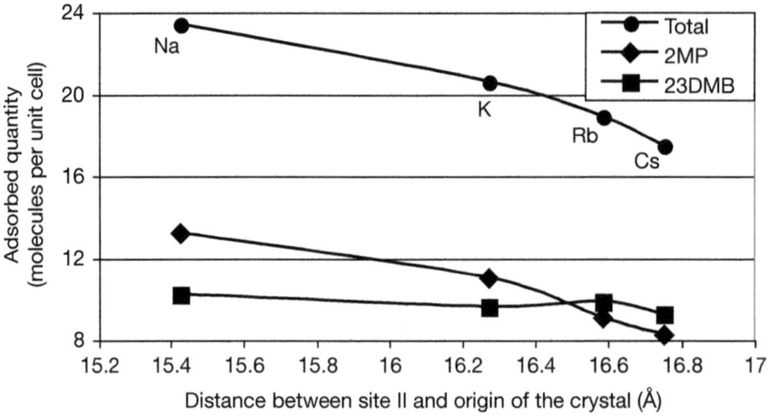

Figure 4.46 Adsorbed quantities of 2-methylpentane (2MP), 2,3-dimethylbutane (23DMB), and total, for equimolar mixtures at 423 K and 20 bar on various Y faujasites with 48 monovalent charge compensating cations, versus distance between cationic site II and the origin of the crystal.

4.5 SEPARATION OF THIOLS FROM NATURAL GAS ON FAUJASITES

Alkanethiols (which were discussed in Section 3.3) are present at low concentrations in natural gases but their strong smell and toxicity makes it important to remove them. In order to extract these molecules selectively at an early stage in production operations, Faujasite zeolites could be used in high pressure conditions (typically 5 MPa). It is therefore important to know the competitive adsorption of alkanethiols with the other components of natural gas (alkanes, aromatics, water, carbon dioxide, hydrogen sulphide). Because of the large number of components involved in such an application, it is essential that a convenient way of estimating the zeolite-guest interaction parameters is available. For this purpose, we will use the effective parameters for the oxygens of the zeolite framework together with combining rules [Pascual *et al.,* 2003], as introduced in Section 4.3.

The investigation of this problem by Monte Carlo simulation will proceed in two main steps. In the first step, we will study the pure component isotherms of a representative alkanethiol, namely ethanethiol, for which recent experimental results are available. In order to know the sensitivity of these simulation results to interaction parameters, we will test the influence of the intermolecular potential of alkanethiols and of the zeolite-guest interaction potential.

In the second step, we will predict the coadsorption of the components of a typical natural gas which is assumed to have been dehydrated because alkanethiol removal is performed at the outlet of a standard dehydration process. This will use a simplified representation of gas composition with nine components. As water turns out to be the major adsorbed species, a preliminary comparison of experimental and simulated water adsorption isotherms will be made to check that its interaction with the zeolite is reasonably reproduced.

4.5.1 Adsorption Isotherms of Alkanethiols

As discussed in Section 3.3, the Anisotropic United Atoms potential of alkanethiols of Delhommelle *et al.* [2000] makes use of atomic charges located on each carbon, sulphur and hydrogen atom. It has been also mentioned that a simplified model with only three electrostatic charges is able to represent reasonably well the properties of alkanethiols. Both potential models are tested here for the prediction of the adsorption isotherms of ethanethiol on NaX zeolite at 298 K and 373 K (Fig. 4.47), as allowed by recently published experimental data [Weber *et al.*, 2005]. In each case, the zeolite-guest interactions are computed with the method of Pascual *et al.* [2003], i.e. using the Lorentz-Berthelot combining rules to obtain unlike interactions parameters between the oxygen framework atoms and the molecular groups.

As can be seen in Figure 4.47, the simulations predict qualitatively the maximum adsorbed amount (15% overshot) and the shift of the isotherm toward higher adsorbed amounts. Also, the isotherm is found with a steeper slope at low temperature, as observed. Given that the model parameters have not been fitted for this purpose, such a result could be considered as perfectly satisfactory. It can be seen that the influence of the intermolecular potential model (nine atomic charges versus three charges) is significant, but not dramatic. Similarly, it has been checked that changing the combining rule from Lorentz-Berthelot to Kong or changing the cation parameters did not result in major changes of the adsorption isotherms. It may be therefore asserted that molecular simulation provides a fair qualitative prediction of the adsorption behaviour in this system.

If a quantitative model is desired, it is possible at this stage to fit slightly one of the model parameters to selected points of the isotherm. However the main goal of the present study is not to provide quantitative predictions, but to investigate the coadsorption of multicomponent mixtures in a qualitative way, as we will see in the next section. We will thus not attempt to optimise further the zeolite-guest potential energy model.

4.5.2 Coadsorption of Alkanethiols with Other Components of Natural Gases

A simplified composition of a natural gas that should be purified from its organic thiols is given in Table 4.11. Methane and the light hydrocarbons (ethane, propane) are the most abundant components. Minute amounts of heavier hydrocarbons (up to eight or nine carbon atoms) are generally present in these gases and here they are represented schematically by *n*-heptane and toluene. The amount of carbon dioxide present is not negligible (2%). Alkanethiols are assumed to be present in minor amounts (50 and 100 ppm) and H_2S in very minor amount (4 ppm). The water content is assumed to be very low (50 ppm) because it is assumed that the gas would be dehydrated before the adsorption process.

The conditions in which the coadsorption is evaluated are 298 K and 5 MPa, i.e. typical of target conditions for the process. In such high pressure conditions, there is a significant departure from ideal gas behaviour. The chemical potential of every component has been thus determined using the biased Widom test insertions in a monophasic *NPT* simulation. Then

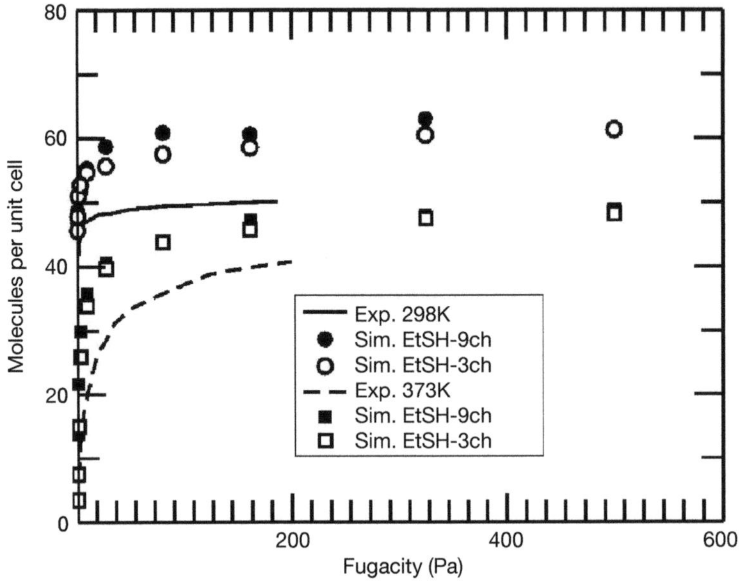

Figure 4.47 Comparison of observed adsorption isotherms of ethanethiol on NaX faujasite at 298 K (circles) and 373 K (squares) by Bellat, Weber *et al.* [2004] with simulation results using two potential models. The first potential model involves nine atomic charges (9ch) [Delhommelle *et al.*, 2000] and the second model involves three charges (3ch).

these chemical potentials have been imposed in a grand canonical simulation with ten components, yielding the composition of the adsorbed phase indicated in Table 4.11.

It may be seen that the most abundant adsorbed species (more than 80%) is water, although it is a minor component in the gas. This is due to the high dipole moment of water which means that it interacts strongly with the zeolite. Organic thiols are adsorbed to a lesser extent but are nevertheless present in significant amounts (10-13% for methanethiol and 3% for ethanethiol). Carbon dioxide is the fourth most abundant which is due to this molecule's high quadrupole moment coupled with its relatively high initial concentration in the gas. The total number of adsorbed molecules (205 molecules per unit cell at 5 MPa) shows that the zeolite is close to being saturated (Faujasites contain approximately 300 water molecules in saturation conditions). This is not surprising if we consider that natural gases are not far from their dew point in such low temperature conditions, as they contain liquid components. The average mole number of every hydrocarbon and H_2S is lower than one molecule per unit cell, which is too small to allow the reliable determination of their concentrations. It may be simply stated that the concentration of these species is lower than one molecule per unit cell, i.e. approximately lower than 0.5%.

At this stage, it is wise to check that the water-zeolite interaction potential is qualitatively correct by testing the prediction of the water adsorption isotherms on Faujasites. This vali-

Table 4.11 Simulation of the adsorption of a dehydrated natural gas containing traces of methanethiol and ethanethiol on NaX Faujasite in representative conditions of an industrial process under investigation (298 K, 0.1 and 5 MPa). Liquid hydrocarbons, carbon dioxide, hydrogen sulphide and water are also minor components in the gas. When less than one molecule is adsorbed per unit cell for a given molecular species, the adsorbed amount is not considered statistically representative and an upper limit is just given.

		***P* = 0.1 MPa**			***P* = 5 MPa**		
	Molar fraction of the gas phase	**Chemical potential/*k* (K)**	**Molecules per unit cell**	**Molar fraction in adsorbed phase**	**Chemical potential/*k* (K)**	**Molecules per unit cell**	**Molar fraction in adsorbed phase**
Methanethiol	50 ppm	– 6 117.5	23	0.130	– 4 951.75	20	0.098
Ethanethiol	100 ppm	– 5 910.98	6	0.034	– 4 745.20	6	0.029
Toluene	100 ppm	– 5 910.98	< 1	< 0.006	– 4 745.2	< 1	< 0.005
n-heptane	1000 ppm	– 5 431.36	–	< 0.006	– 4 265.6	< 1	< 0.005
CO_2	0.02	– 4 332.08	1	0.006	– 3 166.3	< 1	< 0.005
Propane	0.03	4 332.08	< 1	< 0.006	– 3 166.3	< 1	< 0.005
Ethane	0.10	– 4 059.02	< 1	< 0.006	– 2 897.6	< 1	< 0.005
Methane	0.85	– 3 197.69	< 1	< 0.006	– 2 053.84	< 1	< 0.005
H_2O	60 ppm	– 6 063.2	144	0.818	– 4 897.4	177	0.863
H_2S	4 ppm	– 6 870.2	< 1	< 0.006	– 5 704.42	< 1	< 0.005
Total			176			205	

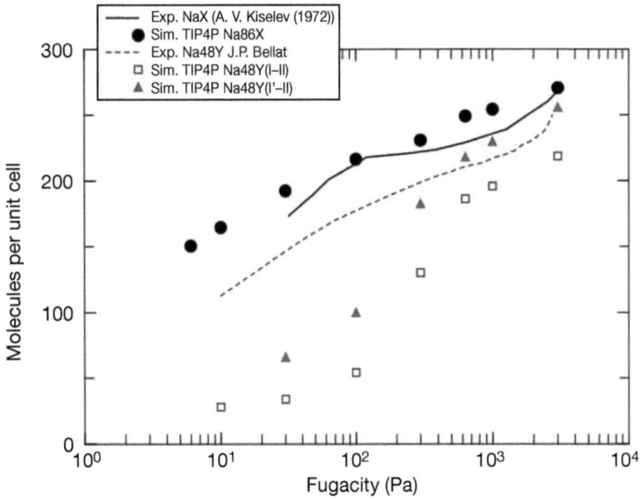

Figure 4.48 Comparison of observed adsorption isotherms of water on NaX and NaY faujasites at 298 K with simulation results using the TIP4P potential [Jorgensen *et al.*, 1983] and the parametrization of guest-host Lennard-Jones interactions by Pascual *et al.* [2003]. Two ways of placing the cations have been tested for NaY faujasite.

dation is illustrated by Figure 4.48. It can be seen that the variation of adsorbed amounts with pressure is reasonably well represented without specific parameter fitting. The stronger adsorption of water in NaX than in NaY Faujasite is qualitatively predicted.

As a conclusion to this work, we can say that molecular simulation has identified water as the most likely competitor of organic thiols in the coadsorption process in sodium-exchanged Faujasites, even if it is assumed that the gas has been subject to a dehydration step. This result should help guide process development in specific directions, e.g.: towards the identification of substitute cations to reduce the adsorption of water compared with that of organic thiols; towards measurement of water/thiol adsorption selectivity; and towards more efficient dehydration of the gas prior to thiol removal. Such a study would not have been possible if a readily applicable zeolite-guest potential had not been available. In this respect, the derivation of zeolite-guest interaction parameters from effective oxygen Lennard-Jones parameters and combining rules has proved to be a key step in the investigation of multicomponent adsorption.

5

Conclusion and Perspectives

For any industry, adopting a new technology like molecular simulation is a step that should only be made with great care because a lot of effort and money could be wasted. Excessive confidence in a new technology inevitably brings disappointment after a few years of fruitless effort – if not wrong decisions – and may undermine its potential industrial application for a decade. In contrast, the industry may lose the opportunity of major savings or breakthroughs if it sticks to conventional technology for too long. What should be the attitude of the oil and gas industry (and more generally the chemical industry) towards the Monte Carlo techniques discussed in this book? Let us list the facts that variously support or argue against the usefulness of Monte Carlo simulation, without hiding negative aspects if any.

As briefly explained in Chapter 2, Monte Carlo simulation is a theory that is based on fundamental principles from the science of physical chemistry, e.g. no approximation is needed to write the laws of statistical mechanics or the Coulomb's law of electrostatic interactions. However, the practical implementation of molecular simulation requires intermolecular potential energy models that are partly empirical. Determining suitable potential parameters for the prediction of thermodynamic properties has made a lot of progress in the last decade but it is still a difficult point that often limits the accuracy of Monte Carlo results. Another problem with Monte Carlo simulation is that it is computer-intensive with typical simulations lasting hours or days rather than seconds or minutes. Also, systems displaying long relaxation times or presenting heterogeneities on a scale larger than 5-10 nanometers are very difficult to simulate at the molecular level because either long simulations are necessary or large systems have to be addressed. Despite these basic limitations, reliable simulations can now be performed on various systems, thanks to the development of numerous ingenious algorithms which save computer time while nevertheless exploring all possible conformations of the systems under investigation.

In Chapters 3 (fluid systems) and 4 (adsorption), a large array of examples of applications were discussed. What kind of practical relevance do these have for the industry and what contributions do they make?

The first contribution relates to the properties of pure components (heavy hydrocarbons, organo-mercuric compounds, etc.) for which basic data acquisition is hampered by toxicity or by the lack of the availability of pure chemicals. This kind of prediction is useful not only for the oil and gas industry but also for the chemical industry as illustrated by the example of aldehydes presented by Kranias *et al.* [2003]. Compared with classical methods of predicting thermodynamic properties involving group contributions, molecular simulation methods can yield better predictions of the equilibrium properties of compounds solely on the basis of their chemical structure. This capacity is mainly due to the transferability of the intermolecular potentials (i.e. force fields) used in these simulations and to predict property differences between isomers. These force fields allow the simultaneous description of volumetric properties, vapour pressure and energy properties over a large range of different temperatures (which is often not the case with engineering equations of state). However, it cannot be expected that molecular simulation provides predictions of similar accuracy to specifically calibrated thermodynamic models for well-known compounds.

A second valuable contribution is in the prediction of fluid properties which are difficult to measure by experimental means under high pressure, including the Joule Thomson coefficients of natural gases, phase diagrams of mixtures containing hydrogen sulphide, and polyethylene swelling upon the absorption of carbon dioxide at high pressure. The various examples of this kind show that existing algorithms and software are up to dealing with complex fluids as we have treated mixtures containing as many as 19 different components and compounds containing as many as 100 carbon atoms with reasonable computer resources. Interestingly, the Gibbs ensemble technique is not only useful for classical liquid-vapour phase equilibrium calculations but also for liquid-liquid equilibria and three-phase equilibria without significant increase in complexity. Phase equilibria with specific constraints can also be handled, as exemplified by bubble point calculations. Liquid-vapour critical regions in binary mixtures can now be simulated in a way that is consistent with near-critical scaling behaviour, which is impossible with analytical equations of state for fundamental reasons.

The third contribution relates to the thermodynamics of adsorption of organic compounds and polar species in microporous solids through grand canonical Monte Carlo simulation. Thanks to the recent extension of the Anisotropic United Atoms force field from vapour-liquid to adsorption in zeolites, it may be asserted that a general, consistent method is now available for obtaining fair predictions of pure component isotherms as well as coadsorption for a very large variety of systems. These methods can account explicitly for the nature and location of cations in zeolites. Here again, today's simulation methods and algorithms are capable of addressing complex systems such as natural gases with ten components including polar compounds. Although predictions are sometimes more qualitative than quantitative, there is no alternative theory to provide comparable results, and molecular simulation can be considered as a real breakthrough technology in this field.

A fourth important aspect of molecular simulation is that it provides an unprecedented understanding of macroscopic properties in terms of molecular structure. In the case of fluid properties, molecular simulation elucidates the influence of polarity upon the behaviour of pure compounds and mixtures. As shown by the examples of the methanol-propane and water-hydrogen sulphide systems, Monte Carlo simulation may be used to characterise associations in mixtures containing associating polar components. Unlike classical association models in which association constants must be correlated, association appears in simulation as the result of interactions occurring at the molecular level. The association constants obtained by simulation could be used in recent equations of state that incorporate explicitly an association term, e.g. CPA [Kontogeorgis *et al.*, 1996]. In the case of adsorption phenomena, molecular simulation indicates the type of site to which the various molecules are preferentially adsorbed. More importantly, it is able to predict qualitatively how zeolite adsorption properties can change with either adsorbent structure, silicon/aluminium ratio or cation type.

What is the actual relevance of these achievements? Do they really work? In the industry, molecular simulation is sometimes considered as an academic discipline which takes too long and costs too much. This opinion might have been justified a decade ago but it is no longer valid today as a result of all the progress made in the software and the hardware. For instance, a commercially available package – the GIBBS code – is able to carry out most of the calculations shown in this book. The availability of this well-tested, multipurpose Monte Carlo software is opening the way to many applications in other industrial sectors, including specialty chemicals and materials science, among others. In addition, two current trends favour efficient applications, namely ongoing increases in computer capacity and speed at constant cost, and the continuing improvement of statistical bias algorithms. For instance, investigating the adsorption of aromatic mixtures in zeolites was a difficult task five years ago (necessitating several days of supercomputer time per data point) but it is now routine (one day on a PC processor) and it is likely to become even easier over the next decade (coming down to just a few hours). Similarly, it is expected that calculations that are considered difficult today (e.g. gas absorption by polymers) will be routine in the future.

Beyond the achievements of molecular simulation hitherto (which might be considered insufficient for the industry to invest in this technology), it is important to evaluate future perspectives for this type of technique. Many new applications may prove successful although some will nevertheless remain problematic.

Using the algorithms proposed by Escobedo [1998, 1999 and 2000], phase equilibrium calculations of real condensate gases are known to be feasible. Using the same kind of approach as with adsorbents, it would be also possible to investigate the interactions of these gases with mineral surfaces in reservoir rocks of high specific surface area – a difficult problem for classical thermodynamic models. However, the explicit modelling of crude oils will remain difficult for two reasons: firstly, it would require intimate knowledge of the chemical structure of the heavy ends (asphaltenes, resins) which is generally not available; and secondly, the minimum size for a representative system is still great as a consequence of the small proportion of very large molecules. Nevertheless, Monte Carlo simulation can provide insight into model systems of interest for gas injection projects by generating phase equilibrium data on binary or ternary systems. This could benefit from histogram reweighting techniques which are reputed to be very efficient for computing some phase equilibria [Potoff *et al.*, 1998 and 1999].

In addition to phase diagrams, simulation data may extend to interfacial tension in the future as techniques are already available for interfacial tension calculations by direct simulation of the liquid-vapour interface [Goujon *et al.*, 2001 and 2002]. As illustrated by recent investigations [Chen *et al.*, 2002], molecular simulation is the only method capable of elucidating and quantifying the most important phenomena that occur at complex interfaces. Among other applications, this would be useful for oil-in-water or water-in-oil emulsions used as drilling fluids, for which the influence of high pressure gases on surfactant behaviour is still poorly known.

As far as polymeric materials are concerned, a feasible challenge is to extend gas solubility and swelling calculations to polymers in real conditions, i.e. to account for semi-crystalline structure (as in polyethylene) or for branched structures. Polymers bearing polar groups (e.g. polymethylmetacrylate) will be difficult to treat because sampling all conformations will require powerful algorithms. There is no reason that enough effort will not make this possible, following the general approach proposed by Theodorou [2004] in a recent review. However, polymer systems are often heterogeneous at a scale that is too great for detailed simulation, even with computers ten or hundred times more powerful than today. In many instances, approximations such as coarse grain modelling will be necessary to treat these systems.

An important field to which the application of Monte Carlo techniques could be extended is solid-fluid phase equilibria. It is already possible to predict the melting properties of molecular systems like low molecular weight n-alkanes [Polson and Frenkel, 1998 and 1999]. Although such computations seem to need more accurate force fields than fluid phase equilibria, it is likely that significant progress will be made in this area as it would be very useful for the pharmaceutical industry to be able to predict crystal structure from molecular structure.

Chemical equilibria and complexation are also possible candidates for elucidation by Monte Carlo techniques, e.g. complexes formed between alkanolamines and acid gases (hydrogen sulphide and carbon dioxide) are intensively used for gas processing. Complexation could be simulated through specific reactive potentials, in the same way as developed for the simulation of complex reaction schemes [van Duin *et al.*, 2001]. It could also require special bias algorithms as already explored by several authors [Lisal *et al.*, 2000]. Further development of gas processes would greatly benefit from such improvements of molecular simulation for the screening of new agents.

Last but not least, significant issues are still open in the simulation of adsorption in microporous adsorbents. Accounting for the mobility of cations and for the flexibility of zeolite structure, developing more general force fields (e.g. for titanium-based adsorbents) are possible avenues to improve the reliability of simulations and to extend the scope of the adsorption problems that can be tackled.

In the light of present and future applications, it would be surprising if an increasing range of industrial Monte Carlo applications were not observed in the next few years. What will be then the status of this discipline? It is impossible that simulation would simply replace classical thermodynamic models like equations of state, activity coefficient models or corresponding state methods. These methods are sufficient for many applications, and their small computing time makes them ideal in large-process simulations of multiphase fluid flow, in which fluid properties must be evaluated billions of times. Monte Carlo simulation does not preclude experimental measurement because the methods still have an empirical basis. As seen many times in this book, reliable experimental measurements are essential to developing intermolecular potentials and to validating predictions in realistic cases. The real place for molecular simulation is an intermediate one somewhere between the classical models and laboratory measurements – being significantly more accurate than classical models and significantly less expensive than measurement. As there are many cases in which measurement is complicated or impossible (due to high pressure conditions, toxic components, microporous adsorbents, etc.), there is no doubt that Monte Carlo simulation will become more widespread in the oil and gas industry, and possibly others.

Appendix

A.1 PARAMETERS OF THE ANISOTROPIC UNITED ATOMS POTENTIAL

The following tables contain the Lennard-Jones parameters, bond lengths, and parameters for bending and torsion potentials (Tables A1.1 to A1.4). Electrostatic charges are given for alcohols in Table A1.5 (results taken from Bourasseau, 2003) and for nitriles in Table A1.6 (results taken from Hadjkali, 2004). In the case of sulphides, thiols, ketones, aldehydes and organomercuric compounds, electrostatic charges are given in the tables of Section 3.3. Atomic units are used throughout : distances in Å, energies (divided by the Boltzmann constant) in K, charges in multiples of the proton electrostatic charge.

A. Lennard-Jones and structural parameters

Table A1.1 Lennard-Jones 6-12 group parameters. δ is the offset of the force center with respect to the main atom of the group.

Group	Molecular mass (g/mol)	ϵ/k (K)	σ (Å)	δ (Å)
CH_3 (sp^3)	15.03	120.15	3.607	0.216
CH_2 (sp^3 – linear)	14.03	86.29	3.461	0.384
CH (sp^3)	13.03	50.98	3.363	0.646
C (sp^3)	12.03	15.04	2.44	0
CH_2 (sp^3 – cyclic)	14.03	90.09	3.461	0.336
CH_2 (sp^2)	14.03	111.1	3.48	0.295
CH (sp^2) in olefins	13.02	90.6	3.32	0.414
C (sp^2)	12.01	61.9	3.02	0
CH (sp^2 aromatic)	13.01	89.4	3.246	0.407
C (sp^2 aromatic)	12.01	42.1 (a) 37.7 (b)	3.065 (a) 3.246 (b)	0 0
SH in R–SH	32.07	220.62	3.655	0.059
S in R'–S–R"	31.07	190	3.60	0
N in R–C ≡ N (nitrile)	14.01	162.41	3.564	0
C in R–C ≡ N (nitrile)	12.01	95.52	3.218	0
C (bonded to an oxygen in ketones – aldehydes)	12.01	36.79	3.60	0
O in ketones R=O and aldehydes R–COH	16.00	103.85	3.04	0
H in R–COH aldehyde	1.01	35.49	1.70	0
Hg in R–Hg–R	200.59	426.11	3.5054	0
OH in alcohols R–OH	17.01	85.267	3.0335	0.0951

(a) Parameters used in the applications of the present book, after Contreras *et al.* [2004b].
(b) Recommended parameters, after Ahunbay *et al.* [2005]

Table A1.2 Bond lengths.

Bond	Length (Å)
C–C (alkanes)	1.535
C=C (olefinic)	1.331
C=C (aromatic)	1.40
C–S (sulphides, thiols)	1.820
C–H (alkyl chains in sulphides and thiols)	1.085
S–H	1.34
S–S	2.10
C=O (ketones, aldehydes)	1.21
H–C(O) (ketones, aldehydes)	1.124
C–Hg	2.09
C–H (alkyl chains in organic mercury compounds)	1.098
C–O (alcohols)	1.425
O–H (alcohols)	0.96
C–CN (nitriles)	1.46
C≡N (nitrile)	1.169

Table A1.3 Bond angles and bending potential.

Atoms involved	θ_o (°)	k_{bend} (K)	Equation form
$C–CH_2–C$	114	62 500	a
(linear alkyl chains and cyclic alkanes)	114	74 900	b
C–CH–C	112	62 500	a
(branched alkyl chain)	112	72 700	b
C=C–C (olefins)	124	–	rigid
C=C=C (aromatics)	120	–	rigid
C=C–C (alkylaromatics)	120	–	rigid
C–C–S	114.7	50 277	b
C–S–C	98.9	62 343	b
C–S–S	103.7	68 376	b
C–S–H	96	43 238	b
C–C–H (alkyl chains)	109.5	–	rigid
C–C≡N (nitriles)	180	24 300	a
C–C–CN (nitriles)	108.8	69 500	a
C–C–C(O) (ketones, aldehydes)	111	63 349	a
C–C=O (ketones, aldehydes)	120.4	80 438	a
C–Hg–C (dimethylmercury)	180	14 100	a
C–Hg–C (diethylmercury)	180	17 140	a
C–C–Hg (diethylmercury)	113.4	23 620	a
C–O–H (alcohols)	108.9	–	rigid
$C–CH_2–O$ (alcohols)	109.5	59 800	b

a: Eq. (2.44); b: Eq. (2.45).

Table A1.4 Parameters of the torsion potential, according to Eq. (2.47), and constraints involving four atoms.

Atoms involved	Parameters (K)
C–CH$_2$–CH$_2$–C (linear alkyl chain or cyclic alkanes C$_5$ and beyond)	a_o= 1 001.35, a_1= 2 129.52, a_2= – 303.06, a_3= – 3 612.27, a_4= 2 226.71, a_5= 1 965.93, a_6= – 4 489.34, a_7= – 1 736.22, a_8= 2 817.37
C–CH$_2$–CH–C (branched alkyl chain or branched cycle)	a_o= 373.05, a_1= 919.04, a_2= 268.15, a_3= – 1 737.21, $a_{4\text{-}8}$= 0
C=CH–CH$_2$–C (alkenes, alkylaromatics)	a_o = 272.37, a_1 = – 938.12, a_2 = 220.73, a_3 = 1 135.06, $a_{4\text{-}8}$ = 0
CH$_2$=CH–CH=CH$_2$ (dienes)	a_o = 3 132.73, a_1 = 898.68, a_2 = – 3 449.75, a_3 = – 1 736.37, a_4 = 1 155.2, $a_{5\text{-}8}$ = 0
C=C=C=C (aromatics)	ϕ = 180° (planar cycle)
C=C=C–X (first substituent of an aromatic group)	ϕ = 0° (the substituent lies in the same plane as the aromatic cycle)
C–C=C–C (olefins)	ϕ = 180° (cis) or ϕ = 0° (trans)
C–CH$_2$–O–H (alcohols)	a_o= 339.41, a_1= 353.97, a_2= 58.34, a_3= – 751.72, $a_{4\text{-}8}$ = 0
C–C–S–C (sulphides)	a_o= 407.6, a_1= 1 373.0, a_2= 868.5, a_3= – 2 319.2, $a_{4\text{-}8}$= 0
C–C–S–H (thiols)	a_o=323 , a_1= 970.7, a_2= 36.87, a_3= – 1 224.9, $a_{4\text{-}8}$= 0
C–S–S–C (disulphides)	a_o= 67.3, a_1= 1 115.4, a_2= 4 352.3, a_3= – 2 016.0, $a_{4\text{-}8}$ = 0
C–C–C=O (ketones, aldehydes)	a_o= 10 565.67, a_1= 0, a_2= – 10 565.67, $a_{3\text{-}8}$ = 0
>C=O (ketones)	The oxygen belongs to the plane defined by the carbon and its two substituents

B. Electrostatic Charges

Table A1.5 Electrostatic charges for methanol and ethanol in the AUA potential model of Bourasseau [2003], in atomic units. q_1 is located on the carbon nucleus, q_2 on the bisector of the COH angle at 0.3 Å from the oxygen nucleus, q_3 on the hydrogen nucleus, and q_4 on the C–O bond, at 0.15 Å from the carbon.

	q_1	q_2	q_3	q_4
Methanol	– 2.05	– 1.49	0.71	2.83
Ethanol	– 1.99	– 1.49	0.70	2.78

Table A1.6 Electrostatic charges of nitriles [HadjKali, 2004].

Atom	Acetonitrile	Propionitrile	Butyronitrile
C_1	– 0.40	– 0.31	– 0.33
C_2	+ 0.15	– 0.25	– 0.16
C_3	–	+ 0.13	– 0.25
C_4	–	–	+ 0.12
$H_1(C_1)$	+ 0.16	+ 0.12	+ 0.11
$H_2(C_1)$	+ 0.16	0.12	+ 0.11
$H_3(C_1)$	+ 0.16	+ 0.12	+ 0.11
$H_4(C_2)$	–	+ 0.15	+ 0.12
$H_5(C_2)$	–	+ 0.15	+ 0.12
$H_6(C_3)$	–	–	+ 0.14
$H_7(C_3)$	–	–	+ 0.14
N	– 0.23	– 0.23	– 0.23
Experimental dipole moment (D)	3.92		
Simulated dipole moment (D)	4.01	4.09	4.21

A.2 IMPLEMENTATION OF MONTE CARLO MOVES WITH THE ANISOTROPIC UNITED ATOMS MODEL

The purpose of this appendix is to provide a guide on how to implement the algorithms mentioned in this book, with particular emphasis on the specific features of the Anisotropic United Atoms (AUA) potential in statistical bias algorithms. The appendix is organised according to the various Monte Carlo moves considered in the present book.

A.2.1 Translation, Rotation, Volume Changes

As the internal conformation of the molecules is unchanged in these Monte Carlo moves, their implementation is exactly the same with AUA as with UA potentials, with the acceptance probabilities provided in Sections 2.3.1 and 2.3.2. This applies also for the concerted volume changes in the Gibbs ensemble discussed in Section 2.3.4.

A.2.2 Flip, Pivot

Eq. (2.56) still applies to the flip and reptation moves when the AUA potential is used, but care has to be taken that more AUA force centres are changed in these moves than with UA potentials. The reason is that the AUA centre attached to an unchanged carbon atom is shifted when the neighbouring carbon atom is changed, because this changes the orientation of the bisector which defines the location of the AUA force centre. The number of modified bending and torsion angles is the same with AUA and UA potentials as these angles are defined from the atomic skeleton, not from the positions of the force centres.

In the case of the flip move, the AUA force centres attached to the two neighbouring atoms (B and C on Fig. 2.18a) are changed so that three force centres (A, B, C) are changed, instead of one with UA potentials (the rotation axis being the axis between the neighbouring carbon nuclei, not the axis between the AUA force centres).

In the case of the pivot move, one additional force centre is moved, namely the AUA force centre attached to the pivot (A on Fig. 2.18c).

A.2.3 CBMC Moves

With Anisotropic United Atoms and All Atoms potentials, the CBMC method is confronted with a special problem because the position of the force centre is still unknown when the related atom is placed at a test location. In consequence, the potential energy increment associated with the new atom cannot be computed. In order to circumvent this problem, the molecule (or part of molecule) is regrown as if a United Atoms potential were being used, and a correction term is introduced into the final acceptance probability once the whole molecule has been regrown [Smit *et al.*, 1995].

In the following section, we will record $\overline{U}_{\text{ext}}$ the Lennard-Jones energy obtained with the UA model during the regrowth process, and U_{ext} the Lennard-Jones energy obtained with the AUA model. Generally, the Lennard-Jones parameters used to compute $\overline{U}_{\text{ext}}$ are the same as those of the AUA model, but this is not necessarily the case.

A. Regrowth

When regrowing the end of the chain from an atom n to the end atom m, the Rosenbluth factors of the new and old configurations are defined exactly as with UA potentials, according to:

$$W_{\text{new}} = \prod_{l=n}^{m} \left(\frac{1}{k_{\max l}} \sum_{k=1}^{k_{\max l}} \exp(-\beta \overline{U}_{\text{ext}\, l}(r_k)) \right) \tag{A.1}$$

$$W_{\text{old}} = \prod_{l=n}^{m} \left(\frac{1}{k_{\max l}} \left(\exp(-\beta \overline{U}_{\text{ext}\, l}(r_o) + \sum_{k=1}^{k_{\max l}-1} \exp(-\beta \overline{U}_{\text{ext}\, l}(r_k)) \right) \right) \tag{A.2}$$

where the subscript o refers to the coordinates of the old configuration and the subscript k refers to test locations. $\overline{U}_{\text{ext}\, l}(r_k)$ is the Lennard-Jones energy increment (intermolecular + intramolecular) associated with one additional force centre in location r_k.

Note that the factor $1/k_{\max l}$ appearing in these expressions is sometimes omitted in the literature (see for instance Smit *et al.*, 1995), as this does not change the acceptance probability of the moves involving a ratio of Rosenbluth factors (Gibbs ensemble transfers, regrowth, displacement, reptation). It is however important to include this factor for grand canonical simulations or Widom test insertions, otherwise the results depends on the numerical parameters $k_{\max l}$ (this remark is not specific to the AUA potential).

The acceptance probability is then given by:

$$p_{\text{acc}} = \min\left(1, \frac{W_{\text{new}} \exp\left(-\beta(\Delta U_{\text{ext}} - \Delta \overline{U}_{\text{ext}})\right)}{W_{\text{old}}}\right) \tag{A.3}$$

where ΔU_{ext} and $\Delta \overline{U}_{\text{ext}}$ are the energy differences between the old and the new configuration, according to the AUA and UA models:

$$\Delta U_{\text{ext}} = U_{\text{ext new}} - U_{\text{ext old}} \tag{A.4}$$

$$\Delta \overline{U}_{\text{ext}} = \overline{U}_{\text{ext new}} - \overline{U}_{\text{ext old}} \tag{A.5}$$

where $\overline{U}_{\text{ext new}}$ and $\overline{U}_{\text{ext old}}$ are obtained by summing the energy increments $\overline{U}_{\text{ext } l}(r_i)$, r_i being the coordinates of the new (resp. old) segment.

Similarly to the pivot move discussed above, care has to be taken in computing ΔU_{ext} that the position of the AUA force centre attached to the last unchanged atom of the chain (*A* on Fig. 2.21) is changed during regrowth. Therefore, $\overline{U}_{\text{ext new}}$ and $\overline{U}_{\text{ext old}}$ are obtained by summing the energy increments $\overline{U}_{\text{ext } l}(r_i)$ from $l = n - 1$ to m.

B. *Grand Canonical Insertions and Deletions*

If we keep the notations of Eqs. (A.1) to (A.5) with the simple difference that the regrowth process concerns the whole molecule ($n = 1$), the acceptance probability of the insertion and destruction moves for species i are given by:

$$p_{\text{acc}}(\text{insertion}) = \min\left(1, \frac{P_o V\ W_{\text{new}} \exp\left(-\beta(U_{\text{ext new}} - \overline{U}_{\text{ext new}})\right)}{kT(N_i + 1)\exp(-\beta\overline{\mu}_i)}\right) \tag{A.6}$$

$$p_{\text{acc}}(\text{deletion}) = \min\left(1, \frac{N_i\, kT \exp(-\beta\overline{\mu}_i)}{P_o V\ W_{\text{old}} \exp\left(-\beta(U_{\text{ext old}} - \overline{U}_{\text{ext old}})\right)}\right) \tag{A.7}$$

where V is the volume of the system, N_i the number of molecules of species i, and $\overline{\mu}_i$ the chemical potential of the species with reference to the ideal gas at pressure P_o and temperature T (see the discussion of the reference state in Sections 2.3.3 and 2.3.5). The subscripts *new* and *old* refer to the inserted molecule and to the deleted molecule.

Note that the factor P_o/kT, which arises from the reference state for the chemical potential, cancels out when the chemical potential is estimated from Widom test insertions according to Eqs. (2.66), (A.9) or (A.15).

C. Gibbs Ensemble Transfer Moves

Keeping the same notations as in Eqs. (A.1) to (A.5), the probability for the transfer of a molecule of species i from box A to box B, using the CBMC algorithm, is given by:

$$p_{\text{acc}} = \min\left(1, \frac{N_i^A V^B W_{\text{new}}^B}{(N_i^B+1)V^A W_{\text{old}}^A} \exp(-\beta(-U_{\text{ext old}}^A + \overline{U}_{\text{ext old}}^A + U_{\text{ext new}}^B - \overline{U}_{\text{ext new}}^B))\right) \quad \text{(A.8)}$$

where the subscripts *old* and *new* refer to the deleted molecule in box A and to the inserted molecule in box B, respectively.

D. Widom Test Insertions

When using CMBC, the chemical potential, with respect to the reference state discussed in Section 2.3.5 (ideal gas under pressure P_o and temperature T), is expressed by:

$$\overline{\mu}_i = -kT \ln\left\langle \frac{P_o V W_{\text{new}}}{kT(N_i+1)} \exp\left(\beta \overline{U}_{\text{ext new}} - \beta U_{\text{ext new}}\right)\right\rangle \quad \text{(A.9)}$$

where the subscript *new* refers to the molecule that is tentatively inserted. Here again n is set to 1, as the whole molecule is regrown.

E. Displacements

The acceptance probability for the displacement move, i.e. the destruction of a molecule and its reconstruction at a randomly selected place of the same box, is given by the same expression as the regrowth move (Eq. A.3), except that the whole molecule is regrown ($n = 1$) with the CBMC algorithm.

F. Reptation

Let us consider a reptation move in which p carbons are added to the end of the molecule and p carbons are deleted at its tail. If the regrowth at the end and the deletion at the tail are both conducted according to the CBMC bias, the acceptance probability is given by the same expression (A.3) as the regrowth move, but the meaning of the variables must be slightly adapted:

- the Rosenbluth factor of the new configuration, W_{new}, must be computed for the new part of the chain, i.e. replacing n by $(m + 1)$ and m by $(m + p)$. The external energy $\overline{U}_{\text{ext}}$ used in this regrowth process must not include the interactions of the newly placed atoms with carbons 1 to p.
- the Rosenbluth factor of the old configuration, W_{old}, must be computed for the deleted part of the chain in a backward regrowth, i.e. replacing n by p and m by 1. The external energy $\overline{U}_{\text{ext}}$ used in the evaluation of W_{old} must not include the interactions of the test atoms with carbons $(m + 1)$ to $(m + p)$.
- the AUA external energy difference ΔU_{ext} is the difference in Lennard-Jones energy (intermolecular + intramolecular) of the chain between its new and old positions. This

calculation must account for the fact that the AUA force centres attached to atoms $(p + 1)$ and m have been changed for the same reason as the pivot or regrowth moves (atoms A and B in Fig. 2.18b).

– the UA external energy difference $\Delta\overline{U}_{\text{ext}}$ may be computed in the same way as the regrowth move, on the basis of the energy increments collected during the growth process at both ends of the chain.

G. Branch Point Sampling

When using the branch point sampling technique [Macedonia and Maginn, 1999] to regrow branched segments in a chain, the sampling of bending and torsion angles is only modified. Therefore this technique does not change the acceptance probabilities from the CBMC moves exposed in Eqs. (A.3) to (A.9). When $k_{\max\,l}$ configurations are tested for a branch, care must be taken that the factor $1/k_{\max\,l}$ is counted only once, although two or three atoms are placed at the same time.

A.2.4 Reservoir Bias

As explained in Section 2.3, the reservoir bias proposed by Bourasseau *et al.* [2002] involves first the test insertion of a Lennard-Jones particle in n locations, then the test insertion of the whole molecule in m conformations in the selected location. In the first step, the Lennard-Jones interaction does not account for the standard long distance correction, and the following Rosenbluth factor is computed:

$$W_1 = \frac{1}{n}\sum_{k=1}^{n} \exp(-\beta\Delta\overline{U}_{\text{LJ}\,k}) \tag{A.10}$$

where $\overline{U}_{\text{LJ}}$ is the interaction energy of the Lennard-Jones test particle.

In the second step, the interaction energy of the whole molecule is considered, and a second Rosenbluth factor is determined:

$$W_2 = \frac{1}{m}\sum_{k=1}^{m} \exp(-\beta\Delta U_{\text{ext}\,k}) \tag{A.11}$$

where U_{ext} comprises the interaction energy with the rest of the system (Lennard-Jones + electrostatic), excluding however the induction (i.e. polarization) and the long distance correction of the Lennard-Jones potential.

A. Gibbs Ensemble Transfers

In the same way as configurational bias, the Rosenbluth factors must be evaluated for the deleted molecule in the donor phase (*old*) and for the inserted molecule in the receiver phase (*new*). The acceptance probability of the Gibbs ensemble transfers with the reservoir bias is given by Bourasseau *et al.* [2002]:

$$P_{\text{acc}} = \min\left(1, \left[\frac{W_1^{\text{new}} W_2^{\text{new}}}{W_1^{\text{old}} W_2^{\text{old}}} \exp\left(-\beta\left(\begin{array}{l} U_{\text{ind}}(\text{new}) + U_{\text{corr}}(\text{new}) - \\ \bar{U}_{\text{LJ}}(\text{new}) - U_{\text{corr}}(\text{old}) - U_{\text{ind}}(\text{old}) + \bar{U}_{\text{LJ}}(\text{old}) \end{array}\right)\right)\right] * \left(\frac{N_i(\text{donor}) V(\text{receiver})}{(N_i(\text{receiver}) + 1) V(\text{donor})}\right)\right) \tag{A.12}$$

where U_{ind} is the induction (i.e. polarization) energy, U_{corr} is the long distance correction. Note that the intramolecular energy does not appear in this expression, because it is already taken into account by picking molecular configurations from the reservoir.

B. Grand Canonical Insertions and Deletions

Keeping the same notations as in Eq. (A.10) to (A.12), the acceptance probability of the insertion and deletion moves are given by:

$$P_{\text{acc}}(\text{insertion}) = \min\left(1, \left[\frac{W_1^{\text{new}} W_2^{\text{new}} V P_{\text{ref}}}{kT(N_i + 1)} \exp\left(-\beta\left(U_{\text{ind}}(\text{new}) + U_{\text{corr}}(\text{new}) - \bar{U}_{\text{LJ}}(\text{new})\right) + \beta\bar{\mu}_i\right)\right]\right) \tag{A.13}$$

$$P_{\text{acc}}(\text{deletion}) = \min\left(1, \left[\frac{kT N_i}{W_1^{\text{old}} W_2^{\text{old}} V P_{\text{ref}}} \exp\left(-\beta\left(-U_{\text{corr}}(\text{old}) - U_{\text{ind}}(\text{old}) + \bar{U}_{\text{LJ}}(\text{old})\right) - \beta\bar{\mu}_i\right)\right]\right) \tag{A.14}$$

where the chemical potential $\bar{\mu}_i$ is expressed with reference to the ideal gas at pressure P_{ref} and temperature *T*, consistently with the Widom test of equation (A.15)

C. Widom Test Insertions

The chemical potential, expressed with respect to the standard reference state (ideal gas at pressure P_{ref} and temperature *T*) is:

$$\bar{\mu}_i = kT \ln\left(\frac{kT}{P_{\text{ref}}}\right) - kT \ln\left\langle \frac{V}{(N_i + 1)} W_1\, W_2 \exp\left(\beta\Delta\bar{U}_{\text{LJ}\,j}{}^{+} - \beta\Delta U_{\text{corr}}{}^{+} - \Delta\bar{U}_{\text{ind}}\right)\right\rangle \tag{A.15}$$

where the same notations are used as above.

References

Abdul-Rehman, H. B., N. A. Hasanain and K. F. Loughlin (1990). “Quaternary, Ternary, Binary, and Pure-Component Sorption on Zeolites. 1. Light Alkanes on Linde S-115 Silicalite at Moderate to High Pressures.” *Ind Eng. Chem. Res.* **29**: 1525-1535.

Ahunbay, M. G., S. Kranias, V. Lachet and P. Ungerer (2004). “Prediction of thermodynamic properties of heavy hydrocarbons by Monte Carlo simulation.” *Fluid Phase Equilibria* **224**: 73-81.

Ahunbay, M. G., P. Ungerer, A. Mackie and J. Perez (2005). “Optimized intermolecular potential for aromatic hydrocarbons based on Anisotropic United Atoms. III. Polyaromatics and naphthenoaromatics.” *J. Phys. Chem. B.* **109**: 2970-2976.

Akkermans, R. L. C. and G. Ciccotti (2004). “On the equivalence of atomic and molecular pressure.” *J. Phys. Chem.* **108**: 6866.

Alberty, R. A. and R. J. Silbey (1997). *Physical Chemistry, 2nd edition.* New York, Wiley.

Allen, M. P. and D. J. Tildesley (1987). *Computer simulation of liquids.* Oxford, Oxford science publications.

Angus, S., B. Armstrong, K. M. d. Reuck, V.V. Altunin, O. G. Gadetskii, G. A. Chapela and J. S. Rowlinson (1973). *International Thermodynamic tables of the fluid state – Carbon dioxide.* Oxford, Pergamon.

Arnaud, J. F. (1995) Caractérisation des propriétés physiques et thermodynamiques des fluides pétroliers à haute pression. Ph.D. thesis, Université de Pau et des pays de l’Adour, Sciences physiques, Pau, France.

Axilrod, B. M. and E. Teller (1943). “Interaction of the van der Waals type between three atoms.” *J. Chem. Phys.* **11**: 299.

Bao, Z., M. Liu, J. Yang and N. Wang (1995). *Hua Hsueh Kung Yeh Kung Cheng* **46**: 230.

Barrat, J. L. and J. P. Hansen (2003). *Basic concepts for simple and complex liquids.* Cambridge, UK, Cambridge University Press.

Beauvais, C., A. A. Boutin and A. H. Fuchs (2004a). “A numerical evidence for nonframework cation redistribution upon water adsorption in Faujasite zeolite.” *ChemPhysChem*(5): 1-7.

Beauvais, C., X. Guerrault, F.-X. Coudert, A. Boutin and A. H. Fuchs (2004b). “Distribution of sodium cations in faujasite-type zeolite: a canonical parallel tempering simulation study.” *J. Phys. Chem. B* **108**: 399-404.

Benson, S. W. (1976). *Thermochemical kinetics.* New York, Wiley.

Berry, V. M. and B. H. Sage (1970). National Bureau of Standards. NSRDS-NBS 32

Bezus, A. G., A. V. Kiselev, A. A. Lopatkin and P. Q. Du (1978). "Molecular statistical calculation of the thermodynamic adsorption characteristics of zeolites using the atom-atom approximation." *J. Chem. Soc. Faraday Trans. II* **74**: 367-379.

Bier, K., J. Kunze and G. Maurer (1976). "Thermodynamic properties of ethane from calorimetric measurements." *J. Chem. Thermodynamics* **8**: 857.

Boulougouris, G. C., I. G. Economou and D. N. Theodorou (1998). "Engineering a molecular model for water phase equilibrium over a wide temperature range." *J. Phys. Chem B* **102**: 1029.

Boulougouris, G. C., I. G. Economou and D.N. Theodorou (1999). "On the calculation of the chemical potential using the particle deletion scheme." *Molecular Physics* **96**: 905-913.

Boulougouris, G. N., I. G. Economou and D.N. Theodorou (2001). "Calculation of the chemical potential of chain molecules using the staged particle deletion scheme." *J. Chem. Phys.* **115**: 8231.

Bourasseau, E., P. Ungerer and A. Boutin (2002a). "Prediction of Equilibrium properties of cyclic alkanes by Monte Carlo simulation – new anisotropic united atoms potential – new transfer bias method." *J. Phys. Chem. B* **106**: 5483-5491.

Bourasseau, E., P. Ungerer, A. Boutin and A. H. Fuchs (2002b). "Monte Carlo simulation of branched alkanes and long chain n-alkanes with anisotropic United Atoms intermolecular potential." *Molecular Simulation* **28**: 317-336.

Bourasseau, E. (2003) Prédiction de propriétés d'équilibre de phases par simulation moléculaire – développement d'algorithmes et optimisation de potentiels. Ph D thesis, Université de Paris Sud, Laboratoire de Chimie Physique, Orsay, France.

Bourasseau, E., M. Haboudou, A. Boutin, A. H. Fuchs and P. Ungerer (2003). "New optimization method for intermolecular potentials – Optimization of a new anisotropic united atoms potential for olefins – Prediction of equilibrium properties." *J. Chem. Phys.* **118**: 3020-3034.

Bourasseau, E., T. Sawaya, I. Mokbel and P. Ungerer (2004). "Measurement and vapour pressures of 2,6,10,14-tetramethylpentadecane (pristane). Experimental and Monte Carlo simulation results." *Fluid Phase Equilibria* **225**: 49-57.

Boutin, A., S. Buttefey, A. H. Fuchs and A. K. Cheetham (2001). "Molecular simulation of adsorption of guest molecules in zeolitic materials: a comparative study of intermolecular potentials." *Molecular Simulation* **27**: 371-385.

Brennan, J. K. and W. G. Madden (2003). "Phase coexistence curves for off-lattice polymer-solvent mixtures: Gibbs-Duhem integration simulations." *Molecular Simulation* **29**: 91.

Buttefey, S. (2002) Modélisation et simulation moléculaire des cations extra-charpente dans les zéolithes de type faujasite. Ph. D. Thesis, Université de Paris Sud, Laboratoire de Chimie Physique, Orsay, France.

Carroll, J. J. and A. Mather (1989). "Phase equilibrium in the system water-hydrogen sulphide: experimental determination of the LLV locus." *Canadian J. of Chem. Engineering* **67**: 468.

Carroll, J. J. and A. E. Mather (1995). "A generalized correlation for the Peng-Robinson interaction coefficients for paraffin-hydrogen sulfide binary systems." *Fluid Phase Equilibria* **105**: 221.

Chacin, A., J. M. Vasquez and E.A. Müller (1999). "Molecular simulation of the Joule-Thomson inversion curve of carbon dioxide." *Fluid Phase Equilibria* **165**: 147-155.

Chen, B., M. G. Martin and J.I. Siepmann (1998). "Thermodynamic properties of the Williams, OPLS-AA, and MMFF94 All-Atom force fields for normal alkanes." *J. Phys. Chem. B* **102**: 2578-2586.

Chen, B. and J.I. Siepmann (1999). "Transferable potentials for phase equilibria: 3. Explicit – hydrogen description of normal alkanes." *J. Phys. Chem.* **103**: 5370-5379.

Chen, B., J. J. Potoff and J. I. Siepmann (2001). "Monte Carlo calculations for alcohols and their mixtures with alkanes, transferable potentials for phase equilibria. 5. United-Atom description of primary, secondary and tertiary alcohols." *J. Phys. Chem. B* **105**: 3093.

Chen, B., J. I. Siepmann and M. L. Klein (2002). "Vapor-liquid interfacial properties of mutually saturated water/1-butanol solutions." *J. Am. Chem. Soc.* **124**: 12232.

Choudhary, V. R. and S. Mayadevi (1993). "Adsorption of Methane, Ethane, Ethylene, and Carbon Dioxide on High-Silica Pentasil Zeolites and Zeolite-Like Materials Using Gas Chromatography Pulse Technique." *Sep. Sci. Technol.* **28**: 2197.

Choudhary, V. R. and S. Mayadevi (1996). "Adsorption of Methane, Ethane, Ethylene, and Carbon Dioxide on Silicalite-I." *Zeolites* **17**: 501.

Colina, C. and E. A. Müller (1997). "Joule-Thomson inversion curves by molecular simulation." *Molecular simulation* **19**: 237.

Colina, C. M., M. Lisal, F. R. Siperstein and K. E. Gubbins (2002). "Accurate CO_2 Joule-Thomson inversion curve by molecular simulations." *Fluid Phase Equilibria* **202**: 253.

Contreras-Camacho, O. (2002) Determinacion del equilibrio liquido-vapor de agua, aromaticos y sus mezclas mediante simulacion molecular. PhD thesis, Universitat Rovira i Virgili, Ingenieria Quimica, Tarragona, Spain

Contreras-Camacho, O., P. Ungerer, A. Boutin, A. H. Fuchs and A. Mackie (2004a). "Optimized intermolecular potential for aromatic hydrocarbons based on Anisotropic United Atoms. I. Benzene." *J. Phys. Chem. B* **108**: 14109-14114.

Contreras-Camacho, O., V. Lachet, M. G. Ahunbay, J. Perez, P. Ungerer, A. Boutin and A. Mackie (2004b). "Optimized intermolecular potential for aromatic hydrocarbons based on Anisotropic United Atoms. II. Alkylbenzenes and styrene." *J. Phys. Chem. B* **108**: 14115-14123.

Cracknell, R. F., D. Nicholson and N.G. Parsonage (1990). "Rotational insertion bias: a novel method for simulating dense phases of structured particles, with particular application to water." *Molecular Physics* **71**: 931.

Culberson, O. L. (1951). "Phase equilibria in hydrocarbon-water systems. The solubility of methane in water at pressures to 10 000 psi." *AIME Petr. Transactions* **192**: 223-226.

Dang, L. X. (1995). "Mechanism and thermodynamics of ion selectivity in aqueous solutions of 18-crown-6-ether – A molecular dynamics study." *J. Am. Chem. Soc.* **117**: 6954-6960.

Delhommelle, J., A. Boutin and A. H. Fuchs (1999). "Molecular simulation of vapour-liquid coexistence curves for hydrogen sulfide-alkane and carbon dioxide-alkane mixtures." *Molecular Simulation* **22**: 351-368.

Delhommelle, J. (2000) Etablissement de potentiels d'interaction pour la simulation moléculaire – Application à la prédiction des équilibres liquide-vapeur de mélanges binaires alcane-molécule multipolaire. Ph. D. thesis, Université de Paris XI, Chimie Physique, Orsay, France

Delhommelle, J., P. Millié and A. H. Fuchs (2000a). "On the role of the definition of potential models in Gibbs ensemble simulations of the H2S-n-pentane mixture." *Molecular Physics* **98**: 1895-1905.

Delhommelle, J., C. Tschirwitz, P. Ungerer, G. Granucci, P. Millié, D. Pattou and A.H. Fuchs (2000b). "Derivation of an optimized potential model for phase equilibria (OPPE) for sulfides and thiols." *J. Phys. Chem. B* **104**: 4745-4753.

Delhommelle, J. and P. Millié (2001). "Inadequacy of the Lorentz-Berthelot combining rules for accurate predictions of equilibrium properties by molecular simulation." *Molecular Physics* **99**: 619-625.

Dhima, A., J. C. d. Hemptinne and G. Moracchini (1998). "Solubility of light hydrocarbons and their mixtures in pure water under high pressure." *Fluid Phase Equilibria* **145**: 129-150.

Dodd, L. R., T. D. Boone and D.N. Theodorou (1993). "A concerted rotation algorithm for atomistic Monte Carlo simulation of polymer melts and glasses." *Molecular Physics* **78**: 961-996.

Du, Z., T. J. H. Vlugt, B. Smit and G. Manos (1998). "Molecular Simulation of Adsorption of Short Linear Alkanes and their Mixtures in Silicalite." *AIChE Journal* **44**: 1756-1764.

Duin, A. C. T. v., S. Dasgupta, F. Lorant and W.A. Goddard (2001). "A reactive force field for hydrocarbons." *J. Phys. Chem. A* **105**: 9396-9409.

Dun, J. J. v. and W. J. Mortier (1988). "Temperature-dependent Cation Distribution in Zeolites. 1. A statistical thermodynamical Model." *J. Phys. Chem. B* **92**: 6740-6746.

Durand, J.-P., A. Fafet and Alain Barreau (1989). "Direct and automatic capillary GC analysis for molecular weight determination and distribution in crude oils and condensates up to C20." *Journal of High Resolution Chromatography* **12**: 230-233.

Dysthe, D., A. H. Fuchs and B. Rousseau (1999). "Fluid transport properties by equilibrium molecular dynamics. I. Methodology at extreme fluid states." *J. Chem. Phys.* **110**: 4047-4059.

Dysthe, D., A. H. Fuchs and B. Rousseau (1999). "Fluid transport properties by equilibrium molecular dynamics. II. Multicomponent systems." *J. Chem. Phys.* **110**: 4060-4067.

Dysthe, D., A. H. Fuchs and B. Rousseau (2000). "Fluid transport properties by equilibrium molecular dynamics. III. Evaluation of united atom interaction potential models for pure alkanes." *J. Chem. Phys.* **112**: 7581.

Economou, I. (2001). "Monte Carlo simulation of phase equilibria of aqueous systems." *Fluid Phase Equilibria* **183-184**: 259-269.

Eder, F. and J. A. Lercher (1997). "Alkane Sorption in Molecular Sieves: The Contribution of Ordering, Intermolecular Interactions, and Sorption on Bronsted Acid Sites." *Zeolites* **8**: 75-81.

Eder, F. and J. A. Lercher (1997). "On the Role of the Pore Size and Tortuosity for Sorption of Alkanes in Molecular Sieves." *J. Phys. Chem. B* **101**: 1273.

Errington, J. R., G. C. Boulougouris, I. G. Economou, A. Z. Panagiotopoulos and D. N. Theodorou (1998). "Molecular simulation of Phase equilibria for water-methane and water-ethane mixtures." *J. Phys. Chem. B* **102**: 8865.

Errington, J. R. and A. Z. Panagiotopoulos (1998). "Phase equilibria of the modified Buckingham exp-6 potential from Hamiltonian scaling grand canonical Monte Carlo." *J. Chem. Phys.* **109**: 1093.

Errington, J. R. and A. Z. Panagiotopoulos (1999). "New intermolecular potential models for benzene and cyclohexane." *J. Chem. Phys.* **111**: 9731.

Errington, J. R. and A. Z. Panagiotopoulos (1999). "A new intermolecular potential model for the n-alkane homologous series." *J. Phys. Chem B* **103**: 6314.

Escobedo, F. (1999). "Tracing coexistence lines in multicomponent fluid mixtures by molecular simulation." *J. Chem. Phys.* **110**: 11999-12010.

Escobedo, F. A. and J. J. de Pablo (1997). "Pseudo-ensemble simulations and Gibbs-Duhem integrations for polymers." *J. Chem. Phys.* **106**: 2911.

Escobedo, F. A. (1998). "Novel pseudoensembles for simulation of multicomponent phase equilibria." *J. Chem. Phys.* **108**: 8761-8772.

Escobedo, F. A. (2000). "Molecular and macroscopic modeling of phase separation." *AIChE Journal* **46**: 2086.

Escobedo, F. A. and Z. Chen (2001). "Simulation of isoenthalps and Joule-Thomson inversion curves of pure fluids and mixtures." *Molecular Simulation* **26**: 395-416.

Eulenberger, G. R., D. P. Shoemaker and J. G. Keil (1967). "The crystal structure of hydrated and dehydrated synthetic zeolites with Faujasite aluminosilicate frameworks. I. The dehydrated sodium, potassium, and silver forms." *J. Phys. Chem* **71**: 1812-1819.

Evans, D. J. and G. P. Morris (1983). "Isothermal-isobaric molecular dynamics." *Chem. Phys.* **77**: 63.

Faller, R., H. Schmitz, O. Biermann and F. Müller-Pathe (1999). "Automatic parametrization of force fields by simplex optimization." *J. Comput. Chem.* **20**: 1009.

Fermeglia, M. and G. Torriano (1999). "Density, viscosity, and refractive index for binary systems of n-C16 and four nonlinear alkanes at 298.15 K." *J. Chem. Eng. Data* **44**: 965.

Fitch, A. N., H. Jobic and A. Renouprez (1986). "Localization of benzene in Sodium-Y zeolite by powder neutron diffraction." *J. Phys. Chem.* **90**: 1311-1318.

Fredenslund, A., J. Gmehling and P. Rasmussen (1977). *Vapour-liquid equilibrium using UNIFAC.* New York, Elsevier.

Frenkel, D. and B. Smit (1996). *Understanding Molecular Simulation.* San Diego, Academic Press.

Friedrich, A. and R. Lustig (2002). "Thermodynamics of fluid benzene from molecular dynamics simulations." *J. Mol. Liq.* **98-99**: 241.

Friend, D. G., H. Ingham and J. F. Ely (1991). "Thermophysical properties of ethane." *J. Phys. Chem. Ref. Data* **20**: 275.

Fuchs, A. H. and A. K. Cheetham (2001). "Adsorption of guest molecules in zeolitic materials: computational aspects." *J. Phys. Chem B* **105**: 7375-7383.

Galivel-Solastiouk, F., S. Laugier and D. Richon (1986). "Vapor-liquid equilibrium data for the propane-methanol and propane-methanol-carbon dioxide system." *Fluid Phase Equilibria* **28**: 73.

Gallezot, P. and B. Imelik (1975). "Structures cristallines de trois zéolithes de type Y partiellement décationisées, (Na26,6H29,4)Y, (Na14,2H41,8)Y, (Na2,2H53,8)Y." *J. Physique Physico-chimie biol.* **68**: 816-821.

Gillespie, P. C. and G. M. Wilson (1982) Vapor-liquid and liquid-liquid equilibria: water-methane, water-carbon dioxide, water-hydrogen sulfide, water-n-pentane, water-methane-n-pentane. Gas Processors Association. Tulsa. Research Report RR 48.

Go, N. and H.A. Sheraga (1976). "On the use of classical statistical mechanics in the treatment of polymer chain conformation." *Macromolecules* **9**: 535.

Goodwin, R. D. (1983) Hydrogen Sulfide provisional thermophysical properties from 188 to 700 K at pressure to 75 MPa. National Bureau of Standards. Boulder, Co.

Goujon, F., P. Malfreyt, A. Boutin and A. H. Fuchs (2001). "Vapour-liquid phase equilibria of n-alkanes by direct Monte Carlo simulations." *Molecular Simulation* **27**: 99-114.

Goujon, F., P. Malfreyt, A. Boutin and A.H. Fuchs (2002). "Direct Monte Carlo simulation of the equilibrium properties of n-pentane liquid-vapor interface." *J. Chem. Phys.* **116**: 8106-8117.

Graham, C., D. A. Imrie and R. E. Raab (1998). "Measurement of the electric quadrupole moments of CO_2, CO, N_2 and BF_3." *Molecular Physics* **93**: 49-56.

Gray, C. G. and K. E. Gubbins (1984). *Theory of molecular fluids.* Oxford, Oxford Science.

Guillot, B. and T. Guissani (2001). "How to build a better pair potential for water." *J. Chem. Phys* **114**: 6720.

Hadjkali, M. (2004). Application de la simulation moléculaire pour le calcul des équilibres liquide-vapeur des nitriles et pour la prédiction des azéotropes. Ph. D. thesis, Institut National Polytechnique de Toulouse, France.

Hansen, J. P. and I. R. McDonald (1986). *Theory of simple liquids.* Oxford, Academic Press.

Harris, J. G. and K. H. Yung (1995). "Carbon dioxide's liquid-vapor coexistence curve and critical properties as predicted by a simple molecular model." *J. Phys. Chem.* **99**: 12021-12024.

Hoover, W. G. (1985). "Canonical dynamics: equilibirum phase-space distributions." *Phys. Rev. A* **31**: 1695.

Hseu, T. D. (1972) The crystal structure of 1,1,4,4-tetramethyl-1,4-diaza-2,5-diboracyclohexane and structural studies of some Faujasite-type zeolites. Ph. D. Thesis, University of Washington, Ann Harbor, Michigan, USA

Hsu, J. J.-C., N. Nagarajan and R. L. Robinson (1985). "Equilibrium phase compositions, phase densities and interfacial tensions for CO2 + hydrocarbon systems. 1. CO2 + n-butane." *J. Chem. Eng. Data* **30**: 485.

Hufton, J. R. and R. P. Danner (1993). "Chromatographic study of alkanes in silicalite – Equilibrium properties." *AIChE Journal* **39**: 954-961.

Huron, M. J. and J. Vidal (1979). "New mixing rules in simple equations of state for representing vapor-liquid equilibria of strongly non-ideal mixtures." *Fluid Phase Equilibria* **1**: 247-265.

Ishihara, K., H. Tanaka and M. Kato (1998). "Phase equilibrium properties of ethane + methanol system at 298.15 K." *Fluid Phase Equilibria* **144**: 131.

Jacobs, P. A., K. K. Beyer and J. Valyon (1981). "Properties of the end members in the pentasil-family of zeolites: characterisation as adsorbents." *Zeolites* **1**: 161-168.

Jaramillo, E. and S. M. Auerbach (1999). "New force field for Na cations in faujasite-type zeolites." *J. Phys. Chem. B* **103**: 9589-9594.

Jedlovszky, P. and R. Vallauri (2001). "Thermodynamic and structural properties of liquid water around the temperature of maximum density in a wide range of pressures: A computer simulation study with a polarizable potential model." *J. Chem. Phys* **115**: 3750.

Jiràk, Z., S. Vratislav and V. Bozacek (1980). *J. Phys. Chem. Solids* **41**: 1089.

Jorgensen, W. L., J. Chandrasekhar and J. D. Madura (1983). "Comparison of simple potential functions for simulating liquid water." *J. Chem. Phys.* **79**: 926-935.

Jorgensen, W. L. and J. D. Madura (1984). "Optimized intermolecular potential functions for liquid hydrocarbons." *J. Am. Chem. Soc.* **106**: 6638.

Jorgensen, W. L. (1986b). "Optimized intermolecular potential functions for liquid alcohols." *J. Phys. Chem.* **90**: 1276-1284.

Jorgensen, W. L. (1986). "Intermolecular potential functions and Monte Carlo Simulations for liquid sulfur compounds." *J. Phys. Chem.* **90**: 6379-6388.

Jorgensen, W. L. and D. L. Severance (1990). "Aromatic-aromatic interactions: free energy profile for the benzene dimer in water, chloroform, and liquid benzene." *J. Am. Chem. Soc.* **112**: 4768.

Jorgensen, W. L., E. R. Laird, T. B. Nguyen and J. Tirado-Rives (1993). "Monte Carlo Simulations of pure liquid substituted benzenes with OPLS potential functions." *J. Comput. Chem.* **14**: 206-215.

Jorgensen, W. L. and T. B. Nguyen (1993). "Monte Carlo simulations of the hydration of substituted benzenes with OPLS potential functions." *J. Comput. Chem.* **14**: 195-205.

Jorgensen, W. L., D. S. Maxwell and J. Tirado-Rives (1996). "Developing and testing of the OPLS-AA force field on conformational energetics and properties of organic liquids." *J. Am. Chem. Soc.* **118**: 11225.

June, R. L., A. T. Bell and D. N. Theodorou (1992). "Molecular dynamics study of butane and hexane in silicalite." *J. Phys. Chem* **96**: 1051-1060.

Kaminsky, R. D. (1994). "Monte Carlo evaluation of ensemble averages involving particle number variations in dense fluid systems." *J. Chem. Phys.* **101**: 4986.

Kellermann, S. J., C. E. Stouffer, P. T. Eubank, J. C. Holste, K. R. Hall, B. E. Gammon and K. N. Marsh (1995) Thermodynamic properties of CO2 + H2S mixtures. Gas Processors Association. Tulsa. Research Report RR-141

Kiyohara, K., K. E. Gubbins and A. Z. Panagiotopoulos (1998). "Phase coexistence properties of polarizable water models." *Mol. Phys.* **94**: 803.

Kofke, D. A. (1993). "Direct evaluation of phase coexistence by molecular simulation via integration along the saturation line." *J. Chem. Phys.* **98**: 4149.

Kofke, D. A. and P. T. Cummings (1997). "Quantitative comparison and optimization of methods for evaluating the chemical potential by molecular simulation." *Molecular Physics* **92**: 973.

Kofranek, M., A. Karpfen and H. Lischka (1992). *Chem. Phys. Lett.* **189**: 281.

Kohn, J. P. and F. Kurata (1959a). "Volumetric behavior of the methane-hydrogen sulfide system at low temperatures and high pressures." *J. Chem. Eng. Data* **4**(1): 33-36.

Kohn, J. P. and F. Kurata (1959b). "Heterogeneous phase equilibria of the methane- hydrogen sulfide system." *AIChE Journal* **4**(2): 211.

Kong, C. L. (1973). "Combining rules for intermolecular potential parameters. II. Rules for the Lennard-Jones (12-6) potential and the Morse potential." *J. Chem. Phys.* **59**: 2464.

Kontogeorgis, G. M., E. C. Voutsas, I. V. Yakoumis and D. P. Tassios (1996). "An Equation of State for Associating Fluids." *Ind. Eng. Chem. Res.* **35**: 4310-4318.

Kortekaas, W. G., C. J. Peters and J. d. S. Arons (1997). "Joule-Thomson expansion of high pressure high temperature gas condensates." *Fluid Phase Equilibria* **139**: 205.

Kranias, S., D. Pattou, B. Levy and A. Boutin (2003). "An optimized potential for phase equilibria computation for ketone and aldehyde molecular fluids." *Phys. Chem. Chem. Phys.* **5**: 4175.

Krishna, R., B. Smit and T. J. H. Vlugt (1998). "Sorption-induced diffusion-selective separation of hydrocarbon isomers using silicalite." *J. Phys. Chem. A* **102**: 7727-7730.

Krishna, R., T. J. H. Vlugt and B. Smit (1999). "Influence of Isotherm Inflection on Diffusion in Silicalite." *Chem. Eng. Science* **54**: 1751.

Krishna, R., S. Calero and B. Smit (2002). "Investigation of Entropy Effects During Sorption of Mixtures of Alkanes in MFI Zeolite." *Chem. Eng. J.* **88**: 81.

Kristof, T. and J. Liszi (1997). "Effective Intermolecular potential for fluid hydrogen sulfide." *J. Phys. Chem. B* **101**: 5480-5483.

Kulprathipanja, S. (1995) US patent, 5382747

Kulprathipanja, S. (1999) US patent, 5900523

Lachet, V., A. Boutin, B. Tavitian and A. H. Fuchs (1998). "Computational study of p-xylene/m-xylene mixtures adsorbed in NaY zeolite." *J. Phys. Chem B* **102**: 9224-9233.

Lachet, V., A. Boutin, B. Tavitian and A. H. Fuchs (1999). "Molecular simulation of p-xylene and m-xylene adsorption in Y zeolites. Single components and binary mixtures study." *Langmuir* **15**: 8678-8685.

Lagache, M., P. Ungerer, A. Boutin and A. H. Fuchs (2001). "Prediction of thermodynamic derivative properties of fluids by Monte Carlo simulation." *Phys. Chem. Chem. Phys.* **3**: 4333-4339.

Lagache, M., P. Ungerer and A. Boutin (2004a). "Prediction of thermodynamic derivative properties of natural condensate gases at high pressure by Monte Carlo simulation." *Fluid Phase Equilibria* **220**: 211-223.

Lagache, M., J. Ridard, A. Boutin and P. Ungerer (2004b). "Force field development for organic mercury compounds." *J. Phys. Chem. B* **108**: 8419-8426.

Landolt, H. and R. Bornstein (1961). *Zahlenwerte und Funktionen aus Physik, Chemie, Astronomie, Geophysik, und Technik, 6th edition*. Berlin, Springer Verlag.

Laugier, S. and D. Richon (1995). "Vapor-liquid equilibria for hydrogen sulfide + hexane, + cyclohexane, + benzene, + pentadecane, and (hexane + pentadecane)." *J. Chem. Eng. Data* **40**: 153-159.

Leblanc, B., B. Braunschweig, H. Toulhoat and E. Lutton (2003). "Improving the sampling efficiency of Monte Carlo molecular simulations: an evolutionary approach." *Mol. Phys.* **101**: 3293-3308.

Lee, B. I. and M.G. Kesler (1975). "A generalized thermodynamic correlation based on three-parameter corresponding states." *AIChE Journal* **21**: 510.

Leeuwen, M. E. v. and B. Smit (1995). "Molecular simulation of the vapor-liquid coexistence curve of methanol." *J. Phys. Chem.* **99**: 1831.

Leu, A. D., J. J. Carroll and D. B. Robinson (1992). *Fluid Phase Equilibria* **72**: 163.

Levelt-Sengers, J. M. H. (1970). "Scaling predictions for thermodynamic anomalies near the gas-liquid critical point." *Ind. Eng. Chem. Fundam.* **9**: 470.

Levelt-Sengers, J. M. H., G. Morrison and R. F. Chang (1983). "Critical behavior in fluids and fluids mixtures." *Fluid Phase Equilibria* **14**: 19.

Lévy, B. and M. Enescu (1998). "Theoretical study of methylene blue: a new method to determine partial atomic charges; investigation of the interaction with guanine." *J. of Molecular Structure* **432**: 235.

Lide, D. R. (1992). *Handbook of Chemistry and Physics*. Boca Raton, CRC Press.

Lievens, J. L., W. J. Mortier and K. J. Chao (1992a). "Cation site energies in high-silica FAU-type zeolites." *J. Phys. Chem. Solids* **53**: 1163-1169.

Lievens, J. L., W. J. Mortier and J. P. Verduijn (1992b). "Influence of the framework composition on the cation-site energy: the structures of dehydrated NaxHGaY (x = 54, 36, and 21) zeolites." *J. Phys. Chem* **96**: 5473-5477.

Lievens, J. L., J. P. Verduijn, A. J. Bons and W. J. Mortier (1992c). "Cation Site Energies in Dehydrated Hexagonal Faujasite (EMT)." *Zeolites* **12**: 698-705.

Lievens, J. L., J. P. Verduijn and W. J. Mortier (1992d). "Cation Site Energies in Dehydrated KFI-type Zeolites: d-NaKFI, d-NaHKFI, and d-KKFI." *Zeolites* **12**: 690-697.

Lisal, M., W. R. Smith and I. Nezbeda (1999). "Accurate computer simulation of phase equilibrium for complex fluid mixtures. Application to binaries involving isobutene, methanol, methyl tert-butyl ether, and n-butane." *J. Phys. Chem. B* **103**: 10496-10505.

Lisal, M., W. R. Smith and I. Nezbeda (2000). "Molecular simulation of multicomponent reaction and phase equilibria in MTBE ternary system." *AIChE Journal* **46**: 4.

Lisal, M., W. R. Smith and I. Nezbeda (2001). "Accurate vapour-liquid equilibrium calculations for complex systems using the reaction Gibbs ensemble Monte Carlo method." *Fluid Phase Equilibria* **181**: 127.

Lisal, M., J. Kolafa and I. Nezbeda (2002). "An examination of the five-site potential (TIP5P) for water." *J. Chem. Phys* **117**: 8892.

Long, L. H. and J. Cattanach (1961). "Antoine vapour-pressure equations and heats of vaporisation for the dimethyls of zinc, cadmiun and mercury"." *J. Inorg. Nucl. Chem.* **112**: 5499.

Lyubartsev, A. P. and A. Laaksonen (1995). "Calculation of effective interaction potentials from radial distribution functions: A reverse Monte Carlo approach." *Phys. Rev. E* **52**: 3730.

Macedonia, M. D. and E. J. Maginn (1999). "A biased grand canonical Monte Carlo method for simulating adsorption using all-atom and branched united atom models." *Molecular Physics* **96**: 1375-1390.

Mackie, A., B. Tavitian, A. Boutin and A. H. Fuchs (1997). "Vapour-liquid phase equilibria predictions of methane-alkane mixtures by Monte Carlo simulation." *Molecular Simulation* **1**: 19.

Maginn, E. J., A. T. Bell and D. N. Theodorou (1995). "Sorption Thermodynamics, siting, and Conformation of Long n-Alkanes in Silicalite as Predicted by Configurational-Bias Monte Carlo Integration." *J. Phys. Chem.* **99**: 2057.

Marra, G. L., A. N. Fitch, A. Zecchina, G. Ricchiardi, M. Salalagio, S. Bordera and C. Lamberti (1997). "Cation Location in Dehydrated NamRbm Zeolite: A XRD and IR Study." *J. Phys. Chem. B* **101**: 10653-10660.

Martin, M. G. and J. I. Siepmann (1998). "Transferable models for phase equilibria 1. United-atom description of n-alkanes." *J. Phys. Chem. B* **102**: 2569.

Martin, M. G. and J. I. Siepmann (1999). "Novel configurational-bias Monte Carlo method for branched molecules. Transferable potentials for phase equilibria. 2. United-atoms description of branched alkanes." *J. Phys. Chem. B* **103**: 4508.

McQuarrie, D. A. (1976). *Statistical Mechanics*. New York, Harper and Collins.

Mehta, M. and D.A. Kofke (1994). "Coexistence diagrams of mixtures by molecular simulation." *Chem. Eng. Science* **49**: 2633-2645.

Mellot, C., D. Espinat, B. Rebours, C. Baerlocher and P. Fischer (1994). "Location of Meta-Xylene in BaX Zeolite and Model for the Filling of the Supercages." *Catalysis Letters* **27**: 159-169.

Mellot-Draznieks, C., S. Buttefey, A. Boutin and A. H. Fuchs (2001). "Placement of cations in NaX faujasite -type zeolite using (N, V, T) Monte Carlo simulations." *Chem. Commun.*: 2200-2201.

Mellot-Draznieks, C., J. Rodriguez-Carvajal, D.E. Cox, A.K. Cheetham (2003). Adsorption of chlorofluorocarbons in nanoporous solids: a combined powder neutron diffraction and computational study of $CFCl_3$ in NaY zeolite. *Phys. Chem. Chem. Phys.* **5**: 1882-1887.

Metropolis, N., A. W. Rosenbluth, M. N. Rosenbluth, A. H. Teller and E. Teller (1953). "Equation of state calculations by fast computing machines." *J. Chem. Phys.* **21**: 1087.

Moïse, J. C. (1999) Equilibres de coadsorption du p-xylène et du m-xylène par les zéolithes X et Y. Effet du cation compensateur, de la température et du taux d'hydratation de la zéolithe sur la sélectivité d'adsorption. Ph. D. Thesis, Université de Bourgogne, Dijon, France

Moldover, M. R. and J. C. Rainwater (1988). "Interfacial tension and vapor-liquid equilibria in the critical region of mixtures." *J. Chem. Phys* **88**: 7772.

Möller, D., J. Oprzynski, A. Müller and J. Fischer (1992). "Prediction of thermodynamic properties of fluid mixtures by molecular dynamics simulations: methane-ethane." *Molecular Physics* **75**: 363.

Möller, D. and J. Fischer (1994). "Determination of an effective intermolecular potential for carbon dioxide using vapoour-liquid phase equilibria from NpT + test particle simulations." *Fluid Phase Equilibria* **100**: 35.

Mortier, W. J., S. K. Ghosh and S. Shankar (1986). "Electronegativity equalization method for the calculation of atomic charges in molecules." *J. Am. Chem. Soc.* **108**: 4315-4320.

Münster, A. (1969). *Statistical Thermodynamics*. Berlin, Springer.

Murthy, C. S., S. F. O'Shea and I. R. McDonald (1983). "Electrostatic interactions in molecular crystals. Lattice dynamics of solid nitrogen and carbon dioxide." *Molecular Physics* **50**: 531.

Narten, A. H. (1977). *J. Chem. Phys.* **67**: 2102.

Nath, S. (2003). "Molecular simulation of vapor-liquid phase equilibria of hydrogen sulfide and its mixtures with alkanes." *J. Phys. Chem. B* **107**: 9498.

Nath, S. A., F. A. Escobedo and J. J. de Pablo (1998). "On the simulation of vapour-liquid equilibria for alkanes." *J. Chem. Phys.* **108**: 9905.

Nath, S. K. and J. J. de Pablo (2000). "Simulation of vapour-liquid equilibria for branched alkanes." *Molecular Physics* **98**: 231-238.

Nath, S. K., B. J. Banaszak and J. J. de Pablo (2001). "A new united Atom force field for alpha-olefins." *J. Chem. Phys.* **114**: 3612.

Neubauer, B., A. Boutin, B. Tavitian and A.H. Fuchs (1999a). "Gibbs ensemble simulation of vapour-liquid equilibria of cyclic alkanes." *Molecular Physics* **97**: 769-776.

Neubauer, B., B. Tavitian, A. Boutin and P. Ungerer (1999b). "Molecular simulations on volumetric properties of natural gas." *Fluid Phase Equilibria* **161**: 45-62.

Neumann, M. (1985). "The dielectric constant of water. Computer simulation experiments with the MCY potential." *J. Chem. Phys.* **82**: 5663.

Neuzil, R. W. (1971) US patents, 3558730, 3558732, 3626020

Neuzil, R. W. (1982) US patent, 4326092

Nicholson, D. and N. G. Parsonage (1982). *Computer simulation and the statistical mechanics of adsorption.* New York, USA, Academic Press.

Nieto-Draghi, C., J. B. Avalos and B. Rousseau (2003). "Dynamic and structural behavior of different rigid nonpolarizable models of water." *J. Chem. Phys.* **118**: 7954-7964.

Nosé, S. (1984). "A unified formulation of the constant temperature molecular dynamics methods." *J. Chem. Phys.* **81**: 511.

Nymand, T. M. and P. Linse (2000). "Ewald summation and reaction field methods for potentials with atomic charges, dipoles, and polarizabilities." *J. Chem. Phys.* **112**: 6152-6160.

Olson, D. H., G. T. Kokotailo, S. L. Lawton and W. M. Meier (1981). "Crystal structure and structure-related properties of ZSM-5." *J. Phys. Chem* **85**: 2238-2243.

Olson, D. H. (1995). "The crystal structure of dehydrated NaX." *Zeolites* **15**: 439-443.

de Pablo, J. J. and J. M. Prausnitz (1989). "Phase equilibria for fluid mixtures from Monte Carlo simulation." *Fluid Phase Equilibria* **53**: 177.

de Pablo, J. J., M. Laso and U. W. Suter (1992a). "Estimation of the chemical potential of chain molecules by simulation." *J. Chem. Phys.* **96**: 6157.

de Pablo, J. J., M. Laso and U.W. Suter (1992b). "Simulation of polyethylene above and below the melting point." *J. Chem. Phys.* **96**: 2395.

Panagiotopoulos, A. Z. (1987). "Direct determination of phase coexistence properties of fluids by Monte Carlo simulation in a new ensemble." *Molecular Physics* **61**: 813-826.

Panagiotopoulos, A. Z. (1994). Molecular Simulation in phase equilibria. *Supercritical fluids – fundamentals for application.* E. K. a. J. M. H. Levelt-Sengers. Dordrecht, Netherlands, Kluwer: 411-437.

Pascal, P. (1957). *Nouveau traité de chimie minérale – Combinaisons organo-mercuriques, tome V.* Paris, Masson.

Paschek, D. and R. Krishna (2001). "Monte Carlo simulations of sorption and diffusion of isobutane in silicalite." *Chem. Phys. Lett.* **342**: 148-154.

Pascual, P., P. Pernot, P. Ungerer, B. Tavitian and A. Boutin (2003). "Development of a transferable guest-host force field for adsorption hydrocarbons in zeolites. Reinvestigation of alkanes adsorption in silicalite by grand canonical Monte Carlo simulation." *Phys. Chem. Chem. Phys.* **5**: 3684.

Pascual, P., A. Boutin, P. Ungerer, B. Tavitian and A. H. Fuchs (2004a). "Adsorption of hydrocarbons in zeolites from molecular simulations. The alkane-ferrierite system revisited." *Chem. Phys. Chem.* **6**: 2015-2017.

Pascual, P., P. Ungerer, B. Tavitian and A. Boutin (2004b). "Development of a transferable guest-host force field for adsorption of hydrocarbons in zeolites. II. Prediction of alkenes adsorption and alkane/alkene selectivity in silicalite." *Journal of Physical Chemistry B* **108**(1): 393-398.

Pedersen, K. S., A. Fredenslund and P. Thomassen (1989). *Properties of oils and natural gases*. Houston, Texas, Gulf Publishing.

Peng, D. Y. and D. B. Robinson (1976). "A new two-constant equation of state." *Ind. Chem. Eng. Fundamentals* **15**: 59-64.

Petrov, A. A. (1984). *Petroleum hydrocarbons*. Berlin, Springer.

Pichon, C., A. Méthivier, M. H. Simonot-Grange and C. Baerlocher (1999). "Location of water and xylene molecules adsorbed on prehydrated zeolite BaX. A low-temperature neutron powder diffraction study." *J. Phys. Chem. B* **103**: 10197-10203.

Pichon, C., A. Méthivier and M. H. Simonot-Grange (2000). "Adsorption of m-xylene on prehydrated zeolite BaX: correlation between temperature-programmed desorption and low-temperature neutron powder diffraction studies." *Langmuir* **16**: 1931.

Pikunic, J., R. J. M. Pellenq, K. T. Thomson, J. N. Rouzaud, P. Levitz and K.E. Gubbins (2001). "Improved molecular models for porous carbons." *Studies in surface science and catalysis* **132**: 647.

Pluth, J. J. (1971) The crystal structures of several ion-exchnaged forms of Faujasite. Ph. D. Thesis, University microfilm n° 71-28459, University of Washington, Ann Harbor, Michigan.

Poling, B. E., J. M. Prausnitz and J. P. O'Connell (2001). *The properties of gases and liquids*. New York, Mc Graw Hill.

Polson, J. M. and D. Frenkel (1998). "Calculation of solid-fluid phase equilibria for systems of chain molecules." *J. Chem. Phys.* **109**: 318-328.

Polson, J. M. and D. Frenkel (1999). "Numerical prediction of the melting curve of n-octane." *J. Chem. Phys.* **111**: 1501-1510.

Porcher, F., M. Souhassou, Y. Dusauvoy and C. Lecomte (1999). "The crystal structure of a low-silica dehydrated NaX zeolite." *Eur. J. Mineral.* **11**: 333-343.

Potoff, J. J. and A. Z. Panagiotopoulos (1998). "Critical point and phase behavior of the pure fluid and a Lennard-Jones mixture." *J. Chem. Phys* **109**: 10914-10920.

Potoff, J. J., J. R. Errington and A. Z. Panagiotopoulos (1999). "Molecular simulation of phase equilibria for mixtures of polar and non-polar components." *Molecular Physics* **97**: 1073.

Prausnitz, J. M., R. N. Lichtenthaler and E. G. d. Azevedo (1986). *Molecular thermodynamics of fluid-phase equilibria*. Englewood Cliffs, New Jersey, USA, Prentice-Hall.

Prodany, N. W. and B. Williams (1971). "Vapor-liquid equilibria in methane-hydrocarbons systems." *J. Chem. Eng. Data* **16**: 1.

Rainwater, J. C. and F.R. Williamson (1986). "Vapor-liquid equilibrium of near-critical binary alkane mixtures." *Int. J. of Thermophysics* **7**: 65-75.

Rao, K. S., B. P. Stoicheff and R. Turner (1960). "High resolution Raman spectroscopy of gases. XIII. Rotational spectra and structures of the zinc-, cadmium- and mercury-dimethyl molecules." *Can. J. Phys.* **38**: 1516.

Reamer, H. H., B. H. Sage and W. N. Lacey (1953). "Phase Equilibria in hydrocarbon systems. Volumetric and phase behavior of n-pentane-hydrogen sulfide system." *Ind. Eng. Chem.* **45**: 1805-1809.

Reamer, H. H. and B. H. Sage (1963). "Phase equilibria in hydrocarbon systems. Volumetric and phase behaviour of the n-decane-CO_2 system." *J. Chem. Eng. Data* **8**: 508.

Reiff, W. E., P. Peters-Gerth and K. Lucas (1987). "A static equilibrium apparatus for (vapour-liquid) equilibrium measurements at high temperatures and pressures. Results for (methane + n-pentane)." *J. Chem. Thermodynamics* **19**: 467.

Renon, H. and J. M. Prausnitz (1967). "On the thermodynamics of alcohol-hydrocarbon solutions." *Chem. Eng. Sci.* **22**: 299-307 and (errata) 1891.

Reuck, K. M. d. and R. J. M. Craven (1993). *International Thermodynamic Tables of the fluid state – Methanol.* London, Blackwell.

Richards, R. E. and L. V. C. Rees (1987). "Sorption and Packing of n-Alkane Molecules in ZSM-5." *Langmuir* **3**: 335.

Rivail, J.-L. (1994). *Eléments de Chimie Quantique à l'usage des chimistes.* Paris, InterEditions/CNRS Editions.

Rosenbluth, M. N. and A. W. Rosenbluth (1955). "Monte Carlo calculation of the average extension of molecular chains." *J. Chem. Phys.* **23**: 356.

Rowley, R. L. (1994). *Statistical mechanics for thermophysical property prediction*, Prentice Hall.

Ryckaert, J. P., I. R. McDonald and M. L. Klein (1989). "Disorder in the pseudo-hexanol rotator phase of n-alkanes: molecular dynamics calculations for tricosane." *Mol. Phys.* **67**: 957.

Sage, B. H. and R. H. Olds (1947). "Volumetric behavior of Oil and Gas from several San Joaquim valley fields." *Trans. AIME* **170**: 156.

Sage, B. H. and V. M. Berry (1971) Phase equilibria in hydrocarbon systems. Behavior of the methane-propane-n-decane system. Monograph on API Research Project 37. American Petroleum Institute.

Slusher, J. T. (1999). "Infinite dilution activity coefficients in hydrogen-bonded mixtures via molecular dynamics: the water/methanol system." *Fluid Phase Equilibria* **154**: 181.

Smit, B. (1995). "Simulating the Adsorption Isotherms of Methane, Ethane, and Propane in the Zeolite Silicalite." *J. Phys. Chem.* **99**: 5597-5603.

Smit, B., S. Karaborni and I. J. Siepmann (1995). "Computer simulation of vapor-liquid phase equilibria of n-alkanes." *J. Chem. Phys.* **102**: 2126.

Smolin, Y. I., Y. F. Shepelev, I. K. Butikova and V. P. Petranovskii (1983). "The crystal structure of Na-X zeolite in hydrated and dehydrated forms." *Soviet Phys. Cristallogr.* **28**: 36-39.

Soave, G. (1972). "Equilibrium constants from a modified Redlich-Kwong equation of state." *Chem. Eng. Science* **27**: 1197-1203.

Soave, G. (1995). "A noncubic equation of state for the treatment of hydrocarbon fluids at reservoir conditions." *Ind. Eng. Chem. Res.* **34**: 3981.

Sportisse, M., A. Barreau and P. Ungerer (1997). "Modeling of gas condensates properties using continuous distribution functions for the characterization of the heavy fraction." *Fluid Phase Equilibria* **139**: 255-276.

Spyriouni, T., I. Economou and D. N. Theodorou (1998). "Phase equilibria of mixtures containing chain molecules predicted through a novel simulation scheme." *Phys. Rev. Letters* **80**: 4466-4469.

Spyriouni, T., I. G. Economou and D.N. Theodorou (1999). "Molecular simulation od alpha-olefins using a new united -atom potential model: vapor-liquid equilibria of pure compounds and mixtures." *J. Am. Chem. Soc.* **121**: 3407-3413.

Stach, H., U. Lohse, H. Thamm and W. Schirmer (1986). "Adsorption equilibria of hydrocarbons on highly dealuminated zeolites." *Zeolites* **6**: 74-90.

Sun, H. (1998). "COMPASS: an ab initio force-field optimized for condensed phase applications – overview with details on alkane and benzene compounds." *J. Phys. Chem. B* **102**: 7338-7364.

Sun, M. S., O. Talu and D. B. Shah (1996). "Adsorption Equilibria of C5-C10 Normal Alkanes in Silicalite Crystals." *J. Phys. Chem.* **100**: 17276.

Sun, M. S., D. B. Shah, H. H. Xu and O. Talu (1998). "Adsorption Equilibria of C1 to C4 Alkanes, CO2, and SF6 on Silicalite." *J. Phys. Chem. B* **102**: 1466.

Sychev, V. V., A. A. Vassrmann, A. D. Kozlov and V. A. Tsymarny (1995). *Thermodynamic properties of butane*, Library of Congress Cataloguing-in-publication data.

Szalay, P. G., H. lischka and A. Karpfen (1989). *J. Phys. Chem* **93**: 6629.

Thamm, H., H. Stach and W. Fiebig (1983). "Calorimetric study of the adsorption of n-butane and but-1-ene on a highly dealuminated Y-type zeolite and on silicalite." *Zeolites* **3**: 95-97.

Theodorou, D. N. and U. W. Suter (1985). "Detailed molecular structure of a vinyl polymer glass." *Macromolecules* **18**: 1467.

Theodorou, D. N. (2004). "Understanding and predicting structure-property relations in polymeric materials through molecular simulations." *Molecular Physics* **102**: 147-166.

Thomson, H. W. and J. W. Linnett (1936). "The vapour pressures and association of some metallic and non-metallic alkyls." *Trans. Faraday Soc.* **32**: 681.

Tissot, B. P. and D.H. Welte (1984). *Petroleum Formation and Occurrence*. Berlin, Springer.

Toth, G. (2001). "Determination of pair-potential parameters from experimental structure factors." *J. Chem. Phys* **115**: 4770.

Tournier, H., A. Barreau, B. Tavitian, D. L. Roux, C. Sulzer and V. d. Beaumont (2000). "Two experimental methods to study adsorption equilibria of xylene isomers in the liquid phase on a Y zeolite." *Microporous and Mesoporous Materials* **39**: 537-547.

Toxvaerd, S. (1990). "Molecular dynamics calculation of the equation of state of alkanes." *J. Chem. Phys.* **93**: 4290.

Toxvaerd, S. (1997). "Equation of state for alkanes II." *J. Chem. Phys.* **107**: 5197.

Ungerer, P., B. Faissat, C. Leibovici, H. Zhou, E. Behar, G. Moracchini and J.P. Courcy (1995). "High pressure – high temperature reservoir fluids: investigation of synthetic condensate gases containing a solid hydrocarbon." *Fluid Phase Equilibria* **111**: 287-311.

Ungerer, P., C. Batut, G. Moracchini, H. B. d. Sant'Ana, J. Carrier and D.M. Jensen (1998). "Measurement and prediction of volumetric and transport properties of reservoir fluids at high pressure." *Revue de l'Institut Français du Pétrole* **53**(3): 265-281.

Ungerer, P., A. Boutin and A.H. Fuchs (1999). "Direct calculation of bubble points by Monte Carlo simulation." *Molecular Physics* **97**: 523-539.

Ungerer, P., C. Beauvais, J. Delhommelle, A. Boutin, B. Rousseau and A.H. Fuchs (2000). "Optimization of the anisotropic United atoms intermolecular potential for n-alkanes." *J. Chem. Phys.* **112**: 5499-5510.

Ungerer, P., A. Boutin and A. H. Fuchs (2001). "Direct calculation of bubble points for alkane mixtures by Monte Carlo simulation." *Molecular Physics* **99**: 1423-1434.

Ungerer, P., A. Wender, G. Demoulin, E. Bourasseau and P. Mougin (2004). "Thermodynamic properties of H_2S-rich systems by Monte Carlo simulation." *Molecular Simulation* **30**: 631-648.

Uytterhoven, L., D. Dompas and W.J. Mortier (1992). "Theoretical investigations on the interaction of benzene with faujasite." *J. Chem. Soc. Faraday Trans.* **88**: 2753-2760.

Vidal, J. (1997). *Thermodynamique – Application au génie chimique et à l'industrie pétrolière*. Paris, Technip.

Vitale, G., C. F. Mellot, L. M. Bull and A.K. Cheetham (1997). "Neutron diffraction and computatinal study of zeolite NaX: influence of SIII' cations on its complex with benzene." *J. Phys. Chem. B* **101**: 4559-4564.

Vlugt, T. J. H., R. Krishna and B. Smit (1999). "Molecular simulations of adsorption isotherms for linear and branched alkanes and their mixtures in silicalite." *J. Phys. Chem. B* **103**: 1102-1118.

Vlugt, T. J. H., C. Dellago and B. Smit (2000). "Diffusion of isobutane in silicalite studied by transition path sampling." *J. Chem. Phys.* **113**: 8791-8799.

Vollhardt, K. P. C. and N. E. Schore (1999). *Organic Chemistry: structure and function, 3d edition*. New York, Freeman.

Vorholz, J., V. I. Harismiadis, B. Rumpf, A. Z. Panagiotopoulos and G. Maurer (2000). "Vapor + liquid equilibrium of water, carbon dioxide, and the binary system, water + carbon dioxide, from molecular simulation." *Fluid Phase Equilibria* **170**: 203.

Vorholz, J., B. Rumpf and G. Maurer (2002). "Prediction of the vapor-liquid phase equilibrium of hydrogen sulfide, and the binary system water-hydrogen sulfide by molecular simulation." *Phys. Chem. Chem. Phys.* **4**: 4449.

Vrabec, J. and J. Fischer (1995). "Vapor-liquid equilibria of mixtures from the NPT plus test particle method." *Molecular Physics* **85**: 781.

Vrabec, J. and J. Fischer (1996). "Vapor-liquid equilibria of binary mixtures containing methane ethane, and carbon dioxide from molecular simulation." *Int. J. of Thermophysics* **17**: 889.

Vrabec, J., J. Stoll and H. Hasse (2001). "A set of molecular models for symmetric quadrupolar fluids." *J. Phys. Chem. B* **105**: 12126-12133.

Wagner, W. and K. M. d. Reuck (1996). *Methane, international thermodynamic tables of the fluid state, Internatinal Union of Pure and Applied Chemistry (IUPAC)*. Oxford, Blackwell Science.

Walter, E. and L. Pronzato (1997). *Identification of parametric models*. Berlin, Springer.

Weber, G., J.P. Bellat, F. Benoit, C. Paulin, S. Limborg-Noetinger, M. Thomas (2005). "Adsorption equilibrium of light mercaptans on faujasites" *Adsorption* (accepted).

Wick, C. D., M. G. Martin and J. I. Siepmann (2000). "Transferable potentials for phase equilibria. 4. United-Atom description of linear and branched alkenes and alkylbenzenes." *J. Phys. Chem B* **104**: 8008.

Widom, B. (1963). "Some topics in the theory of fluids." *J. Chem. Phys.* **39**: 2808-2812.

Wilding, N. (1995). "Critical point and coexistence curve properties of the Lennard-Jones fluid: a finite-size scaling study." *Phys. Rev. E* **53**: 926.

Yan, Q. and J. J. de Pablo (1999). "Hyper-parallel tempering Monte Carlo: Application to the Lennard-Jones fluid and the restricted primitive model." *J. Chem. Phys.* **111**: 9509.

Zervopoulou, E., V. G. Mavrantsas and D. N. Theodorou (2001). "A new Monte Carlo simulation approach for the prediction of sorption equilibria of oligomers in polymer melts: solubility of long alkanes in linear polyethylene." *J. Chem. Phys* **115**: 2860.

Zhu, L. and K. J. Seff (1999). "Reinvestigation of the crystal structure of dehydrated sodium zeolite X." *J. Phys. Chem. B* **103**: 9512-9518.

Zhu, W., J. M. Graaf, L. J. P. Broeke, F. Kapteijn and J. A. Moulijn (1998). "TEOM: a unique technique for measuring adsorption properties. Light alkanes in Silicalite-1." *Ind Eng. Chem. Res.* **37**: 1934-1942.

Zhu, W., F. Kapteijn, B. Linden and J. A. Monlijn (2001). "Equilibrium adsorption of linear and branched C6 alkanes on silicalite-1 studied by the tapered element oscillating microbalance." *Phys. Chem. Chem. Phys.* **3**: 1755-1761.

Zhuravlev, N. D. and J. I. Siepmann (1997). "Exploration of the vapour-liquid phase equilibria and critical points of triacontane isomers." *Fluid Phase Equilibria* **134**: 55.

Zhuravlev, N. D., M. G. Martin and J. I. Siepmann (2002). "Vapor-liquid phase equilibria of triacontane isomers: deviations from the principle of corresponding states." *Fluid Phase Equilibria* **202**: 307.

Index